Bond Graph Modelling for Control, Fault Diagnosis and Failure Prognosis

Wolfgang Borutzky

Bond Graph Modelling for Control, Fault Diagnosis and Failure Prognosis

Springer

Wolfgang Borutzky
Bonn-Rhein-Sieg University of Applied
Sciences
Sankt Augustin
Germany

ISBN 978-3-030-60966-5 ISBN 978-3-030-60967-2 (eBook)
https://doi.org/10.1007/978-3-030-60967-2

This Springer imprint is published by the registered company Springer Nature Switzerland AG
The registered company address is: Gewerbestrasse 11, 6330 Cham, Switzerland

To my wife, Heidrun

Preface

Bond graph methodology has been successfully applied in various engineering fields all over the world since bond graphs (BGs) were devised by Professor H. Paynter back in 1959 at the Massachusetts Institute of Technology (MIT) in Cambridge, Massachusetts, USA, and were elaborated into a methodology by his former PhD students and leading pioneers of bond graph modelling Professor D. Karnopp and Professor D. Margolis (University of California at Davis) and Professor R. Rosenberg (Michigan State University, East Lansing, Michigan). Since then, remarkable progress has been achieved in almost all engineering fields. The advances are still ongoing as it is quite evident by looking, for instance, at the interplay of information technology with engineering systems in smart manufacturing, or all kinds of intelligent autonomously operating systems.

These developments are also reflected in bond graph-related publications. First, the graphical, step-by-step approach of multidisciplinary bond graph methodology based on an understanding of physics was successfully used mainly for modelling, analysis, and simulation of mechatronic systems in a wide range of application fields. BG methodology has become quite popular as the development of a model, in general, does not start at the level of mathematics. Equations needed for simulation can rather be derived automatically from a bond graph.

In the beginning, bond graph methodology was confined to the development of continuous time models for simulation as the exchange of energy between system components takes place continuously. For systems with fast state transitions in some components, e.g. systems with fast switching devices, the abstraction of discontinuous state changes can help avoid problems in the computation of a time continuous model and can save computational time. Accordingly, various extensions of bond graph methodology have been reported in the literature that enable to represent hybrid models by bond graphs. As a result, many publications on bond graph modelling and simulation in a variety of application areas have come up.

As the concurrent design of an engineering system with components from different disciplines goes along with the design of a control, it is no surprise that control engineers over the decades have increasingly taken an interest in the structural and computational properties of bond graphs. In fact, quite some control-relevant information can be deduced from a bond graph, for instance, whether a system is structurally observable and structurally controllable.

Nowadays, engineering systems of ever-increasing complexity are designed. In various areas, the notion *system of systems* has become in use. Accordingly, safety, reliability, and availability become more and more important. The equipment of engineering systems with an increasing number of all kinds of networked sensors and embedded systems in which algorithms can process the information delivered by the sensors enable not only the control of a system but also fault diagnosis, fault tolerant control in cooperation with the algorithms implemented in a supervision system, and failure prognosis. As to fault detection and isolation (FDI) based on time continuous models as well as on the so-called *hybrid* bond graphs, numerous articles, PhD theses, and some textbooks have been published in the course of the last decade. More recently, there are also some approaches to use bond graph modelling for failure prognosis, which is of still ongoing interest with less publications in comparison to the body of literature on bond graph modelling and on bond graph-based FDI.

Four decades ago, my interest in bond graph methodology also first started with modelling and simulation of closed loop engineering systems, especially hydraulic systems and mechatronic systems. Over time, the focus and my research interest have shifted and also include the exciting field of fault diagnosis and failure prognosis. My first book is a comprehensive presentation of bond graph methodology. The second book addresses bond graph modelling of hybrid systems and concludes with a short chapter on failure prognosis based on bond graphs. In addition, two compilation texts I edited with co-authors from all over the world include some few chapters on model-based fault diagnosis and prognosis.

The aim of this book is a presentation of bond graph modelling with a focus on how the methodology can contribute to model-based control, fault diagnosis, fault tolerant control, and failure prognosis by reviewing the state of the art and by taking into account the results of latest research work compiled in some PhD theses and articles. In online fault diagnosis, measurements are used as inputs into a so-called *diagnostic bond graph* model. Its evaluation provides information on the health state of a system that can be used to detect faults and to assess a degradation trend in the case of an incipient fault. However, measurements are corrupted with noise, and, in general, fault detection based on diagnostic BGs also needs the time derivatives of some of the measured signals. In addition, disturbances, model uncertainties, and parameter uncertainties have an effect on the estimation of states and parameters and on the projection of the estimated current state into the future, which has led to the use of Kalman and particle filters in failure prognosis. For that reason, the book is not confined to bond graph methodology but provides a more comprehensive presentation that puts BG modelling into a broader context by considering physical model-based as well as data-driven approaches to fault diagnosis and prognosis and their integration into a hybrid bond graph model-based, data-driven approach to failure prognosis.

Over the decades of my professional life, I had the opportunity to meet in person with leading personalities in the bond graph modelling community and to collaborate with many colleagues from all over the world. Their research activities, discussions with them, and their publications have inspired my own research direction of which this book is one result.

As to teaching, my experience has been that bond graph modelling and model-based control have been well received by students due to the graphical, rule-based step-by-step model development starting from a system schematic on the one hand side, and due to the available software support on the other hand for the automatic generation of equations, for the simulation, and for the design of a control of a system of which the dynamic behaviour is studied.

This book has been written for students specialising in the overlap of engineering and computer science as well as for researchers, and engineers in industry dealing with modelling, simulation, control, fault diagnosis, and failure prognosis in various application fields and who might be interested to see how bond graph modelling can support their work. For convenience, the book provides appendices with a list of definitions of key notions, an introduction into bond graph modelling, some mathematical background, and a glossary.

I would like to express my sincere thanks to Professors A. Fakri, Université Paris-Est, France, Y. Merkuryev, Riga Technical University, Latvia, and D. Murray-Smith, University of Glasgow, Scotland, UK, for their encouragement and their support of this project.

Furthermore, I would like to thank the Editorial Team with Springer in New York, NY, USA, especially Mary James, and Zoe Kennedy for their invitation to this book project and their kind support during this book project. Their good cooperation is much appreciated during a year in which a new pandemic has spread all over the world and in which New York City among many other places in the world was badly affected. My thanks also go to Mr. Pandian, Project Coordinator, with Springer Nature in Chennai, India and to Ms. Pearly Percy, production manager at SPi Global in Puducherry, India.

Last but not the least, I wish to express my sincere thanks and appreciation to my wife for her support and her patience with me when I was not available sometimes for many hours while working on my computer.

Sankt Augustin, Germany
August 2020

Wolfgang Borutzky

Contents

Abbreviations

AFTC	Active fault tolerant control
AR	Auto-regressive
ARMA	Auto-regressive moving average
ARR	Analytical redundancy relation
BD	Block diagram
BG	Bond graph
BGI	Bond graph in preferred integral causality
BGD	Bond graph in preferred derivative causality
BN	Bayesian network
BPNN	Backward-propagation neural network
CBM	Condition-based maintenance
DAE	Differential-algebraic equation
DBG	Diagnostic bond graph
DM	Degradation model
EoL	End of life
FDI	Fault detection and isolation
FSM	Fault signature matrix
FT	Failure threshold
FTC	Fault tolerant control
HI	Health indicator
HMM	Hidden Markov model
incBG	Incremental bond graph
IVP	Initial value problem
KF	Kalman filter
LFT	Linear fractional transformation form
LTI	Linear time-invariant system
MIMO	Multiple-input multiple-output system
NLSP	Nonlinear least square problem
NN	Neural network
ODE	Ordinary differential equation
pdf	Probability density function

PF	Particle filter
PFTC	Passive fault tolerant control
PHM	Prognostic and health management
PID	Proportional, integral, and derivative
PoF	Maximum allowable probability of failure
RLS	Recursive least square
RSME	Root mean square error
RUL	Remaining useful life
SCAP	Sequential causality assignment procedure
SISO	Single-input single-output system
SoH	State of health
SPBG	Sensitivity pseudo bond graph
SVF	State variable filter
TTF	Time to failure
UIO	Unknown input observer

Chapter 1
Introduction

1.1 Motivation

The equipment of more and more engineering systems with embedded systems in conjunction with numerous networked sensors and actors enable them not only to process raw measurement data and to use filtered signals as inputs into controllers, but also to determine health indicators (HIs) and to perform fault diagnosis and failure prognostic in situ by algorithms implemented in local embedded systems, and to communicate results with collaborating subsystems and operators via remote smart human interfaces.

Beyond fault diagnosis, failure prognosis is of major importance for safety critical engineering systems and processes such as nuclear power plants or chemical plants, where a failure may have catastrophic consequences, for supervision, automation, and condition based maintenance (CBM) of industrial processes, predictive maintenance, and for all kinds of emerging autonomous intelligent operating mobile systems such as unmanned aerial vehicles.

Based on continuous monitoring of the state of health (SoH) of an engineering system and by estimating the progressive degradation of some component behaviour, failure prognostic or prognosis from the Greek word $\pi\rho o\gamma\nu\omega\sigma\iota\zeta$ is the core of predictive maintenance of industrial processes enabling a longer lifetime of process components, increased safety, a more efficient use of resources, and a reduction of costs. Predictive anomaly identification based on real-time data, monitoring degradation, detecting precursors to failure, and predicting the remaining useful life (RUL) of components and of subsystems becomes even more important for complex systems in industry 4.0 smart manufacturing as the increased range of interaction between intelligent autonomous machines and system interdependencies has an influence on process faults and failures.

In mobile autonomous systems such as unmanned aerial vehicles, prediction of a failure of an actuator or a critical state of charge (SoC) of the battery in the near

W. Borutzky, *Bond Graph Modelling for Control, Fault Diagnosis and Failure Prognosis*, https://doi.org/10.1007/978-3-030-60967-2_1

future can be used for changing the control or even the mission to avoid all kind of possible damage.

Continuous monitoring of the system health and a repeated prediction of the time to failure (TTF) as of the current time, i.e., failure prognostic is clearly of technical and economical importance. Fault detection and isolation (FDI) as a prerequisite for failure prognostic has been a subject of research with regard to various applications. Data-driven as well as physics model-based approaches to FDI and failure prognosis are the two most commonly used approaches in industry and academia and have been reported in the literature over the last past decades. The increasing importance of failure prognosis is well reflected by quite a number of latest PhD theses such as [1, 5, 12, 13, 15, 18, 21] to name a few. More recently, combinations of both approaches are of growing interest. In comparison to the field of fault diagnosis with various established approaches, the combined use of model-based and data-driven methods for failure prognosis is a still rather young, a still developing research subject with contributions from various fields [10]. Latest reviews may be found in [7, 9].

Bond Graphs (BGs) introduced by Prof. H. Paynter at Massachusetts Institute of Technology, Cambridge, MA, USA back in 1959, have proven as a well suited and powerful tool for modelling, analysis, and simulation of a wide range of engineering systems and processes, especially for the concurrent design of closed loop mechatronic systems. The methodology intuitively starts from considering the energy exchange between system components from possibly different energy domains and conversions of energy from one form into another. As the exchange and conversion or transformation respectively of energy takes place continuously with respect to time, the graphical methodology was initially confined to the representation of continuous time models but has been extended in various ways so that hybrid models can be represented by BGs as well.

Beyond the graphical, physics-based systematic development of models for the purpose of model analysis and simulation of the dynamic behaviour of engineering systems, application of BG methodology has extended to model-based fault diagnosis over the last ten years, which has been documented in numerous conference and journal articles, PhD theses, and in some textbooks [2, 14, 19, 20], [3, Chap. 6]. More recently, there is some interest in using bond graphs also in hybrid model-based, data-driven approaches to failure prognosis. So far, this subject has only recently been addressed in some articles [4, 6, 8, 16, 17, 23], PhD theses [11, 22], and in some few book chapters [3, Chap. 7].

The aim of this book is to demonstrate how bond graph methodology can contribute to the tightly interrelated tasks of model-based control, fault diagnosis, fault tolerant control, and failure prognosis. Research works reported in the literature address these engineering tasks from various angles. Methods from quite diverse fields are used and are combined for fault diagnosis and failure prognosis. In order to put BG methodology into a broader context, the contribution of physics-based BG modelling is embedded in a comprehensive presentation that comprises to some extent also purely mathematical methods operating on measured data such as signal preprocessing, filters, and data-driven failure prognosis. Therefore, one focus of this text is on a combined BG model-based, data-based approach to failure prognosis.

1.2 Organisation of the Book

One objective and the focus of this book is to show how bond graph methodology can be used for the offline physics model-based part of a hybrid approach to fault diagnosis and failure prognosis. Accordingly, the next chapter first summarises and points out how bond graphs can be useful to check a model for structural properties such a structural observability. Clearly, observability is a prerequisite for the detection of abnormal system behaviour. In addition, it is necessary to identify a faulty component and to map the faulty behaviour to parametric faults. Therefore, Chap. 3 presents bond graph model-based fault detection and isolation and proposes a graphical method for the placement of sensors so that a maximum number of potential faulty elements can be isolated.

Chapter 4 continues with a brief presentation of data-based and model-based failure prognosis methods, proposes a hybrid bond graph model-based approach, and concludes by discussing various types of uncertainties in failure prognostic.

Once a component fault has been detected, isolated, and assessed with respect to its magnitude and the component's remaining useful life has been estimated, it might become necessary to change the system's control in order to accommodate for the fault if the feedback controller cannot cope sufficiently with the fault and to ensure a safe, possibly limited functionality for the remaining time until failure preventing actions must be taken.

For instance, if an actuator of a multicopter fails, it may continue its mission with a changed control. If the state of charge of the battery is approaching a critical level, the decision will be to change the mission and to safely land the aerial vehicle. Chapter 5 shows how bond graph-based fault diagnosis, BG-based model inversion can serve active as well as passive fault tolerant control.

Chapter 6 considers some software packages with a focus on open source software as to how it can be used for fault diagnosis, failure prognosis, and fault tolerant control.

The last but one chapter illustrates bond graph modelling, fault diagnosis, failure prognosis, and FTC by application to some small example systems.

The concluding chapter summarises the contributions of bond graph methodology regarding model-based control, fault diagnosis, an integration of a bond graph modelling with failure prognosis in a hybrid bond graph model-based, data-driven approach and considers some topics of further ongoing research.

Appendix A compiles some definitions of key terms used throughout this book that are in accordance with the outcome of a standardisation effort of the IFAC SAFEPROCESS Technical Committee. Appendix B provides a short introduction into bond graph methodology. For more details, the reader may find some textbooks on bond graph modelling in the list of references. Appendix C briefly recalls some mathematical background. Finally, a glossary lists explanations of some key notions.

References

1. Bektas, O. (2018). An Adaptive Data Filtering Model for Remaining Useful Life Estimation. Ph.D. thesis, University of Warwick.
2. Borutzky, W. (2015). *Bond Graph Model-Based Fault Diagnosis of Hybrid Systems*. Cham: Springer International Publishing.
3. Borutzky, W. (Ed.) (2016). *Bond Graphs for Modelling, Control and Fault Diagnosis of Engineering Systems*. Cham: Springer International Publishing.
4. Borutzky, W. (2019). A Hybrid Bond Graph Model-Based – Data Driven Method for Failure Prognostic. In *Proceedings of the International Conference on Industry 4.0 and Smart Manufacturing* (ISM 2019).
5. Butler, S. (2012). Prognostic Algorithms for Condition Monitoring and Remaining Useful Life Estimation. Ph.D. thesis, National University of Ireland, Maynooth.
6. Danes, M., Ramezani, A., & Moghaddam, J. Z. (2015). Enhanced prognosis of hybrid systems with unknown mode changes. *Modares Journal of Electrical Engineering, 15*(2), 21–26.
7. Dawn, A., Kim, N. H., & Joo-Ho, C. (2015). Practical options for selecting data-driven or physics-based prognostics algorithms with reviews. *Reliability Engineering & System Safety, 133*, 223–236.
8. Djeziri, M., Thi-Bich-Lien, N., Benmoussa, S., M'Sirdi, N. (2016). Fault prognosis based on physical and stochastic models. In *2016 European Control Conference (ECC)*, pp. 2269–2274.
9. Ekanayake, T., Dewasurendra, D., Abeyratne, S., Lin, M., Yarlagadda, P. (2019). Model-based fault diagnosis and prognosis of dynamic systems: a review. *Procedia Manufacturing, 30*, 435–442, 14th Global Congress on Manufacturing and Management (GCMM-2018).
10. Eker, Ö, F. (2015). A Hybrid Prognostic Methodology and its Application to Well-Controlled Engineering Systems. Ph.D. thesis, Cranfield University.
11. Jha, M. (2015). Diagnostics and prognostics of uncertain dynamical systems in a bond graph framework. Ph.D. Thesis, École Centrale de Lille, Université Lille Nord-de-France.
12. Kimotho, J. K. (2016). Development and performance evaluation of prognostic approaches for technical systems. Ph.D. thesis, Universität Paderborn.
13. Kulkani, C. S. (2013). A physics-based degradation modeling framework for diagnostic and prognostic studies in electrolytic capacitors. Ph.D. thesis, Graduate School of Vanderbilt University.
14. Merzouki, R., Samantaray, A., Pathak, P., Ould Bouamama, B. (2013). *Intelligent Mechatronic Systems*. London: Springer.
15. Mosallam, A. (2014). Remaining useful life estimation of critical components based on Bayesian approaches. Ph.D. thesis, Université de Franche-Comté.
16. Lu, N., Zhou, D., & Jiang, B. (2014). A fault prognosis scheme for chemical reaction process using Pseudo-Bond Graph based Bayesian network. In *Proceeding of the 11th World Congress on Intelligent Control and Automation* (pp. 5869–5874). Shenyang: IEEE. https://doi.org/10.1109/WCICA.2014.7053723.
17. Prakash, O., Samantaray, A., Bhattacharyya, R., & Ghoshal, S. (2018). Adaptive prognosis for a multi-component dynamical system of unknown degradation modes. In *Proceedings of 10th IFAC Symposium on Fault Detection, Supervision and Safety for Technical Processes*. Warsaw: IFAC.
18. Rodriguez Obando, D. J. (2018). From Deterioration Modeling to Remaining Useful Life Control: A comprehensive framework for post-prognosis decision-making applied to friction drive systems. Ph.D. thesis, Université Grenoble Alpes.

19. Samantaray, A., & Ould Bouamama, B. (2008). Model-based process supervision – a bond graph approach. *Advances in Industrial Control*. London: Springer.
20. Wang, D., Yu, M., Low, C., & Arogeti, S. (2013). *Model-Based Health Monitoring of Hybrid Systems*. New York: Springer.
21. Wileman, A. (2016). An Investigation into the Prognosis of Electromagnetic Relays. Ph.D. Thesis, Cranfield University.
22. Ming, Y. (2012). Fault Diagnosis and Prognosis of Hybrid Systems using Bond Graph Models and Computational Intelligence. Ph.D. thesis, Nanyang Technological University.
23. Wu, Y., Jiang, B., Lu, N., & Zhou, Y. (2015). Bayesian Network Based Fault Prognosis via Bond Graph Modeling of High-Speed Railway Traction Default. Mathematical Problems in Engineering Article ID 321872. http://dx.doi.org/10.1155/2015/321872.

Chapter 2
Structural Properties of Bond Graphs for Model-Based Control

This chapter compiles some features of Bond Graph methodology developed over the past decades by various authors that have proven useful in model-based control. The final step in a rule-based step-by step development of a bond graph model from a schematic of an engineering system is the assignment of computational causality either manually for medium size models or automatically. It is the check of causal paths in a bond graph that is the basis for various tasks in control.

Clearly, prior to the design of a control law for a system, a developed model should be checked whether the system with given sensors and actuators is structurally, i.e. independently from numerical values of system parameters, state observable and structurally state controllable. If this is not the case, then the system behaviour cannot be controlled. Additional sensors first must be placed in appropriate places if physically feasible. The question then, however, is how many sensors should be placed in which locations.

As block diagrams are a standard, well known graphical model representation in control engineering and well supported by a number of widely used commercial and open source software packages such as Matlab®/Simulink® or Scilab/Xcos, it is briefly recalled that bond graphs may be transformed into block diagrams.

The computation of transfer functions from the matrices of a linear state space model is costly and requires the support of software packages such as Matlab® or GNU Octave. Back in 1973, Brown has shown that for small to medium size bond graph models, transfer functions can directly be obtained from a bond graph by following causal paths [1].

Following causal paths in an extension of bond graphs introduced by Gawthrop and called *bicausal bond graphs* [6] has proven useful for model-based control tasks such as parameter estimation and system inversion. Parameter estimation is of importance for fault diagnosis and online failure prognosis, while system inversion is a part in fault tolerant control. Both subjects are addressed in later chapters. The presentation in this chapter is less theoretical than illustrated in detail by various easy to survey examples.

W. Borutzky, *Bond Graph Modelling for Control, Fault Diagnosis and Failure Prognosis*, https://doi.org/10.1007/978-3-030-60967-2_2

2.1 Structural Observability and Structural Controllability

For fault detection, it is necessary that at least some physical quantities of a system are observable. If the system's dynamic behaviour can be described by a linear time-invariant (LTI) state space model

$$\dot{\boldsymbol{x}}(t) = \mathbf{A}\boldsymbol{x}(t) + \mathbf{B}\boldsymbol{u}(t)\,, \quad \boldsymbol{x}(0) = \boldsymbol{x}_0 \tag{2.1a}$$

$$\boldsymbol{y}(t) = \mathbf{C}\boldsymbol{x}(t) + \mathbf{D}\boldsymbol{u}(t) \tag{2.1b}$$

where $\boldsymbol{x}(t) \in \mathbb{R}^n$ denotes the vector of states, $\boldsymbol{u}(t) \in \mathbb{R}^m$ the vector of inputs, and where $\mathbf{A}, \mathbf{B}, \mathbf{C}, \mathbf{D}$ are constant coefficient matrices of appropriate dimensions, then Kalman's well known numerical observability criterion can be used to check whether the system is completely state observable by computing the rank of the observability matrix

$$\mathscr{O} = \begin{bmatrix} \mathbf{C} \\ \mathbf{CA} \\ \mathbf{CA}^2 \\ \vdots \\ \mathbf{CA}^{n-1} \end{bmatrix} \tag{2.2}$$

A prerequisite for the design of a controller is that a system is controllable. A well known necessary and sufficient criterion for complete state controllability is that the rank of Kalman's controllability matrix $\mathscr{C}(\mathbf{A}, \mathbf{B})$ is equal to the number n of system states.

$$\text{rank}\mathscr{C}(\mathbf{A}, \mathbf{B}) = \text{rank}\left[\mathbf{B}\middle|\mathbf{AB}\middle|\mathbf{A}^2\mathbf{B}\middle|\ldots\middle|\mathbf{A}^{n-1}\mathbf{B}\right] = n \tag{2.3}$$

The use of the observability matrix and the controllability matrix, respectively, however, is not robust as the entries in the matrices depend on the system's physical parameters. The actual numerical values of the latter may deviate from their known nominal values and are therefore uncertain. To get rid of this parameter dependency, the property of *structural* observability has been considered in the literature. To that end, entries in the state space matrices different from zero are replaced by a symbol regardless of their actual numerical value. The resulting matrices capture the matrix structure that is the same for all members of a class of models.

If the structure of a mathematical state space model is known, structural dependencies between input, state, and output variables are represented in classical control theory by a digraph. A necessary condition for structural observability is that there are direct or indirect connections between the state variables and the output variables, which can be checked on the graph. The second part of a necessary and sufficient criterion for complete structural state observability is that the *structural rank* of the structural observability matrix is equal to the number of states. Let $\mathbf{A}^*$

and $\mathbf{C}^*$ be the structural matrices corresponding to matrices $\mathbf{A}$ and $\mathbf{C}$, and n the number of states, then

$$s - \text{rank}[\mathcal{O}^*] := s - \text{rank}[\mathbf{A}^*|\mathbf{C}^*] = n \tag{2.4}$$

The s-rank can be checked by appropriately rearranging columns and rows. Note that from $\text{rank}(\mathcal{O}^*) < n$ it cannot be concluded that $\mathcal{O}$ does not have full rank, i.e. that the system is not observable.

In the following, the notion of a *causal path* in a BG is used.

Definition 2.1 (Causal path) A causal path p between power variables in a causal BG is a series of power variables that are successively connected according to the assignment of causality.

Definition 2.2 (power line) In contrast to a causal path, a power line denotes a series of bonds and junctions structure elements in an *acausal* bond graph connecting the ports of two components.

2.1.1 Bond Graph-Based Analysis of Structural State Observability

Bond graph methodology enables to develop rule based on a causal bond graph without prior writing of equations. A state space model can be rather derived either manually for medium size models by following causal paths or automatically by software. As to a check of *structural* observability, there is no need to derive state space equations in order to determine structural matrices. Their entries can be directly determined by following causal paths in a BG. For instance, if there is a causal path between two storage elements associated with the j-th and the i-th state variable that contains the least number of storage elements and does not pass through any other storage element in integral causality and if none of its bonds is activated, then coefficients a_{ij}, a_{ji} of matrix $\mathbf{A}$ are different from zero and are replaced by a wildcard, e.g. $*$. Likewise, coefficient c_{ij} in matrix $\mathbf{C}$ is replaced by $*$, if there is a causal path from the ith state to the jth output that does not pass through any other storage element.

Another necessary and sufficient criterion for structural observability even avoids the setup of structural matrices and the determination of the structural rank of the structural observability matrix. It is enough to check for certain causal paths in a BG with storage elements in integral causality, then convert to preferred derivative causality and check again for certain causal paths.

It is assumed that

- energy sources, energy storage elements, and dissipators in a bond graph are 1-port elements,
- there are only power bonds in the bond graph, no signals, and
- all elements are linear.

Theorem 2.1 (Sueur and Dauphin-Tanguy, 1991 [3, 17]) *A linear time-invariant MIMO system with matrices* **A** *and* **C** *is structurally completely state observable if and only if the following two conditions are satisfied.*

1. *Given that the preferred causality applied to the energy storage elements in the bond graph is integral causality, then every energy storage element in integral causality must have at least one causal path to a sensor.*
2. *In a bond graph with preferred integral causality, all energy storage in integral causality must take derivative causality if derivative causality is the preferred causality. If this condition is not directly met, then it is achieved by replacing some sensors in appropriate places by their dual.*

Definition 2.3 (Bond Graph rank) Let $1 \leq k < n$ denote the number of storage ports that must take integral causality when derivative causality is the preferred causality, then the difference $n - k$ is called the bond graph rank of a $n \times n$ state matrix **A** and is denoted as BG − rank**A**. (This difference is also equal to the number of non-zero eigenvalues of **A**.)

Remark 2.1 If the two conditions of Theorem 2.2 are met, then rank$\mathscr{O} = n$. If BG − rank**A** $= n$, then one observer, suitably placed, is sufficient to assure complete observability. Otherwise, if BG − rank**A** $= q < n$, then $n - q$ observers are needed to guarantee observability. They are to be placed such that the first condition of Theorem 2.2 is satisfied.

2.1.2 Bond Graph-Based Analysis of Structural State Controllability

There is a similar bond graph-based necessary and sufficient criterion for structural controllability.

Theorem 2.2 (Sueur and Dauphin-Tanguy, 1991 [3, 17]) *A linear time-invariant MIMO system with the $n \times n$ state matrix* **A** *and the $n \times m$ matrix* **B** *is structurally completely state controllable if and only if the following two conditions are satisfied.*

1. *Given that the preferred causality applied to the energy storage elements in the bond graph is integral causality, then for each energy storage element in integral causality there exists a causal path from a controlled source to the energy storage element. In other words, all states are* input-reachable.
2. *In a bond graph with preferred integral causality, all energy storage elements in integral causality take derivative causality if derivative causality is the preferred causality. If this condition is not met directly, then it is achieved by replacing some controlled sources in appropriate places by their dual.*

Remark 2.2 (Number of control sources)

- If $BG - rank\mathbf{A} = n$, then one properly located input source is sufficient to control the system.
- If $BG - rank\mathbf{A} = n - k$, then at least k input sources properly located are necessary to control the system.
- Proper location means that if the sources are replaced by their duals, then all storage ports are in derivative causality, which means that the system is state controllable by the added sources [3, 17].

Example: Masses-Spring Oscillator

Consider the simple mechanical oscillator in Fig. 2.1 consisting of two masses connected by a spring with an excitation force F moving the masses against an external force F_e. Figure 2.2 shows the BG in preferred integral causality developed from the schematic.

Check for Structural Controllability on a Bond Graph

As can be seen from the BG in Fig. 2.2, there are two causal paths from the excitation force source $MSe : F$ to the storage elements highlighted by added signals in red

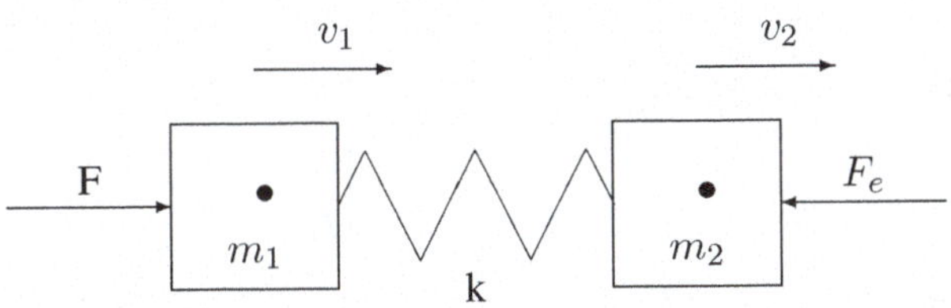

Fig. 2.1 Mechanical oscillator with two masses

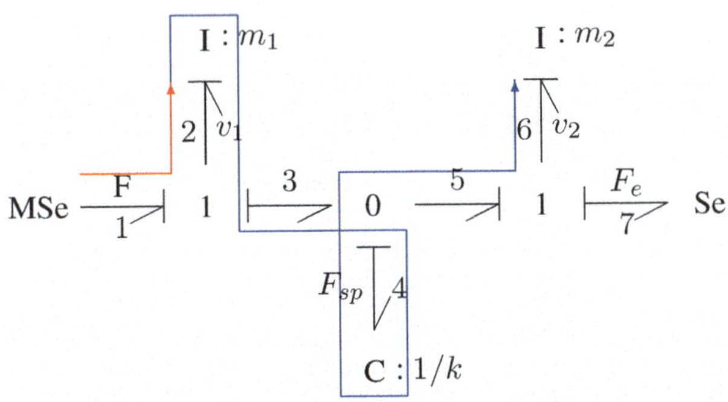

Fig. 2.2 Bond graph of the mechanical oscillator with two masses in preferred integral causality

and blue, respectively. That is, the necessary attainability condition of Theorem 2.2 is fulfilled. However, if preferred derivative causality is assigned to the BG, then the storage element $I : m_2$ must retain integral causality. Derivative causality at all storage elements can be achieved if the modulated effort source MSe is replaced by a flow source (Fig. 2.3). The sufficient condition of Theorem 2.2 is also met. Accordingly, the mechanical oscillator is structurally state controllable.

In order to see whether the oscillator is also numerically controllable, Kalman's controllability matrix is computed. From the BG in preferred integral causality the following state space model is derived.

$$\underbrace{\begin{bmatrix} \dot{v}_1 \\ \dot{v}_2 \\ \dot{F}_{sp} \end{bmatrix}}_{\dot{\boldsymbol{x}}(t)} = \underbrace{\begin{bmatrix} 0 & 0 & -\frac{1}{m_1} \\ 0 & 0 & \frac{1}{m_2} \\ k & -k & 0 \end{bmatrix}}_{\mathbf{A}} \underbrace{\begin{bmatrix} v_1 \\ v_2 \\ F_{sp} \end{bmatrix}}_{\boldsymbol{x}(t)} + \underbrace{\begin{bmatrix} \frac{1}{m_1} & 0 \\ 0 & -\frac{1}{m_2} \\ 0 & 0 \end{bmatrix}}_{\mathbf{B}} \underbrace{\begin{bmatrix} F \\ F_e \end{bmatrix}}_{\boldsymbol{u}(t)} \tag{2.5a}$$

$$y(t) = v_2 = \underbrace{\begin{bmatrix} 0 & 1 & 0 \end{bmatrix}}_{\mathbf{C}} \begin{bmatrix} v_1 \\ v_2 \\ F_{sp} \end{bmatrix} \tag{2.5b}$$

With these state space model matrices the controllability matrix reads

$$\mathscr{C}(\mathbf{A}, \mathbf{B}) := [\mathbf{B}|\mathbf{AB}|\mathbf{A}^2\mathbf{B}] = \left[\begin{array}{cc|cc|cc} \frac{1}{m_1} & 0 & 0 & 0 & -\frac{k}{m_1^2} & \frac{k}{m_1 m_2} \\ 0 & -\frac{1}{m_2} & 0 & 0 & \frac{k}{m_1 m_2} & \frac{k}{m_2^2} \\ 0 & 0 & \frac{k}{m_1} & -\frac{k}{m_2} & 0 & 0 \end{array}\right] \tag{2.6}$$

As $\operatorname{rank}\mathscr{C}(\mathbf{A}, \mathbf{B}) = n = 3$, the oscillator is also numerically completely state controllable. Mass m_1 is directly controllable, m_2 indirectly via the spring.

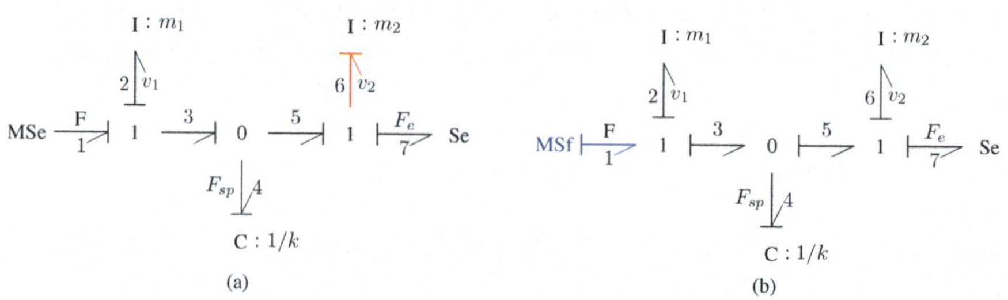

Fig. 2.3 Check for structural controllability on the BG of the mechanical oscillator. (**a**) BG in preferred derivative causality. (**b**) BG after replacement of the modulated effort source MSe by its dual

Check for Structural Observability on a Bond Graph

As to structural state observability of the oscillator, it can be seen from Fig. 2.4a that there is a direct causal path from the inertia element $I : m_2$ to the flow detector and an indirect causal path from $I : m_1$ via the C-element and $I : m_2$ to the detector. That is, the attainability condition of Theorem 2.1 is fulfilled. Note that the sufficient condition of Theorem 2.1 can be met if the flow detector takes inverse causality or is replaced by an effort detector as depicted in Fig. 2.4b.

The computation of Kalman's observability matrix reveals that the mechanical oscillator is also numerically completely state observable with the single chosen flow detector. With the system matrix **A** and the sensor matrix **C** one obtains for the observability matrix

$$\operatorname{rank}\mathcal{O}(\mathbf{A}, \mathbf{C}) := \operatorname{rank}\begin{bmatrix} \mathbf{C} \\ \mathbf{CA} \\ \mathbf{CA}^2 \end{bmatrix} = \operatorname{rank}\begin{bmatrix} 0 & 0 & 1 \\ \hline 0 & 1 & \dfrac{1}{m_2} \\ \hline \dfrac{k}{m_2} & -\dfrac{k}{m_2} & 0 \end{bmatrix} = 3 \tag{2.7}$$

If instead of the flow sensor Df : v_2 an effort sensor De is used and attached to the 0 junction, then the system is not structurally state observable. There are causal paths from all storage elements to the effort detector as can be seen in Fig. 2.5a so that the necessary attainability condition is met.

However, the inertia element $I : m_2$ retains integral causality when preferred derivative causality is assigned to the BG as depicted by Fig. 2.5b. Derivative causality at $I : m_2$ cannot be achieved if the effort detector is exchanged by its dual. That is, the sufficient condition of Theorem 2.2 is not fulfilled. Hence, the oscillator with an effort sensor De attached to the 0 junction is not structurally observable.

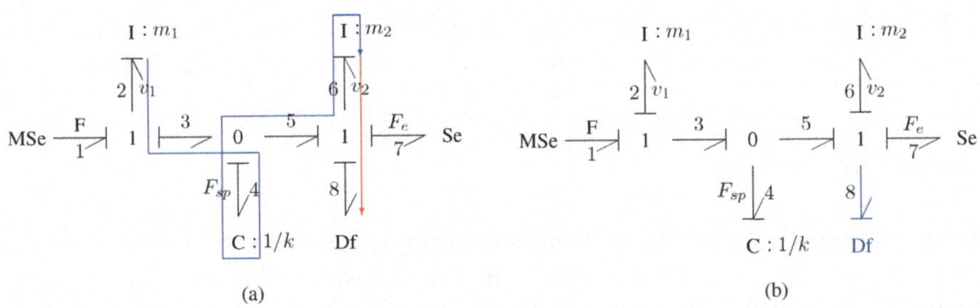

Fig. 2.4 Check for structural observability on the BG of the mechanical oscillator. (**a**) BG in integral causality with flow detector. (**b**) BG in preferred derivative causality with flow detector in derivative causality

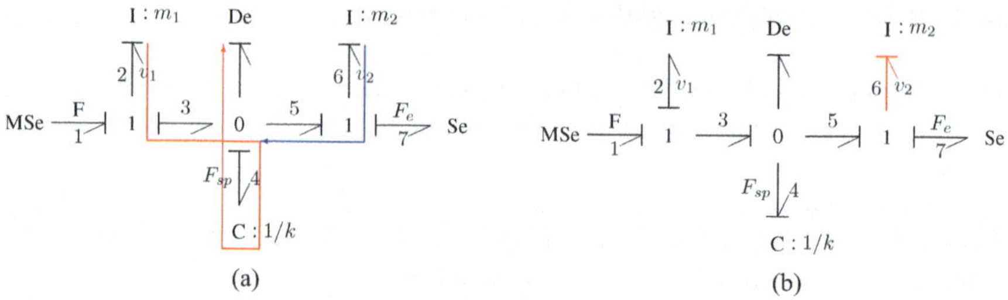

Fig. 2.5 Check for structural observability on the BG of the mechanical oscillator with an effort detector attached to the 0 junction. (**a**) BG in integral causality with effort detector. (**b**) BG in preferred derivative causality with effort detector

Computation of the observability matrix confirms that the oscillator with this sensor is also not numerically observable.

$$\mathscr{O}(\mathbf{A}, \mathbf{C}) := \begin{bmatrix} \mathbf{C} \\ \mathbf{CA} \\ \mathbf{CA}^2 \end{bmatrix} = \left[\begin{array}{ccc} 0 \;\; 0 & 1 \\ \hline k \;\; -k & 0 \\ \hline 0 \;\; 0 & -\dfrac{k}{m_1} - \dfrac{k}{m_2} \end{array}\right] \tag{2.8}$$

As can be seen, rank$\mathscr{O} = 2 < n = 3$.

Remark 2.3 (Structural state observability – numerical state observability)

- If a system is numerically completely state observable (controllable), then it is structurally completely state observable (controllable).
- If a system is structurally completely state observable (state controllable), then almost all systems having a state space model with the same structural matrices are also numerically completely state observable (state controllable).
- If a system is *not* structurally completely state observable (state controllable), then it is *not* numerically completely state observable (state controllable).

Example: RC Network

Figure 2.6 shows a simple RC network with a single capacitor.

Checks on the corresponding BGs in Fig. 2.7 show that the conditions of Theorems 2.1 and 2.2 are fulfilled. Hence, the circuit is structurally observable and structurally controllable. However, whether the circuit is also numerically controllable depends on the parameter values of the resistances.

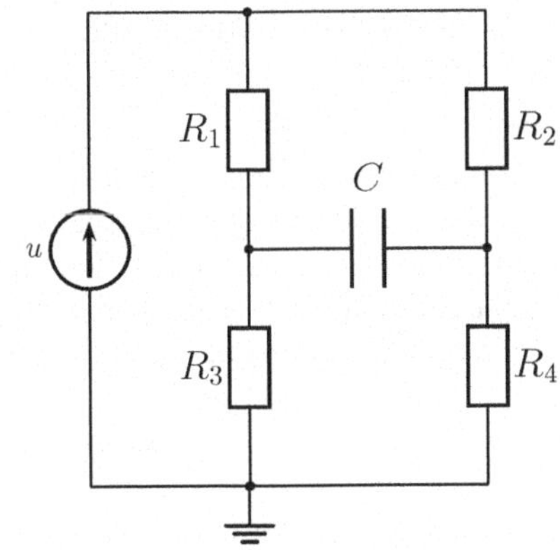

Fig. 2.6 Circuit schematic of a RC bridge

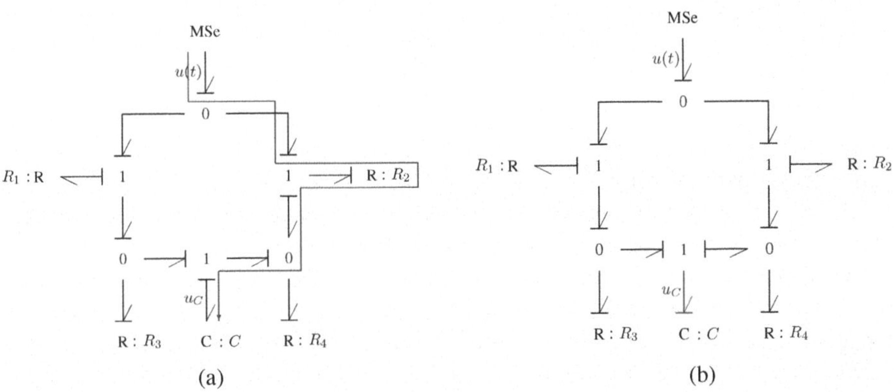

Fig. 2.7 Check for structural controllability on the BG of the RC network in Fig. 2.6. (**a**) Attainability. (**b**) Storage element in derivative causality

From the equations derived from the causal BG, the following state space model can be obtained after some steps.

$$u_C(t) = \left(\frac{1}{\frac{R_1}{R3}+1} - \frac{1}{\frac{R_2}{R_4}+1} \right) u(t) - \left(\frac{R_1}{\frac{R_1}{R3}+1} + \frac{R_2}{\frac{R_2}{R4}+1} \right) C\dot{u}_C(t) \quad (2.9a)$$

$$y(t) = u_C(t) \quad (2.9b)$$

As can be seen, if $R_1/R_3 = R_2/R4$, then the capacitor voltage u_C vanishes from the state equation (2.9). The voltage u_C across the capacitor cannot be controlled by the voltage source $MSe : u(t)$.

2.2 Transfer Functions

Once it has been checked that a multivariable system with given sources and sensors is completely state controllable and observable, transfer functions of interest can be determined. To this end, various approaches can be used. One option is to use the matrices of the state space model. Given vanishing initial values, Laplace transform of the state space equation and the observation equation yields for the transfer function matrix $\mathbf{G}(s)$ the well known expression

$$\mathbf{G}(s) = \mathbf{C}\,(s\mathbf{I} - \mathbf{A})^{-1}\,\mathbf{B} + \mathbf{D} \tag{2.10}$$

Any entry g_{ij} of the matrix $\mathbf{G}$ relating the jth input to the ith output is a transfer function of the form

$$g_{ij} = \frac{N_{ij}(s)}{D(s)} \tag{2.11}$$

where $N_{ij}(s)$ is the numerator polynomial in $s \in \mathbb{C}$ and $D(s) = \det(s\mathbf{I} - \mathbf{A})$ the denominator polynomial.

The determination of transfer functions via the matrices of the state space model may be quite costly with regard to the computational effort.

Another approach is to represent the equations of the state space by means of a block diagram and to simplify the structure of the graph by merging rule-based step-by-step signal blocks, or to have a software program generating required transfer functions from a block diagram or even from an iconic model representation.

An alternative known since a long time is to establish a signal flow graph and to apply Mason's loop rule on the graph [9].

In contrast to block diagrams and signal flow diagrams, bond graph modelling does not start from equations but from the consideration of energy flows and physical phenomena. Various forms of mathematical models can be derived from a causal bond graph which is the result of a rule-based step-by-step modelling approach. As to transfer functions, F. Brown [1] showed as early as 1972 that there is no need to transform a causal bond graph into a block diagram or into a signal flow graph in order to determine transfer functions. Mason's loop rule based on signal flow graphs can be applied directly on a causal bond graph. This demonstrates the usefulness of bond graphs for model-based control and shall be illustrated by means of two example systems.

Example: LC Network

Consider the simple LC circuit in Fig. 2.8a. A check of the corresponding BG in Fig. 2.8b shows that the circuit with the effort source and the flow sensor is structurally completely state controllable and structurally completely state observable.

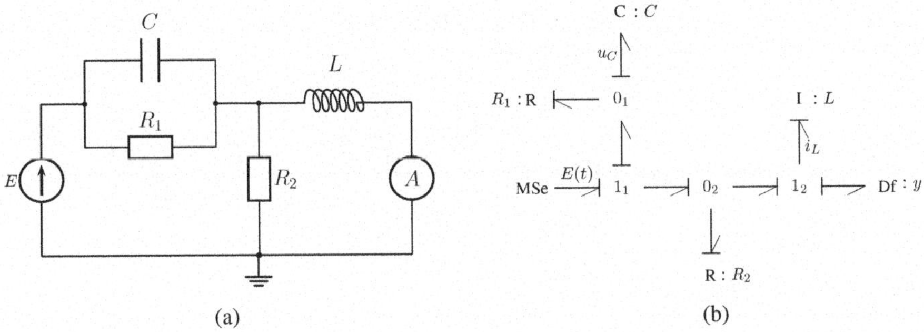

Fig. 2.8 LC circuit and corresponding BG. (**a**) Schematic of a LC circuit. (**b**) BG of the LC circuit in preferred integral causality

From the causal BG in preferred integral causality the following state space model can be derived.

$$\frac{d}{dt}\begin{bmatrix} u_c \\ i_L \end{bmatrix} = \underbrace{\begin{bmatrix} -\frac{1}{RC} & \frac{1}{C} \\ -\frac{1}{L} & 0 \end{bmatrix}}_{\mathbf{A}} \begin{bmatrix} u_c \\ i_L \end{bmatrix} + \underbrace{\begin{bmatrix} \frac{1}{R_2C} \\ \frac{1}{L} \end{bmatrix}}_{\mathbf{B}} E \tag{2.12a}$$

$$y = i_L = \underbrace{\begin{bmatrix} 0 \; 1 \end{bmatrix}}_{\mathbf{C}} \begin{bmatrix} u_c \\ i_L \end{bmatrix} \tag{2.12b}$$

Application of Kalman's criteria shows that the circuit is also numerically completely state controllable as well as numerically completely state observable.

$$\mathscr{C} = [\mathbf{B}|\mathbf{AB}] = \left[\begin{array}{c|c} \frac{1}{R_2C} & -\frac{1}{RC}\cdot\frac{1}{R_2C}+\frac{1}{LC} \\ \frac{1}{L} & -\frac{1}{L}\cdot\frac{1}{R_2C} \end{array}\right] \tag{2.13}$$

$$\mathscr{O} = \begin{bmatrix} \mathbf{C} \\ \mathbf{CA} \end{bmatrix} = \left[\begin{array}{cc} 0 & 1 \\ \hline -\frac{1}{L} & 0 \end{array}\right] \tag{2.14}$$

Both matrices $\mathscr{C}$, $\mathscr{O}$ are of rank two.

The classical computation of the transfer function $\mathscr{L}y/\mathscr{L}E$ based on the matrices of the state space model yields

$$\frac{\mathscr{L}y}{\mathscr{L}E} = \mathbf{C}\,(s\mathbf{I}-\mathbf{A})^{-1}\,\mathbf{B} = \frac{\frac{1}{L}\left(s+\frac{1}{R_1 C}\right)}{s\left(s+\frac{1}{RC}\right)+\frac{1}{LC}} = \frac{\frac{1}{L}\left(s+\frac{1}{R_1 C}\right)}{\det(s\mathbf{I}-\mathbf{A})} \tag{2.15}$$

where $1/R := 1/R_1 + 1/R_2$.

2.2.1 Mason's Loop Rule

Mason's formula for the transfer function g_{ij} between the jth input signal and the ith output signal of a linear time-invariant multivariable system reads

$$g_{ij} = \frac{1}{\Delta}\sum_k G_k \Delta_k \tag{2.16}$$

where Δ denotes the so-called graph determinant or the system determinant, which is equal to the characteristic polynomial $\det(s\mathbf{I}-\mathbf{A})$ divided by its highest power of s.

The symbol G_k represents the gain of the kth signal path between two nodes in the signal flow graph representing an input and an output variable, respectively. The signal path gain is the product of all influence coefficients along the path.

The influence coefficient of a linear 1-port resistor is either its parameter R or $1/R$ depending on its causality. Likewise, the influence coefficient of a linear 1-port energy storage element of parameter θ is either $1/(s\theta)$ for integral causality or $s\theta$ in the case of derivative causality after Laplace transform of its constitutive equation.

The symbol Δ_k denotes the determinant of the reduced signal flow graph that results if the kth signal path is expunged with all its nodes. The sum extends over all signal paths from the node of the jth input to the ith output. Given the signal paths gains, the graph determinant Δ is obtained as the sum

$$\Delta = 1 - \sum_i G_i + \sum_{i,j} G_i G_j - \sum_{i,j,k} G_i G_j G_k + \ldots \tag{2.17}$$

In this sum, the term $\sum_i G_i$ is the sum of gains of all individual loops in the signal flow graph. The sum $\sum_{i,j} G_i G_j$ extends over all products of loop gains of two loops that do not touch, in other words, that do not share a node. Since loops in a signal flow graph usually often touch one another, there are only few pairs of non-touching loops among all loops. Similarly, the term $\sum_{i,j,k} G_i G_j G_k$ is the sum of all products of loop gains for sets of three non-touching loops (Different sets of three loops may share two loops but must have a distinct third loop). In practice, such sets are rather rare. For the kth path between the nodes of an input and an output variable, the reduced signal graph is obtained by expunging all loops that touch the kth path. For computation of the denominator, Δ, *all* loops must be considered. Among these, all loops not touching the kth path between the nodes of the input and the output variable contribute to the determinant of the *reduced* signal flow graph, Δ_k, in the numerator.

2.2.2 Application of Mason's Loop Rule Directly on a Causal Bond Graph

The gain $G^{(0)}$ of a *direct* causal path from an input source to a detector via transformers and/or gyrators is

$$G^{(0)} = (-1)^{n_0+n_1} \prod_i m_i^{k_i} \prod_j r_j^{l_j} \tag{2.18}$$

where n_0 and n_1 denote the total number of changes of the reference power flow direction at 0 junctions and 1 junctions, respectively along the causal path, m_i is the modulus of the ith transformer, and r_j the ratio of the rth gyrator. Depending on the causality at these two port elements, the value of the exponents k_i and l_j is either $+1$ or -1.

The gain $G^{(1)}$ of an *indirect* causal path from an input source to a detector has to take into account the gain G_e of all elements passed on the way from the input signal to the output signal.

$$G^{(1)} = G^{(0)} \cdot \prod_e G_e \tag{2.19}$$

The gain $G^{(2)}$ of an indirect causal path between a resistor and a storage element or between two storage elements of different type reads

$$G^{(2)} = (-1)^{n_0+n_1} \prod_i (m_i^{k_i})^2 \prod_j (r_j^{l_j})^2 \prod_e G_e \tag{2.20}$$

The above result (2.15) may be obtained directly from the BG in Fig. 2.8b by identifying the causal path from the effort source MSe : E to the flow detector Df : y, causal paths between a resistor and a storage element, and between two storage elements of different type. The second step is to compute the gain for each of the identified causal paths. Finally, the gains are used in Mason's loop rule.

First, there is an indirect causal path p_0 from the input source to the detector via the inductor:

$$p_0 : \quad MSe : E(t) \rightharpoonup 1_1 \rightharpoonup 0_2 \rightharpoonup 1_2 \rightharpoonup I : L \rightharpoonup Df : y$$

Its gain is equal to $1/sL$. Furthermore, there are the following causal paths:

$$\begin{aligned}
p_1 : &\quad C : C \rightharpoonup R : R_1 \\
p_2 : &\quad C : C \rightharpoonup 0_1 \rightharpoonup 1_1 \rightharpoonup 0_2 \rightharpoonup R : R_2 \\
p_3 : &\quad C : C \rightharpoonup 0_1 \rightharpoonup 1_1 \rightharpoonup 0_2 \rightharpoonup 1_2 \rightharpoonup I : L
\end{aligned}$$

Accordingly, the graph determinant reads

$$\begin{aligned}
\Delta &= 1 - \sum G_i \\
&= 1 - \left(-\frac{1}{R_1 sC}\right)\left(-\frac{1}{R_2 sC}\right)\left(-\frac{1}{sCsL}\right) \\
&= \frac{1}{s^2}\left[s\left(s + \frac{1}{RC}\right) + \frac{1}{LC}\right] = \frac{1}{s^2}\det(s\mathbf{I} - \mathbf{A})
\end{aligned} \tag{2.21}$$

As the causal paths p_2 and p_3 touch the causal path p_0 from the input source to the detector, they are not taken into account in the computation of the *reduced* graph determinant Δ_1.

$$\Delta_1 = 1 - \left(-\frac{1}{R_1 sC}\right) = \frac{1}{s}\left(s + \frac{1}{R_1 C}\right) \tag{2.22}$$

Inserting all partial results into Mason's formula yields

$$\frac{\mathscr{L}y}{\mathscr{L}E} = \frac{\dfrac{1}{sL}\dfrac{1}{s}\left(s + \dfrac{1}{R_1 C}\right)}{\dfrac{1}{s^2}\det(s\mathbf{I} - \mathbf{A})} \tag{2.23}$$

Example: DC Motor Drive

From the schematic of a DC motor drive in Fig. 2.9, the causal BG in Fig. 2.10 can be obtained.

The BG model in Fig. 2.10 assumes that the shaft has a finite stiffness k and that the motor drives the mechanical load with an inertia J_l against an external torque $T(t)$ and friction denoted by the coefficient r. The motor parameters are the armature resistance R_a, the inductance I_a, the torque constant k_m, the rotor inertia I_m, and the friction coefficient r_m. The torsion moment on the shaft is denoted by M_s.

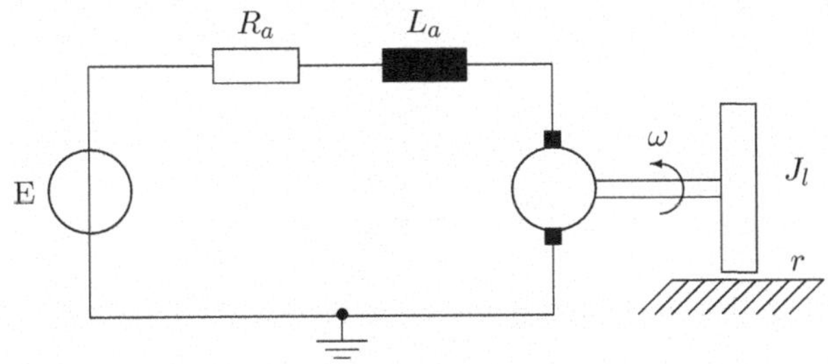

Fig. 2.9 Schematic of a DC motor drive

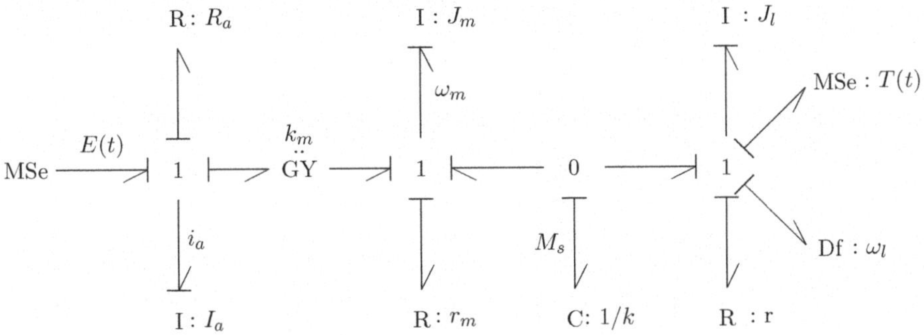

Fig. 2.10 BG of a DC motor drive

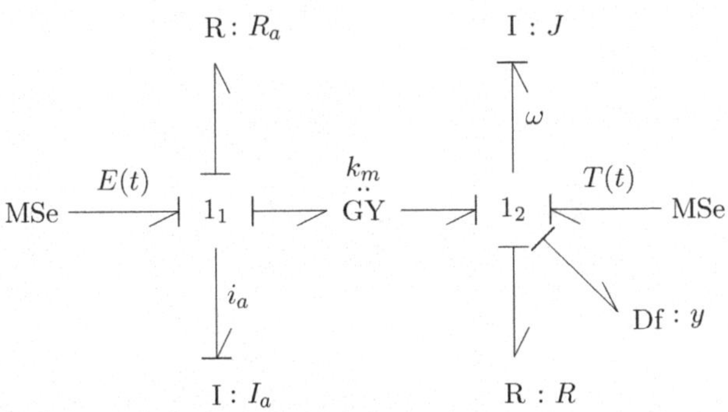

Fig. 2.11 Simplified BG model of a DC motor drive

The model can be simplified if the shaft is considered rigid. Then the two mechanical inertia can be merged into one. Likewise, the two friction effects can be represented by a single resistance. Figure 2.11 depicts the simplified model.

In the following, Mason's loop rule shall be used to determine the response to setpoint changes, i.e. the transfer function $\mathscr{L}y/\mathscr{L}E$. From the BG in Fig. 2.11 it can be seen that there is an indirect causal path p_0 from the voltage source MSe : E to the flow detector Df : y.

$$p_0: \ MSe : E \rightharpoonup 1_1 \rightharpoonup I : I_a \rightharpoonup 1_1 \rightharpoonup GY : k_m \rightharpoonup 1_2 \rightharpoonup I : J \rightharpoonup 1_2 \rightharpoonup Df : y$$

Its gain is

$$G^{(0)} = k_m \frac{1}{sI_a} \frac{1}{sJ} \tag{2.24}$$

Furthermore, the following causal paths are to be taken into account:

$$\begin{aligned} p_1: &\quad R : R_a \rightharpoonup 1_1 \rightharpoonup I : I_a \\ p_2: &\quad I : I_a \rightharpoonup 1_1 \rightharpoonup GY : k_m \rightharpoonup 1_2 \rightharpoonup I : J \\ p_3: &\quad R : R \rightharpoonup 1_2 \rightharpoonup I : J \end{aligned}$$

Accordingly, the computation of the graph determinant in this case yields

$$\begin{aligned} \Delta &= 1 - \sum_i G_i + \sum_{i,j} G_i G_j \\ &= 1 - \left(-\frac{R_a}{sI_a}\right) - \left(-\frac{R}{sJ}\right) - \left(-\frac{k_m^2}{s^2 I_a J}\right) + \frac{R_a}{sI_a} \cdot \frac{R}{sJ} \end{aligned} \tag{2.25}$$

As all above causal paths touch the causal path from the input source to the detector, the reduced graph determinant Δ_1 is just equal to 1. Hence, the transfer function to be determined becomes

$$\begin{aligned} g_{11}(s) = \frac{\mathscr{L}y}{\mathscr{L}E} &= \frac{\left(\frac{k_m}{I_a J s^2}\right) \cdot 1}{1 + \frac{k_m^2}{I_a J s^2} + \frac{R_a}{I_a s} + \frac{R}{Js} + \frac{R_a}{I_a s} \cdot \frac{R}{Js}} \\ &= \frac{\frac{k_m}{I_a J}}{s^2 + \left(\frac{R_a}{I_a} + \frac{R}{J}\right) s + \frac{R_a R}{I_a J} + \frac{k_m^2}{I_a J}} \\ &= \frac{\frac{k_m}{I_a J}}{\left(s + \frac{R_a}{I_a}\right)\left(s + \frac{R}{J}\right) + \frac{k_m^2}{I_a J}} \end{aligned} \tag{2.26}$$

The same result can be obtained by means of the state space model matrices **A**, **B**, **C**. From the BG in Fig. 2.11, the following state space model equations can be derived:

$$\underbrace{\frac{d}{dt}\begin{bmatrix} i_a \\ \omega \end{bmatrix}}_{\boldsymbol{x}(t)} = \underbrace{\begin{bmatrix} -\frac{R_a}{I_a} & \frac{k_m}{I_a} \\ \frac{k_m}{J} & -\frac{R}{J} \end{bmatrix}}_{\mathbf{A}} \underbrace{\begin{bmatrix} i_a \\ \omega \end{bmatrix}}_{\boldsymbol{x}(t)} + \underbrace{\begin{bmatrix} \frac{1}{I_a} & 0 \\ 0 & -\frac{1}{J} \end{bmatrix}}_{\mathbf{B}} \underbrace{\begin{bmatrix} E \\ T \end{bmatrix}}_{\boldsymbol{u}(t)} \tag{2.27a}$$

$$y(t) = \underbrace{\begin{bmatrix} 0 & 1 \end{bmatrix}}_{\mathbf{C}} \underbrace{\begin{bmatrix} i_a \\ \omega \end{bmatrix}}_{\boldsymbol{x}(t)} \tag{2.27b}$$

$$\begin{aligned} \mathbf{G}(s) &= \mathbf{C}\,(s\mathbf{I} - \mathbf{A})^{-1}\,\mathbf{B} \\ &= \begin{bmatrix} 0 & 1 \end{bmatrix} \frac{1}{(s + \frac{R_a}{I_a})(s + \frac{R}{J}) + \frac{k_m^2}{I_a J}} \begin{bmatrix} s + \frac{R}{J} & -\frac{k_m}{I_a} \\ \frac{k_m}{J} & s + \frac{R_a}{I_a} \end{bmatrix} \begin{bmatrix} \frac{1}{I_a} & 0 \\ 0 & -\frac{1}{J} \end{bmatrix} \\ &= \frac{1}{\det(s\mathbf{I} - \mathbf{A})} \begin{bmatrix} \frac{k_m}{I_a J} & -\frac{1}{J}\left(s + \frac{R_a}{I_a}\right) \end{bmatrix} = \begin{bmatrix} G_{11} & G_{12} \end{bmatrix} \end{aligned} \tag{2.28}$$

The application of Mason's loop rule directly on a causal bond graph is practically limited to bond graph models of small to medium size. For large bond graph models which may have been composed of sub-models from a library, the derivation of transfer functions directly from the causally completed bond graph according to Mason's loop rule requires the identifications and examination of many causal paths so that software support will be needed.

2.3 Bond Graphs and Block Diagrams

In control engineering, the traditional graphical model representation are block diagrams. Commonly used software such as Matlab® or Scilab has a graphical user interface that supports the development of block diagrams. The previous sections have demonstrated that model properties such as structural controllability and structural observability can be checked directly on a causal bond graph and that transfer functions can be directly derived from a causal bond graph. Once a causal bond graph model has been developed by starting from a system schematic and by following the steps of a graphical procedure, the bond graph may be systematically transformed into a block diagram, if further model processing shall be performed by software such as Matlab® or Scilab that does not accept bond graphs.

A systematic transformation of causal BGs is feasible because each causal bond can be replaced by an effort signal and a flow signal in opposite direction and the constitutive equations of each BG element can be represented by a block diagram. This shall be illustrated by a small example. Figure 2.12 shows a simple LC circuit and its corresponding BG. Figure 2.13 indicates the step-by-step transformation of the BG in Fig. 2.12b into a block diagram.

As early as 1977, van Dixhoorn [4] noticed that an alphanumeric description of a block diagram in the form of a list of assignment statements can be directly obtained from a causal BG if all storage elements are in integral causality and if there are no causal paths between resistors. To that end, all bonds are enumerated. For each BG element different from a 0- or 1-junctions, the inputs are traced back along causal paths until they can be expressed by the outputs of other BG elements. By that

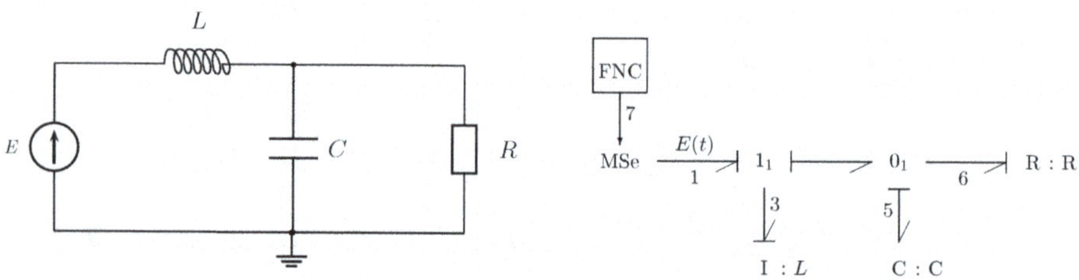

(a) Simple LC circuit (b) BG of the LC circuit

Fig. 2.12 Simple LC circuit and corresponding BG. (**a**) Simple LC circuit. (**b**) BG of the LC circuit

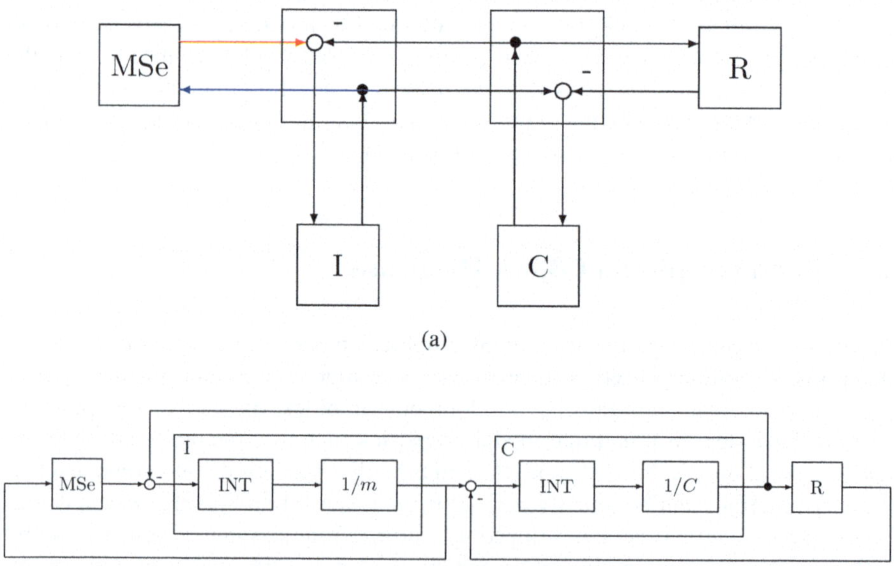

(a)

(b)

Fig. 2.13 Transformation of the BG in Fig. 2.12b into a block diagram

way, the summation of efforts or flows at a junction is transferred into the element that causally dominates the junction. At that time, the assignment statements were written in the short form

```
<Block number>  <Block type>  <Input_1>,...,<Input_n>.
```

and the result was called a *structure table*.

From the causal BG in Fig. 2.12b the structure table in Fig. 2.14 can be derived. Its graphical representation is the block diagram in Fig. 2.15.

If the available software does not provide functional blocks for the I- and C-element, their functionality has to be built by means of standard signal blocks as depicted in Fig. 2.13b.

If there is a causal path between two resistors as in the modified BG in Fig. 2.16, then the corresponding block diagram contains an *algebraic signal loop* as depicted

Fig. 2.14 Structure table of the causal bond graph in Fig. 2.12b

1	MSE	7	
3	I	1	-5
5	C	3	-6
6	R	5	
7	FNC		

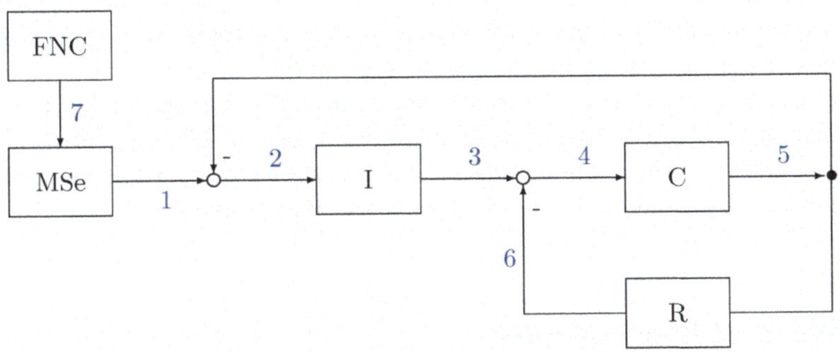

Fig. 2.15 Block diagram given by the structure table in Fig. 2.14

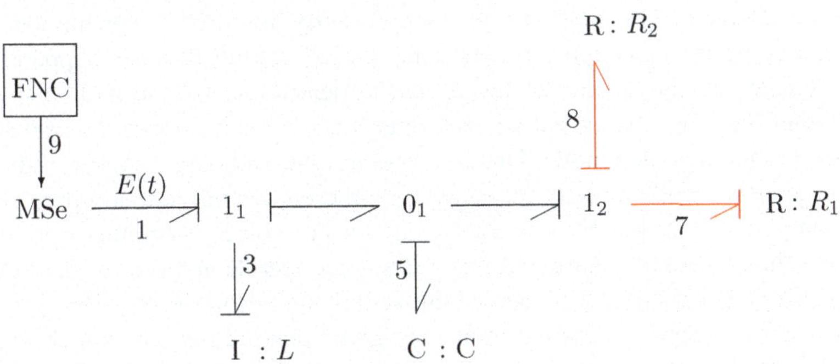

Fig. 2.16 BG with a causal path between two resistors

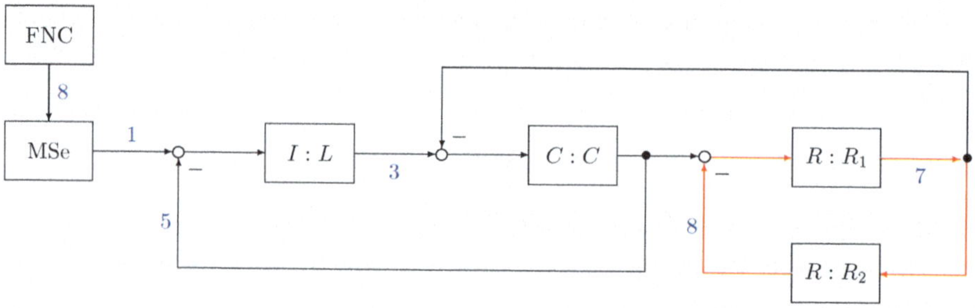

Fig. 2.17 Block diagram obtained from the BG in Fig. 2.16

in Fig. 2.17. The causal path R_1 — R_2 in the BG highlighted in red and its corresponding signal loop in the block diagram represent two algebraic equations for the unknowns f_7, e_8.

$$f_7 = \frac{1}{R_1}(e_5 - e_8) \tag{2.29}$$

$$e_8 = R_2\, f_7 \tag{2.30}$$

Today's modelling and simulation programs solve such subsystems of algebraic equations automatically either symbolically if they are linear with respect to the unknowns or numerically by iteration at the present time instant.

Note that a causal BG can be transformed into a block diagram, but not every block diagram can be transformed into a causal BG. The reason is that block diagrams may represent any functional relation between signals, while the equations derived from BGs should comply with first physical principles.

2.4 Bicausal Bond Graphs

The following chapters of this book consider three major engineering tasks that are closely related. The first task is to constantly monitor the behaviour of a controlled engineering system, to detect and isolate a fault that has happened due to disturbances, or due to tear and wear and to determine its kind and its severity. Once a fault has been diagnosed an important question is how long the system can still operate in a degraded mode. Due to costs, weight, and other reasons, redundant hardware may be available only for some system components so that it may happen that a component that has become faulty cannot be replaced. Another option may be to alter the control to ensure that the system can operate at reduced functionality and can still produce sufficiently acceptable output signals in the presence of a fault. This last task means to determine inputs for given desired outputs, which requires an inversion of the system model. An aim of this book is to show how bond graph modelling can support these tasks.

As early as 1995, Gawthrop [6] extended the computational causality concept of bond graph modelling by introducing the so-called *bicausal* bond graphs. They may be used for

- state estimation,
- parameter estimation,
- fault detection, and
- system inversion.

These tasks are essential in various applications of model-based control, for instance, in feedforward control.

The assignment of computational causality means that the signals associated with the two conjugate power variables effort and flow do have opposite orientation. This constraint is lifted. Effort and flow at *bicausal* bonds may have the *same* direction. This is represented by splitting the perpendicular stroke at one bond end and by attaching half causal strokes at both bond ends as shown in Fig. 2.18. In addition, Gawthrop also introduced a so-called *Source Sensor* element denoted by SS that can take bicausality. In bicausal BGs a source and detector at the ends of a bicausal path are replaced by SS elements. By following a bicausal path inverse system behaviour can be determined. Table 2.1 lists the possible causal patterns at SS elements. The admissible causal patterns at resistor and storage element ports, and at the ports of the junction structure elements in a bicausal BG are given in Table 2.2. Note that at 0- and 1-junctions only two bonds are allowed to take bicausality.

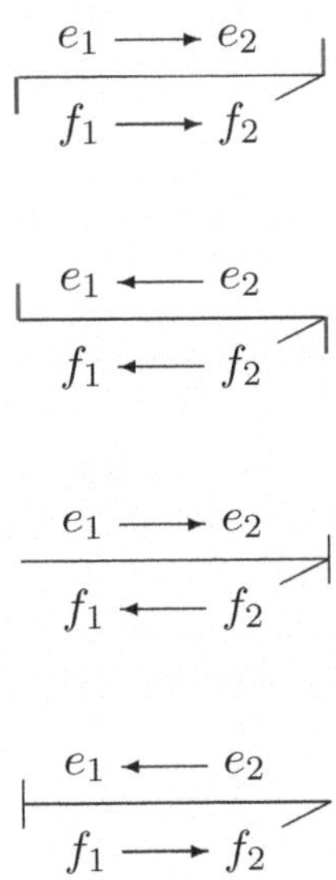

$e_2 := e_1$, $f_2 := f_1$

$e_1 := e_2$, $f_1 := f_2$

$e_2 := e_1$, $f_1 := f_2$

$e_1 := e_2$, $f_2 := f_1$

Fig. 2.18 Bicausal bonds

Table 2.1 Causal patterns for the source sensor element SS (cf. Ngwompo and Gawthrop, 1999)

Causal pattern	Nature of the SS element	
SS 	Effort source, flow sensor	Se element
SS 	Flow source, effort sensor	Sf element
SS 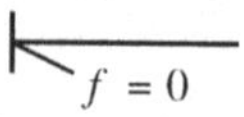	Zero flow source, effort sensor	De element: Effort detector
SS 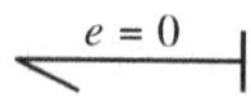	Zero effort source, flow sensor	Df element: Flow detector
SS 	Flow source, effort source	
SS 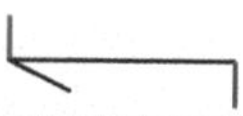	Flow sensor, effort sensor	

In the section on bicausal bond graph-based system inversion, the notion of the length of a causal path is needed.

Definition 2.4 (Length of a causal path) The length $l(p)$ of a causal path p from a source, a storage element, or a resistor to a detector, a storage element, or a resistor is the number $n_I(p)$ of storage elements in integral causality minus the number $n_D(p)$ of storage elements in derivative causality encountered on the causal path.

Definition 2.5 (Length associated with a set P of disjoint causal paths) [13] The length associated with a set P of disjoint causal paths p_i denoted as $\omega(P)$ is the sum of the lengths of all causal paths in the set:

$$\omega(P) = \sum_i l(p_i) \tag{2.31}$$

2.5 Parameter Estimation Based on Bicausal Bond Graphs

As parameter estimation is one way to detect and to isolate faults, parameter estimation by means of a bicausal bond graph is illustrated on the example of the often considered two-tank system and is used in Sect. 4.5.1 for providing values of an unknown degradation trend in failure prognosis. For other applications of bicausal bond graphs it is referred to [6].

Table 2.2 Admissible causal patterns at storage elements, resistors, and junction structure elements

Bicausal pattern	Assignment statements
	$C := q/e$
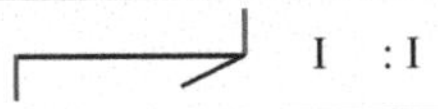	$I := p/f$
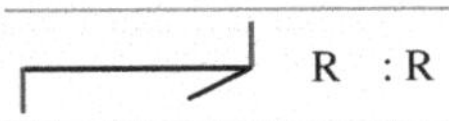	$R := e/f$
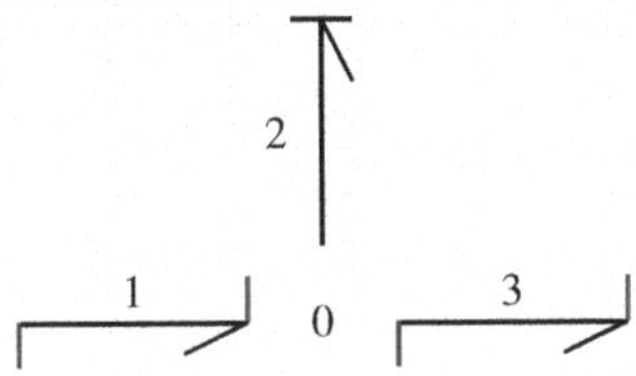	$e_2 := e_1$ $e_3 := e_1$ $f_3 := f_1 - f_2$
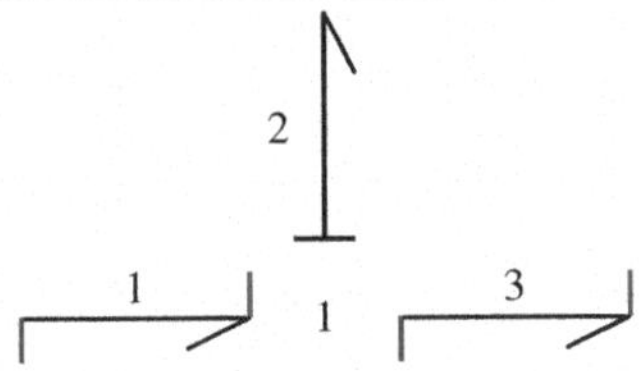	$f_2 := f_1$ $f_3 := f_1$ $e_3 := e_1 - e_2$
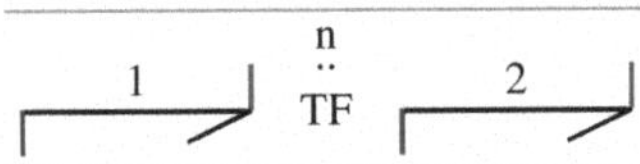	$e_2 := e_1/n$ $f_2 := n \times f_2$
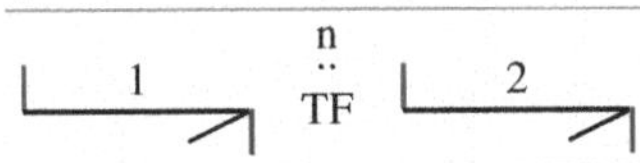	$e_1 := n \times e_2$ $f_1 := f_2/n$
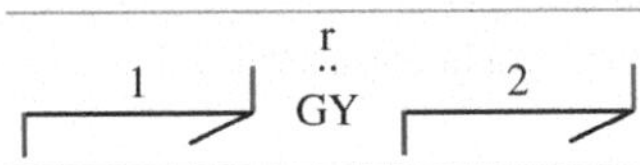	$e_2 := r \times f_1$ $f_2 := e_1/r$
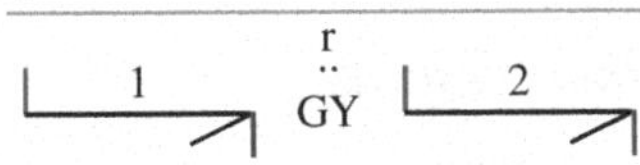	$e_1 := r \times f_2$ $f_1 := e_2/r$

Example 1: Parameter Estimation Applied to a Two-Tank System

Figure 2.19 displays a schematic of a two-tank system.

A constant flow pump delivers a volume flow Q_p into Tank 1 while the outflow Q_o from Tank 2 is measured. Suppose that the valve between the two tanks is going to be partially blocked due to contamination and that its resistance shall be estimated. All other elements are assumed to be fault free.

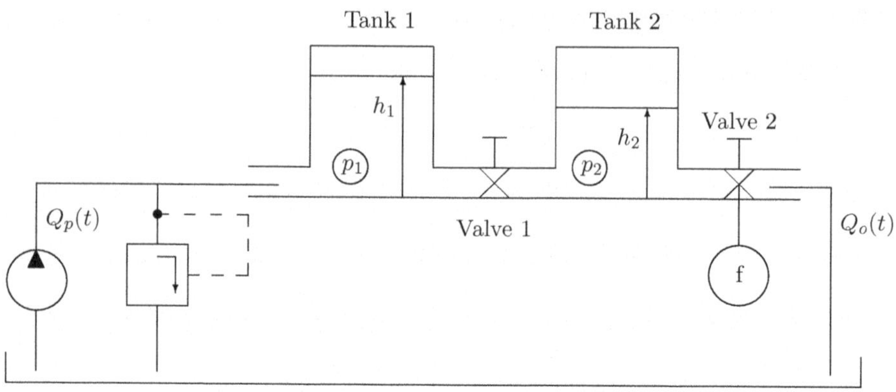

Fig. 2.19 Schematic of a two tanks system

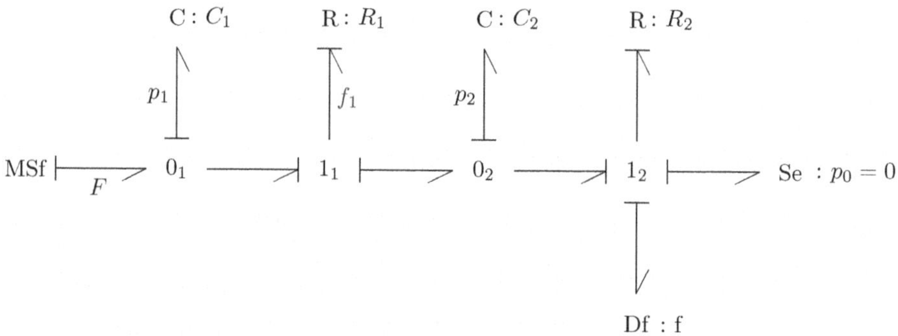

Fig. 2.20 BG of the two tanks system in Fig. 2.19

Controllability and Observability

First, it is checked that the system with the flow pump and the flow sensor is completely state controllable and completely state observable. Figure 2.20 displays a BG with preferred integral causality of the system.

For simplicity, linear resistors are assumed for the two valves. From the causal BG in Fig. 2.20 the following state space model can be directly derived.

$$\begin{bmatrix} \dot{p}_1 \\ \dot{p}_2 \end{bmatrix} = \underbrace{\begin{bmatrix} -\dfrac{1}{R_1 C_1} & \dfrac{1}{R_1 C_1} \\ \dfrac{1}{R_1 C_2} & -\dfrac{1}{R C_2} \end{bmatrix}}_{\mathbf{A}} \underbrace{\begin{bmatrix} p_1 \\ p_2 \end{bmatrix}}_{\boldsymbol{x}} + \underbrace{\begin{bmatrix} 1 \\ 0 \end{bmatrix}}_{\mathbf{B}} F \tag{2.32}$$

$$f = \underbrace{\begin{bmatrix} 0 & \frac{1}{R_2} \end{bmatrix}}_{\mathbf{C}} \boldsymbol{x} \tag{2.33}$$

where $1/R := 1/R_1 + 1/R_2$, $F = Q_p$, and $f = Q_o$.

The controllability and the observability matrix read

$$\mathscr{C} = [\mathbf{B}|\mathbf{AB}] = \left[\begin{array}{c|c} 1 & -\frac{1}{R_1 C_1} \\ 0 & -\frac{1}{R_1 C_2} \end{array}\right] \tag{2.34}$$

$$\mathscr{O} = \begin{bmatrix} \mathbf{C} \\ \mathbf{CA} \end{bmatrix} = \begin{bmatrix} 0 & \frac{1}{R_2} \\ -\frac{1}{R_1 R_2 C_2} & -\frac{1}{R_2 R C_2} \end{bmatrix} \tag{2.35}$$

Both matrices are of rank two.

Estimating the Resistance of Valve 1

Now, Fig. 2.21 shows a BG with a bicausal path from the flow detector $\mathrm{Df} : f$ replaced by a source sensor SS element to the resistor $R : R_1$. The flow f of the source sensor is propagated by junction 1_2. The sum of efforts at 1_2 is propagated by junction which results in derivative causality at capacitor $C : C_2$, while the sum of flows at junction 0_2 is propagated by junction 1_2 and thus an input into the resistor $\mathrm{R} : R_1$. Finally, the difference of the effort p_1 and the effort propagated by junction 0_2 is also an input into the resistor $R : R_1$. Given the effort and the flow into that element, its parameter R_1 can be estimated.

From the bicausal BG in Fig. 2.21, the following equations can be derived:

$$1_2 : \quad p_2 = 0 + p_{R_2} + p_0 \tag{2.36}$$

$$R : R_2 : \quad f = \frac{1}{R_2} p_{R_2} = \frac{1}{R_2} p_2 \tag{2.37}$$

$$C : C_2 : \quad f_2 = C_2 \dot{p}_2 = C_2 R_2 \dot{f} \tag{2.38}$$

$$0_2 : \quad f_{R_1} = f_2 + f = C_2 R_2 \dot{f} + f \tag{2.39}$$

$$R : R_1 : \quad f_{R_1} = \frac{1}{p_{R_1}} \tag{2.40}$$

$$1_1 : \quad p_{R_1} = p_1 - p_2 \tag{2.41}$$

$$0_1 : \quad \dot{p}_1 = \frac{1}{C_1}(F - f_{R_1}) \tag{2.42}$$

Differentiation of Equations 2.36, 2.40, and 2.41 with respect to time gives

$$\dot{f}_{R_1} = \frac{1}{R_1}\dot{p}_{R_1} = \frac{1}{R_1}(\dot{p}_1 - \dot{p}_2) \tag{2.43}$$

$$= \frac{1}{R_1}\left[\frac{1}{C_1}(F - f_{R_1}) - R_2\dot{f}\right] \tag{2.44}$$

Substituting f_{R_1} in (2.44) finally yields an equation that relates the known pump flow F and the measured output flow f and enables to estimate resistance R_1:

$$C_2R_2\ddot{f} + \dot{f} = \frac{1}{R_1}\left[\frac{1}{C_1}(F - C_2R_2\dot{f} - f) - R_2\dot{f}\right] \tag{2.45}$$

As can be seen from (2.45), time derivatives of the measured outflow f are needed to estimate the resistance R_1, which requires to smooth the noisy measurement by means of a low pass filter before it is differentiated with respect to time.

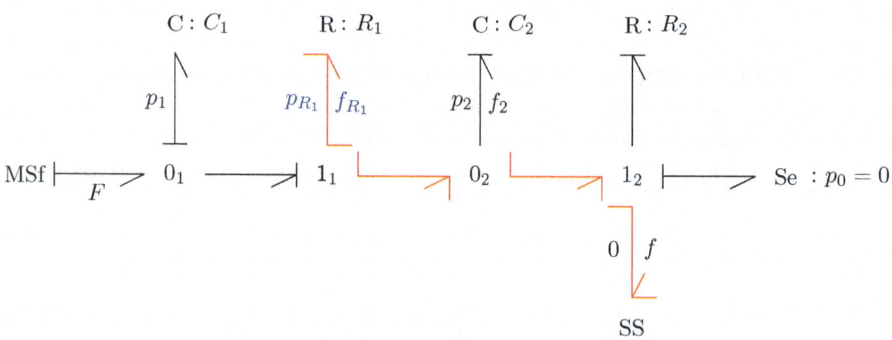

Fig. 2.21 Bicausal BG of the two tanks system in Fig. 2.19

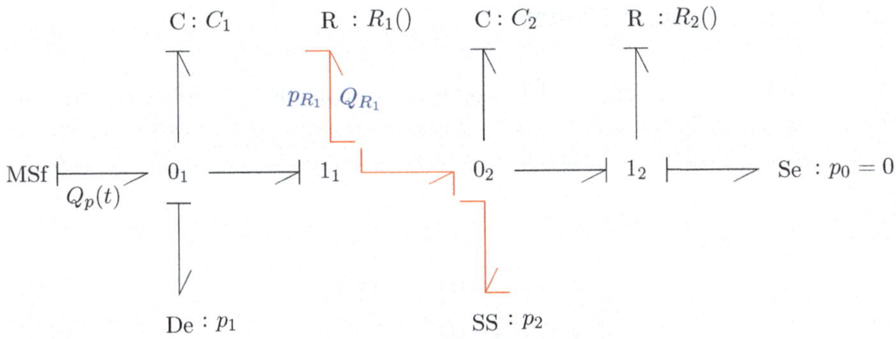

Fig. 2.22 Bicausal BG of the two tanks system with two pressure sensors

Example 2: Two-Tank System with Two Pressure Sensors

Example 2 assumes that for each of the two tanks a pressure (level) sensor is available. Again, it is assumed that Valve 1 connecting the two tanks is partially blocked and that all other components are fault free. From the bicausal BG in Fig. 2.22, an equation can be easily deduced that determines the discharge coefficient c_{d_1} of Valve 1.

In the bicausal BG in Fig. 2.22, two effort detectors represent the two pressure sensors. Detector De : p_1 is in inverted causality. It delivers the measured pressure p_1 into the model. Detector De : p_2 has been replaced by source sensor (SS) which is the starting point of a bicausal path to the nonlinear resistor R : $R_1()$ of Valve 1. From the bicausal BG, effort and flow into the resistor R : $R_1()$ can be deduced.

$$\begin{aligned} Q_{R_1} &= C_2 \dot{p}_2 + Q_{R_2} \\ &= C_2 \dot{p}_2 + c_{d_2} A_{20} \sqrt{\frac{2}{\rho} p_2} \end{aligned} \tag{2.46}$$

$$p_{R_1} = p_1 - p_2 \tag{2.47}$$

Substituting the expressions for both power variables into the constitutive equation of the valve gives a relation for the discharge coefficient c_{d_1} to be determined.

$$C_2 \dot{p}_2 + c_{d_2} A_{20} \sqrt{\frac{2}{\rho} p_2} = c_{d_1} A_{12} \operatorname{sign}(p_1 - p_2) \sqrt{\frac{2}{\rho} |p_1 - p_2|} \tag{2.48}$$

Inputs into (2.48) are the appropriately filtered measurements of pressures p_1 and p_2. If Valve 1 gets partially clogged, the value of its discharge coefficient will be different from the one of the fully open valves as of some time instant.

2.6 Inverse System Models

Inverse system models are used in application areas such as robot control, for actuator sizing, in fault diagnosis, or fault tolerant control. There is a whole body of mathematical literature on system inversion, for instance, [2, 15, 16, 18, 19] to name a few.

Let

$$\dot{\boldsymbol{x}}(t) = \mathbf{A}\boldsymbol{x}(t) + \mathbf{B}\boldsymbol{u}(t) \tag{2.49a}$$

$$\boldsymbol{y}(t) = \mathbf{C}\boldsymbol{x}(t) \tag{2.49b}$$

be a state space model with $\boldsymbol{x}(t) \in \mathbb{R}^n$, $\boldsymbol{u}(t) \in \mathbb{R}^m$, $\boldsymbol{y}(t) \in \mathbb{R}^p$, and constant coefficient matrices $\mathbf{A} \in \mathbb{R}^{n\times n}$, $\mathbf{B} \in \mathbb{R}^{n\times m}$, $\mathbf{C} \in \mathbb{R}^{p\times n}$.

Definition 2.6 (Forward bond graph model) A bond graph in preferred integral causality representing the state space model (2.49), i.e. the dynamics of the states $\boldsymbol{x}(t)$ and the outputs $\boldsymbol{y}(t)$ in terms of the inputs $\boldsymbol{u}(t)$ is called a *forward bond graph* model.

Assume that the system is numerically completely state controllable and completely state observable and that the matrix $(s\mathbf{I}-\mathbf{A})$ is non-singular. Let $\mathbf{G}(s)$, $s \in \mathbb{C}$, denote the $p \times m$ transfer function matrix $\mathbf{G}(s) := \mathbf{C}(s\mathbf{I} - \mathbf{A})^{-1}\mathbf{B}$.

Definition 2.7 (left inverse) A *left inverse* is a matrix $\mathbf{G}_L(s)$ for which

$$\mathbf{G}_L(s)\mathbf{G}(s) = \mathbf{I}_m \tag{2.50}$$

with $p \geq m$ and $s \in \mathbb{C}$.

Definition 2.8 (right inverse) A *right inverse* is a matrix $\mathbf{G}_R(s)$ that is defined in a similar manner.

$$\mathbf{G}(s)\mathbf{G}_R(s) = \mathbf{I}_p \tag{2.51}$$

with $m \geq p$ and $s \in \mathbb{C}$.

Let $\mathbf{M}$ be the matrix

$$\mathbf{M}(s) := \begin{bmatrix} \mathbf{A} - s\mathbf{I} & \mathbf{B} \\ \mathbf{C} & \mathbf{0} \end{bmatrix} \tag{2.52}$$

Theorem 2.3 (Moylan [10]) *The forward model* (2.49) *is invertible if and only if rank*$\mathbf{M}(\lambda) = n + m$ *for some* $\lambda \in \mathbb{R}$.

Theorem 2.4 (Moylan [10]) *There exists an asymptotically stable left inverse for the above system* (2.49) *if and only if rank*$M(\lambda) = n + m$ *for all* $\lambda \in \mathbb{C}$ *with* $Re(\lambda) \geq 0$.

Another necessary and sufficient invertibility criterion has been given by Sain and Massey [14].

Let

$$\dot{x}(t) = \mathbf{A}x(t) + \mathbf{B}u(t) \quad x(0) = \mathbf{0} \tag{2.53a}$$

$$y(t) = \mathbf{C}x(t) + \mathbf{D}u(t) \tag{2.53b}$$

a state space model with $x(t) \in \mathbb{R}^n$, $u(t) \in \mathbb{R}^m$, $u(t) \in \mathbb{R}^m$, and constant coefficient matrices $\mathbf{A}, \mathbf{B}, \mathbf{C}, \mathbf{D}$ of appropriate dimensions. Furthermore, let $\mathbf{M}_0 := [\mathbf{D}]$ and

$$\mathbf{M}_k = \begin{bmatrix} \mathbf{D} & \mathbf{0} & \mathbf{0} & \dots & \mathbf{0} \\ \mathbf{CB} & \mathbf{D} & \mathbf{0} & \dots & \mathbf{0} \\ \mathbf{CAB} & \mathbf{CB} & \mathbf{D} & \dots & \mathbf{0} \\ \vdots & \vdots & \vdots & \ddots & \vdots \\ \mathbf{CA}^{k-1}\mathbf{B} & \mathbf{CA}^{k-2}\mathbf{B} & \mathbf{CA}^{k-3}\mathbf{B} & \dots & \mathbf{D} \end{bmatrix} \tag{2.54}$$

Theorem 2.5 (Sain and Massey [14]) *The system* (2.53) *is invertible if and only if*

$$\text{rank}\mathbf{M}_n \; - \; \text{rank}\mathbf{M}_{n-1} = m \tag{2.55}$$

Now, assume $p = m$ and that the above forward model (2.49) is invertible. Then an inverse model is given by the time domain equations

$$\begin{bmatrix} \mathbf{I} & \mathbf{0} \\ \mathbf{0} & \mathbf{0} \end{bmatrix} \dot{z} = \begin{bmatrix} \mathbf{A} - \mathbf{BC} & \mathbf{B} \\ -\mathbf{C} & \mathbf{0} \end{bmatrix} z + \begin{bmatrix} \mathbf{B} \\ \mathbf{I} \end{bmatrix} y \tag{2.56a}$$

$$u = \begin{bmatrix} -\mathbf{C} \; \mathbf{I} \end{bmatrix} z + y \tag{2.56b}$$

where $z^T := [x^T \, u^T]^T$.

Let $\hat{\mathbf{G}}(s)$ be the transfer function matrix of the inverse model. Then

$$\hat{\mathbf{G}}(s) = \mathbf{G}^{-1}(s) \tag{2.57}$$

Laplace transform of the equations of the inverse model yields

$$\begin{bmatrix} s\mathbf{I} - \mathbf{A} + \mathbf{BC} & -\mathbf{B} \\ \mathbf{C} & \mathbf{0} \end{bmatrix} Z(s) = \begin{bmatrix} \mathbf{B} \\ \mathbf{I} \end{bmatrix} \mathbf{Y} \tag{2.58a}$$

$$U(s) = \begin{bmatrix} -\mathbf{C} \; \mathbf{I} \end{bmatrix} \mathbf{Z} + Y \tag{2.58b}$$

Substituting (2.58) into (2.58) gives

$$U(s) = \underbrace{\left\{ [-\mathbf{C}\ \mathbf{I}] \begin{bmatrix} s\mathbf{I} - \mathbf{A} + \mathbf{BC} & -\mathbf{B} \\ \mathbf{C} & \mathbf{0} \end{bmatrix}^{-1} \begin{bmatrix} \mathbf{B} \\ \mathbf{I} \end{bmatrix} + \mathbf{I} \right\}}_{\hat{\mathbf{G}}(s)} \mathbf{Y} \tag{2.59}$$

$$\begin{aligned}
\hat{\mathbf{G}}(s) &= [-\mathbf{C}\ \mathbf{I}] \left\{ \begin{bmatrix} s\mathbf{I} - \mathbf{A} & -\mathbf{B} \\ \mathbf{C} & \mathbf{0} \end{bmatrix} \begin{bmatrix} \mathbf{I} & \mathbf{0} \\ -\mathbf{C} & \mathbf{I} \end{bmatrix} \right\}^{-1} \begin{bmatrix} \mathbf{B} \\ \mathbf{I} \end{bmatrix} + \mathbf{I} \\
&= [-\mathbf{C}\ \mathbf{I}] \begin{bmatrix} \mathbf{I} & \mathbf{0} \\ \mathbf{C} & \mathbf{I} \end{bmatrix} \begin{bmatrix} s\mathbf{I} - \mathbf{A} & -\mathbf{B} \\ \mathbf{C} & \mathbf{0} \end{bmatrix}^{-1} \begin{bmatrix} \mathbf{B} \\ \mathbf{I} \end{bmatrix} + \mathbf{I} \\
&= [\mathbf{0}\ \mathbf{I}] \begin{bmatrix} \mathbf{0} & \mathbf{C}^{-1} \\ -\mathbf{B}^{-1} & \mathbf{G}^{-1} \end{bmatrix} \begin{bmatrix} \mathbf{B} \\ \mathbf{I} \end{bmatrix} + \mathbf{I} \\
&= [-\mathbf{B}^{-1}\ \mathbf{G}^{-1}] \begin{bmatrix} \mathbf{B} \\ \mathbf{I} \end{bmatrix} + \mathbf{I} \\
&= -\mathbf{I} + \mathbf{G}^{-1} + \mathbf{I} = \mathbf{G}^{-1}
\end{aligned} \tag{2.60}$$

Due to the low pass behaviour of physically realisable engineering systems, the entries in their transfer function matrix $\mathbf{G}(s)$ are proper rational functions, i.e. the degree of the numerator polynomial is less or equal to the degree of the denominator polynomial. As result, some entries in the transfer function matrix $\hat{\mathbf{G}}(s) = \mathbf{G}^{-1}(s)$ of the inverse system model are *not* proper which precludes a direct physical realisation of its mathematical model so that the inverse must be concatenated with a pre- or post low pass filter.

2.7 System Inversion Based on Bicausal Bond Graphs

Bicausal BGs may not only be used for parameter estimation but can be used for system inversion as well [7, 11–13]. To that end, first, some notions are recalled from [13].

Definition 2.9 (Power line) A power line between two components is a series of bonds and junctions structure elements connecting the two elements.

Remark 2.4 Power lines are different from causal paths. They are independent of the concept of computational causality and can be identified in *acausal* BGs.

Definition 2.10 (I/O causal path, I/O power line) A causal path from a source to a detector is called an input-output (I/O) causal path.

An input-output (I/O) power line is associated with an I/O causal path if the causal path goes through one variable of each bond of the power line.

For illustration, in the BG of Fig. 2.23, two disjoint I/O causal paths are highlighted by additional signal lines along the bonds of the paths. The bonds of the I/O power lines from E to y_1 and from $I(t)$ to y_2 associated with them are drawn as thick green lines.

Proposition 2.1 (Structural invertibility Ngwompo, 2001 **[13])** *Let $\mathcal{M}$ be a bond graph model with m inputs and m outputs. If there is a unique set of m disjoint I/O power lines, then the model is* structurally *invertible.*

Remark 2.5 Structural invertibility considers only the structure of a BG and the type of its elements. Constitutive relationships of the latter and parameter values are not taken into account.

Now, for simplicity, it is assumed that all n storage elements in a BG of a forward model have integral causality. That means that the order of the forward model is n. To obtain a bicausal BG of the inverse system, the following steps are carried out.

1. First, a BG in integral causality of the forward model is checked, whether there are disjoint input-output (I/O) power lines. If there are none, then the system is structurally not invertible.
2. If there are m disjoint I/O power lines associated with a minimal length I/O causal path in the forward bond graph model, then the source and the sensor at the ends of each power line are replaced by source sensor elements SS.
3. Now, bicausality on the *acausal* BG is assigned to the bond at an output SS element according to the type of the sensor replaced by the SS element. The information of both effort and flow is propagated through the BG junction structure along a power line associated with one of the disjoint I/O causal paths of minimal length identified in step 1 towards the input SS element at the other end of the power line replacing a source. As the aim of these steps is not parameter estimation but the inversion of the forward model, no bond impinged to a storage element or a resistor receives bicausality.
4. The causal implications of bicausal assignment along a power line between SS elements are extended into the BG as far as possible.
5. If elements remain without causality, the classical sequential causality assignment procedure (SCAP) is applied until the BG is causally completed.

Remark 2.6

1. Propagation of bicausality from an output to an input in the *acausal* BG follows a power line associated with an I/O causal path in the forward BG model. That is, only bonds of the power line are bicausally assigned.
2. Bicausal bonds always impose the common variable at a junction. They are said to have a *strong causal determination*. As a consequence, a storage element

attached to a junction on a bicausal path receives the causality imposed at the junction and is enforced to take derivative causality. That is, the order of the inverse model is lower than the one of the forward model. The order n of a BG in preferred integral causality is the number of storage elements that are in integral causality. For the inverse model this number is reduced by the number of storage elements on a shortest causal path that must take derivative causality as a result of bicausality assignment. Bicausal BG model-based inversion yields an inverse model of reduced order.

3. Bond graph elements of a bicausal path are not causally affected. Especially storage elements that are not on a considered shortest causal path keep their causality they have received when preferred integral causality was assigned to the forward BG. Therefore, if bond graph elements remain causally unassigned after bicausality has been propagated along all power lines associated with disjoint I/O causal paths of minimal length identified in the forward BG model, classical SCAP is used to complete causality assignment.
4. Propagation of bicausality along power lines associated with I/O causal paths means that both effort and flow information is propagated. In order to make sure that no causal conflicts result, the power lines associated with the I/O causal paths identified in step 1 must be disjoint.

After these steps, equations of the inverse model can be obtained from the bicausal BG by following causal paths from a SS element associated with an output signal to a source sensor associated with an input signal.

Example: Bicausal Bond Graph-Based System Inversion Applied to a RLC Circuit

Bicausal bond graph-based system inversion is illustrated on the example of the simple passive RLC circuit depicted in Fig. 2.23.

Figure 2.24 displays a BG in integral causality of the RLC circuit in Fig. 2.23.

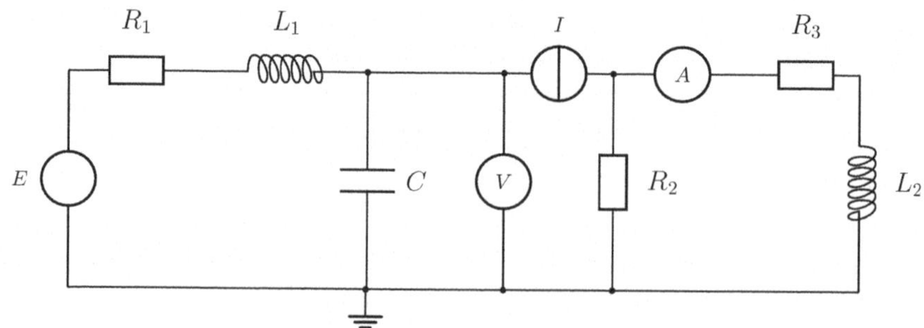

Fig. 2.23 RLC Circuit with two sources and two sensors

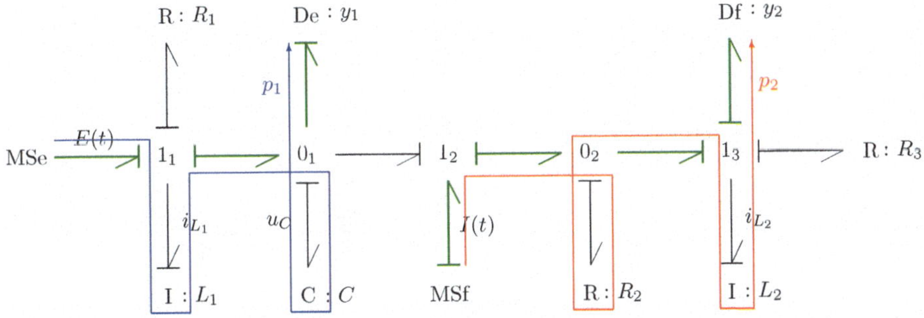

Fig. 2.24 Forward BG model of the RLC Circuit in Fig. 2.23 with two disjoint I/O causal paths of minimal length

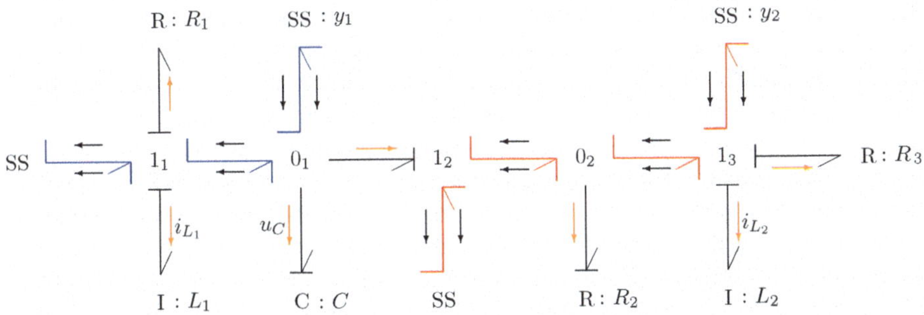

Fig. 2.25 Bicausal BG of the RLC Circuit in Fig. 2.23 with two disjoint bicausal paths

As can be seen from the BG in Fig. 2.24, there is a causal path p_1 between MSe : E and De : y_1 and another causal path p_2 disjoint from p_1 between MSf : I and Df : y_2. The power line associated with the O/I causal path p_1 is

$$p_1 : \quad \text{Df} : y_1 \leftharpoondown 0_2 \leftharpoondown 1_1 \leftharpoondown \text{MSe} : E$$

The power line associated with the causal path p_2 is

$$p_2 : \quad \text{Df} : y_2 \leftharpoondown 1_3 \leftharpoondown 0_2 \leftharpoondown 1_2 \leftharpoondown \text{MSf} : I$$

Figure 2.25 shows a bicausal BG of the inverse model. Model inversion on a BG does not change the structure of the BG in contrast to block diagrams in which structural changes generally result.

From the BG in Fig. 2.25 it can be seen that one of the two signals carried by a bicausal bond always determines the common variable at the junction to which the bicausal bond is attached. For instance, the effort imposed on junction 0_1 by the source sensor SS : y_1 determines the causality at all other bonds attached to 0_1. The sum of flows at 0_1 is also propagated to junction 1_1 where it determines the flow at all other bonds attached to 1_1. The propagation of both information from

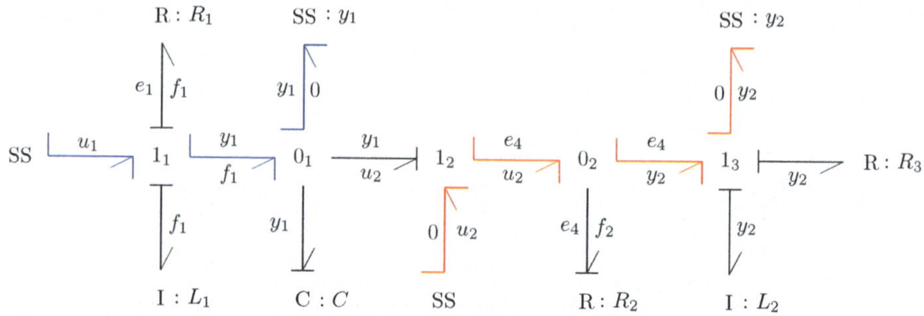

Fig. 2.26 Bicausal BG of the RLC Circuit in Fig. 2.23

the source sensor SS : y_1 along the power line to the source sensor replacing the modulated effort source is highlighted by additional signal pairs. Single additional signals indicate the result of a strong causality determination at a junction by a bicausal bond on the power line. As all storage elements are linked to a junction on one of the two power lines, they receive the causality imposed on the junction by a bicausal bond. That is, their integral causality in the forward BG model is turned into derivative causality in the bicausal BG of the inverse model. That is, while the order of the forward model is equal to three, the inverse model is stateless. Moreover, the storage elements in derivative causality indicate that time derivatives of the outputs are required. No states and time derivatives of the outputs mean that the transfer functions are not proper so that the inverse mathematical model cannot be implemented without adding an appropriate low pass filter.

Since time derivatives of measured signals are not desired, I/O causal paths of minimal length are chosen among I/O causal paths identified in the forward BG model in order to minimise the number of storage elements in integral causality linked to a junction in an output-input power line and to minimise the number of storage elements that must take derivative causality in the bicausal BG of the inverse model. In the case of the considered example circuit, there are no alternative output-input power lines.

From the bicausal BG in Fig. 2.26 the following equations can be derived:

$$1_1: \quad e_1 = u_1 - y_1 - L_1 \dot{f}_1 \tag{2.61}$$

$$R: R_1: \quad e_1 = R_1 f_1 \tag{2.62}$$

$$0_1: \quad f_1 = 0 + C\,\dot{y}_1 + u_2 \tag{2.63}$$

$$1_2: \quad e_4 = y_1 + 0 \tag{2.64}$$

$$0_2: \quad u_2 = f_2 + y_2 \tag{2.65}$$

$$R: R_2: \quad f_2 = \frac{1}{R_2} e_4 \tag{2.66}$$

$$1_3: \quad e_4 = 0 + L_2 \dot{y}_2 + R_3 y_2 \tag{2.67}$$

Substituting auxiliary variables finally yields the equations of the inverse model.

$$u_1 = R_1 C \dot{y}_1 + R_1 [\frac{1}{R_2}(R_3 y_2 + L_2 \dot{y}_2) + y_2] + y_1 + L_1 C \ddot{y}_1$$

$$+ L_1 [\frac{1}{R_2}(R_3 \dot{y}_2 + L_2 \ddot{y}_2) + \dot{y}_2] \tag{2.68}$$

$$u_2 = \frac{1}{R_2}(R_3 y_2 + L_2 \dot{y}_2) + y_2 \tag{2.69}$$

As there are no storage elements in integral causality in the bicausal BG, the inverse model is stateless. The inputs u_1, u_2, in fact, only depend the outputs y_1, y_2 and their time derivatives.

Let $R_1 = R_3 = 0$ to simplify the expressions. The transfer function matrix $\hat{\mathbf{G}}(s)$ of the inverse model then reads

$$\begin{bmatrix} U_1 \\ U_2 \end{bmatrix} = \underbrace{\begin{bmatrix} L_1 C s^2 + 1 & \frac{L_1 L_2}{R_2} s^2 + L_1 s \\ 0 & \frac{L_2}{R_2} s + 1 \end{bmatrix}}_{\hat{\mathbf{G}}(s)} \begin{bmatrix} Y_1 \\ Y_2 \end{bmatrix} \tag{2.70}$$

The transfer function matrix $\hat{\mathbf{G}}(s)$ is not proper. Thus, the inverse model cannot be directly implemented. To that end, a concatenation with a low pass filter is needed.

The result for $\hat{\mathbf{G}}(s)$ obtained from a bicausal BG of the inverse model can be checked by following the classical matrix-based approach. That is, the transfer function matrix $\mathbf{G}(s)$ is computed from the matrices of the forward model and then inverted.

For the illustrating example with the simplification $R_1 = R_3 = 0$, it is sufficient to derive the equations from the BG of the forward model (Fig. 2.24), to apply a Laplace transform and to solve the resulting algebraic equations for the inputs $U_1(s)$, $U_2(s)$. For $R_1 = R_3 = 0$, the time domain equations derived from the BG of the forward model (Fig. 2.24) read

$$I : L_1 : \quad \dot{f}_1 = \frac{1}{L_1}(E - e_1) \tag{2.71}$$

$$C : C : \quad \dot{e}_1 = \frac{1}{C}(f_1 - I) \tag{2.72}$$

$$I : L_2 : \quad \dot{f}_2 = \frac{1}{L_2} R_2 (I - f_2) \tag{2.73}$$

$$u_1 = E \tag{2.74}$$

$$u_2 = I \tag{2.75}$$

$$y_1 = e_1 \tag{2.76}$$

$$y_2 = f_2 \tag{2.77}$$

From Equation (2.71)–(2.77), one obtains

$$U_1 = \left(L_1 C s^2 + 1\right) Y_1 + \left(\frac{L_1 L_2}{R_2} s^2 + L_1 s\right) Y_2 \tag{2.78}$$

$$U_2 = \left(\frac{L_1}{R_2} + 1\right) Y_2 \tag{2.79}$$

in accordance with (2.70).

2.8 Bond Graph-Based Stability Analysis

A major task in control engineering is to analyse the stability of a system and to design an appropriate controller to ensure system stability. For LTI MIMO systems, a common approach is to determine the solutions of the characteristic equation

$$0 = \det(s\mathbf{I} - \mathbf{A}) \quad s \in \mathbb{C} \tag{2.80}$$

where $\mathbf{A}$ denotes the system matrix and $\mathbf{I}$ the identity matrix of appropriate dimension. $\det(s\mathbf{I}-\mathbf{A})$ is a polynomial in s which is the denominator of each entry in the transfer function matrix of a MIMO system. For small to medium scale models, transfer functions can be directly deduced from a causal BG according to Mason's loop rule (Sect. 2.2). Alternatively, equations can be deduced from a causal BG and formulated as LTI state space model. Once its matrices are available, open source software such as GNU Octave can be used to determine the poles by calling the function `pole()`, or `isstable()` to check an LTI system for stability.

Another option is to make use of Lyapunov's second method, which is also applicable for nonlinear models without having to solve the state space equations. As has been shown by Junco in [8], stability according to Lyapunov can be checked directly on a causal BG even without the need of state equations. This will be illustrated in the following. First, some definitions are provided.

Definition 2.11 (equilibrium point) Let $\dot{\boldsymbol{x}}(t) = \mathbf{f}(\boldsymbol{x}(t)), \boldsymbol{x}(0) = \boldsymbol{x}_0$ be a nonlinear time-invariant system, where $\mathbf{f} : \mathscr{D} \subseteq \mathbb{R}^n \rightarrow \mathbb{R}^n$. A point $\boldsymbol{x}_e \in \mathscr{D}$ is called an *equilibrium* point for $\boldsymbol{x}_e \in \mathscr{D}$ if

$$\mathbf{f}(t; \boldsymbol{x}_e) = \mathbf{0} \quad \forall\, t \geq t_0$$

$\boldsymbol{x}_e$ is an *isolated* equilibrium point if there is a $R > 0$ such that the neighbourhood $B_R(\boldsymbol{x}_e) := \{\boldsymbol{x} : ||\boldsymbol{x} - \boldsymbol{x}_e|| < R\}$ contains no other equilibrium point.

Definition 2.12 (Lyapunov stability) Let $\dot{x}(t) = \mathbf{f}(x(t))$, $x(0) = x_0$ be a nonlinear time-invariant system, where $\mathbf{f} : \mathscr{D} \subseteq \mathbb{R}^n \to \mathbb{R}^n$ and x_e an *equilibrium* point, i.e. $\mathbf{f}(x_e) = \mathbf{0}$.

A system (an equilibrium point x_e) is *stable* in the sense of Lyapunov if

$$\forall \epsilon > 0 \quad \exists \delta > 0 \text{ such that } ||x_0 - x_e|| < \delta \Rightarrow ||x(t; x_0) - x_e|| < \epsilon \quad \forall t \geq 0$$

Definition 2.13 (local asymptotic stability) A system is said to be *locally asymptotically* stable (l.a.s) if it is stable and if a δ can be chosen such that

$$||x_0 - x_e|| < \delta \Rightarrow \lim_{t \to \infty} ||x(t; x_0) - x_e|| = 0$$

Definition 2.14 (global asymptotic stability) A system (an equilibrium point x_e) is *globally asymptotically stable* (g.a.s) if it is stable and

$$x(t; x_0) \to x_e \quad \text{as} \quad t \to \infty \quad \forall x_0 \in \mathscr{D}$$

Remark 2.7

1. A nonlinear system may have a number of equilibrium points.
2. *Global* asymptotic stability implies that x_e is the *unique* equilibrium point.
3. Often a coordinate change $\tilde{x} := x - x_e$ is performed so that $\tilde{x}_e = \mathbf{0}$.
4. A linear system $\dot{x} = \mathbf{A}x$ is g.a.s with $x = \mathbf{0}$ if and only if all n eigenvalues λ_i of the system matrix $\mathbf{A}$ have a negative real part, i.e. $\text{Re}\{\lambda_i\} < 0, i = 1, \ldots, n$.
5. A *linear* system $\dot{x} = \mathbf{A}x$ is l.a.s near $x = \mathbf{0}$ if and only if $\text{Re}\{\lambda_i\} < 0, i = 1, \ldots, n$. That is, local and global asymptotical stability are equivalent for LTI systems.
6. $\text{Re}\{\lambda_i\} < 0 \Leftrightarrow -\mathbf{Q} := \mathbf{A}^T\mathbf{P} + \mathbf{P}\mathbf{A} < 0$ for some positive definite matrix $\mathbf{P} = \mathbf{P}^T$.

Definition 2.15 (Positive definite functions) Let $x_e = \mathbf{0}$ be an equilibrium point of $\dot{x}(t) = \mathbf{f}(x(t))$, $x(0) = x_0$. A function $V : \Omega \subseteq \mathbb{R}^n \to \mathbb{R}$ is said to be *positive definite* with respect to x_e if

1. $V(x) \geq 0 \quad \forall x \in \Omega$
2. $V(x) = 0 \quad$ if and only if $x = \mathbf{0}$
3. $V(x) \to \infty$ as $x \to \infty$

Remark 2.8

1. A function V that fulfils the third the property is said to be radially unbounded.
2. Let $\mathbf{P}$ be a symmetric $n \times n$ matrix. Then $V(x) := x^T\mathbf{P}x$ is a positive definite function if and only if $\mathbf{P} > 0$, i.e. is positive definite.

Theorem 2.6 (Lyapunov) *If there exists a positive definite function* $V : \mathbb{R}^n \to \mathbb{R}$ *such that for its time derivative* $\dot{V}$ *along the trajectory* $x(t)$ *holds*

1. $\dot{V}(\boldsymbol{x}) = (\nabla V(\boldsymbol{x}))^T \mathbf{f}(\boldsymbol{x}) < 0 \quad \forall \boldsymbol{x} \neq \mathbf{0}$
2. $\dot{V}(\mathbf{0}) = 0$

then every trajectory of $\dot{\boldsymbol{x}}(t) = \mathbf{f}(\boldsymbol{x}(t))$ *converges to zero as* $t \to \infty$, *i.e. the system equilibrium point* $\boldsymbol{x}_e = \mathbf{0}$ *is globally asymptotically stable.*

Remark 2.9

1. If such a function V exits, it is called a *Lyapunov function* for the system.
2. Note: The existence of a Lyapunov function is a sufficient condition for global asymptotical stability.
3. V may be considered a positive definite generalised energy function.
4. Lyapunov's theorem allows to check for g.a.s without knowing the trajectories, i.e. without having to solve the differential equations.
5. The key problem is to find a Lyapunov function. An approach is to select a parametrised function assumed to be a Lyapunov function, e.g. a quadratic function, and to determine its parameters so that the assumption made is justified. Such a function is called a *candidate Lyapunov function*.
6. For an LTI system, the opposite conclusion of Lyapunov's theorem can be proven: If the system $\dot{\boldsymbol{x}} = \mathbf{A}\boldsymbol{x}$ is g.a.s, then here exists a quadratic Lyapunov function that proves g.a.s.

In a bond graph framework, it is a natural approach to choose the energy ϵ stored in the modelled system as a candidate Lyapunov function. Its time derivative is the power that flows into the storage elements of the BG and can be determined directly on the BG. Let $\boldsymbol{q}$ be the vector of all n_C energy variables of the C elements and let $\boldsymbol{p}$ denote the vector of all n_I energy variables of the I storage elements in integral causality. The energy stored in the system is then expressed as

$$V(\boldsymbol{q}, \boldsymbol{p}) = \epsilon = \sum_{i=1}^{n_C} \frac{q_i^2}{2C_i} + \sum_{j=1}^{n_I} \frac{p_i^2}{2I_j} \tag{2.81}$$

where C_i is the capacitance of the ith C element and I_j the inertia of the jth I element. Let $\boldsymbol{x} = [\boldsymbol{q}^T \ \boldsymbol{p}^T]^T$. The transformation $\Delta\boldsymbol{x} := \boldsymbol{x} - \boldsymbol{x}_e^n$, where $\boldsymbol{x}_e^n$ is the equilibrium point of the system with nominal parameters, then shifts the equilibrium into the origin [5]. Evidently, $V(\Delta\boldsymbol{x}) > 0$. The time derivative of V as a function of the new coordinates then reads

$$\dot{V}(\Delta\boldsymbol{q}, \Delta\boldsymbol{p}) = \sum_{i=1}^{n_C} \frac{\Delta q_i \Delta\dot{q}_i}{C_i} + \sum_{j=1}^{n_I} \frac{\Delta p_i \Delta\dot{p}_i}{I_i} \tag{2.82}$$

According to the energy conservation principle, the energy into the storage element $\dot{V}$ equals the energy supplied into the system ϵ_{supply} minus the energy ϵ_R converted into heat by the dissipators of the system.

$$\dot{V}(\Delta\boldsymbol{q}, \Delta\boldsymbol{p}) = \Delta\epsilon_{\text{supply}} - \Delta\epsilon_R \tag{2.83}$$

That is, $\dot{V}(\Delta\boldsymbol{q}, \Delta\boldsymbol{p}) < 0$ if $\Delta\epsilon_R > \Delta\epsilon_{\text{supply}}$, which means that the system is passive.

Illustrative Example 1

The simple circuit in Fig. 2.27a is easily converted into the causal BG in Fig. 2.27b.
The state equations derived from the BG in Fig. 2.27b read

$$\dot{q} = I - \frac{1}{L}p \tag{2.84a}$$

$$\dot{p} = \frac{1}{C}q + R_1\left(I - \frac{1}{L}p\right) - R_2\frac{1}{L}p \tag{2.84b}$$

This gives for the time derivative of the candidate Lyapunov function

$$\begin{aligned}\dot{V}(\Delta q, \Delta p) &= \frac{\Delta p}{L}\Delta\dot{p} + \frac{\Delta q}{C}\Delta\dot{q} \\ &= \frac{\Delta p}{L}\left[\frac{1}{C}\Delta\dot{q} - \frac{1}{L}R\Delta\dot{p}\right] + \frac{\Delta q}{C}\left[-\frac{1}{L}\Delta p\right] \\ &= -R\frac{(\Delta p)^2}{L^2} \\ &= \frac{R}{L^2}[\Delta p\ \Delta q]^T \underbrace{\begin{bmatrix}-1 & 0\\ 0 & 0\end{bmatrix}}_{-\mathbf{P}}\begin{bmatrix}\Delta p\\ \Delta q\end{bmatrix} \leq 0\end{aligned} \tag{2.85}$$

where $R := R_1 + R_2$.

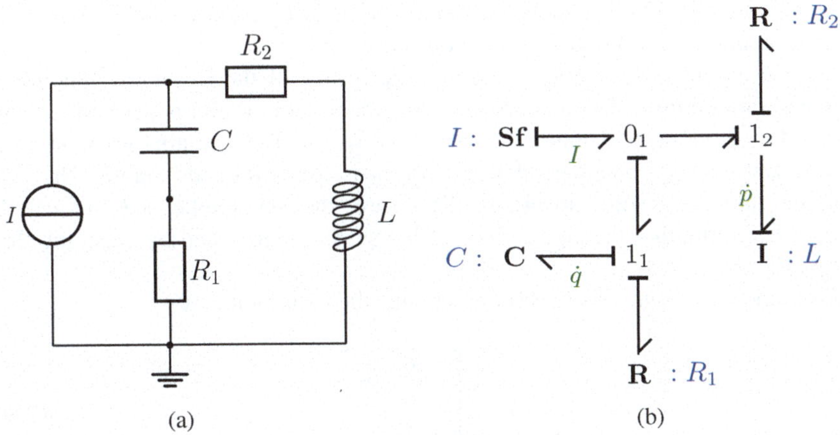

Fig. 2.27 Schematic and BG of a simple RLC circuit. (**a**) RLC circuit. (**b**) BG of the RLC circuit

The symmetric matrix $-\mathbf{P}$ is negative semi-definite, its eigenvalues are $\lambda = 0$ and $\lambda = -1$. That is, the candidate Lyapunov function is negative semi-definite. The origin is stable. However, nothing can be said about asymptotical stability.

This result can be obtained without any equations by just inspecting the causal BG due to the following two propositions recalled from [8].

Proposition 2.2 (Junco 2001) *If the origin is asymptotically stable, then* each *integral storage element imposes causality on at least one dissipative element.*

Proposition 2.3 (Junco 2001) Lyapunov stability*: If each R element in a BG is strictly dissipative and if its causality is imposed by only one state variable, then the origin is at least Lyapunov-stable.*

Asymptotic stability*: Suppose that the necessary condition in Proposition* 2.2 *holds. Further, suppose that each R element is strictly dissipative and that its causality is imposed by only one state variable, then the origin is asymptotically stable.*

In the above Example 1, the causality on the two resistors is imposed solely by the I : L storage element. The C element does not affect the causality on any of the resistors. That is, the condition of Proposition 2.2 is not fulfilled.

If a candidate Lyapunov function $V : \mathbb{R}^n \rightarrow \mathbb{R}$ for $\dot{\boldsymbol{x}}(t) = \mathbf{f}(\boldsymbol{x}(t))$ on $\boldsymbol{x}(t)$ is only negative semi-definite, then according to LaSalle's invariance principle (Appendix C), the origin is asymptotically stable, if the only solution of

$$\dot{z}(t) = \mathbf{f}(z(t)), \quad \dot{V}(z(t)) = 0$$

is $z(t) = \mathbf{0} \quad \forall\, t$.

In [8], Junco has used LaSalle's invariance principle to conclude that an isolated equilibrium point of a *subclass* of BG models without any modulated elements is asymptotically stable if the equilibrium point is Lyapunov stable even if the necessary condition of Proposition 2.2 does not hold.

For the circuit in Example 1, the time derivative of the Lyapunov function is negative semi-definite. The causality of two resistors in the BG is determined only by the I : L storage element. Nevertheless, as the BG contains no modulated elements, it can be concluded that the origin is asymptotically stable. The state equations are linear time invariant. Therefore, the asymptotic stability can be verified by computing the eigenvalues of the system matrix and by checking their real part.

From the state equations (2.84) one obtains the system matrix

$$\mathbf{A} = \begin{bmatrix} -\frac{R}{L} & \frac{1}{C} \\ -\frac{1}{L} & 0 \end{bmatrix} \tag{2.86}$$

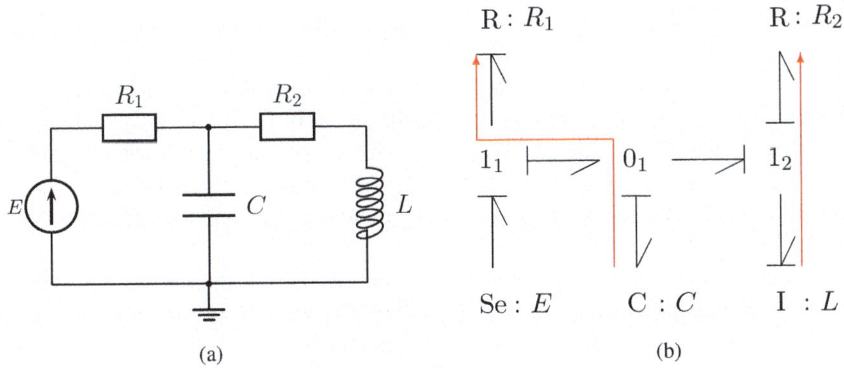

Fig. 2.28 Schematic and BG of a simple RLC circuit. (**a**) RLC circuit. (**b**) BG of the RLC circuit

Their eigenvalues are both negative.

$$s_{1;2} = -\frac{R}{2L} \pm \frac{1}{L}\sqrt{\frac{R^2}{4} - \frac{L}{C}} < 0 \tag{2.87}$$

Illustrative Example 2

The BG of the second simple RLC circuit in Fig. 2.28 indicates that the causality on $\mathrm{R} : R_1$ is imposed by the integral causality of the capacitor, while the causality on $\mathrm{R} : R_2$ is an immediate consequence of the integral causality of the inductor. Given that the two resistors are strictly dissipative, then, according to Proposition 2.3, it can be directly concluded by inspection of the causal BG that the origin is asymptotically stable.

The conclusion obtained by inspection of the BG is easily verified. A computation as in Example 1 yields that the time derivative of the Lyapunov function, in fact, is negative definite.

$$\dot{V}(\Delta p, \Delta q) = -\frac{R_2}{L^2}(\Delta p)^2 - \frac{1}{R_1 C}(\Delta q)^2 < 0 \tag{2.88}$$

$$= -R_2(\Delta i_L)^2 - \frac{1}{R_1}(\Delta u_C)^2 = -P_R \tag{2.89}$$

2.9 Summary

An outstanding feature of Bond Graphs is that the methodology enables a systematic graphical, rule-based development of models for multidisciplinary systems guided by physical insight. A causal BG as the result of the step-by-step graphical modelling procedure can serve as a basis for the automatic generation of various mathematical models depending on the application area and the purpose of the model.

Beyond the development of models for the purpose of simulating the dynamic behaviour of an engineering system, BG methodology can also support various tasks that are of major concern in control. This chapter has shown that structural state observability and structural state controllability as a necessary prerequisite for the design of a controller can be checked directly on a causal bond graph of the model.

If a multivariable system under consideration is numerically state observable and state controllable, transfer functions of interest can be directly derived manually from a causal bond graph of medium complexity according to Mason's loop rule. For large scale bond graphs, one might use modelling and simulation software to derive state space model equations from the bond graph and pass the obtained state space matrices to a mathematical software such as GNU Octave or Matlab® .

Moreover, bicausal BGs introduced by Gawthrop as an extension of classical BGs can support a number of control engineering tasks. This chapter shows how they can serve the determination of an inverse model directly from a bicausal BG and how bicausal bond graphs can be used for parameter estimation which is one possible approach to fault detection and isolation (FDI). The following chapter shows in detail how bond graphs can support model-based FDI. Failure prognosis considered in Chap. 4 builds up on FDI.

Finally, it is illustrated that an inspection of the causalities of a BG enables to draw conclusions with regard to asymptotical stability.

References

1. Brown, F. (1992). Direct application of the loop rule to bond graphs. *Transactions of the ASME Journal of Dynamic Systems, Measurement and Control, 94*(3), 253–261.
2. Buchholz, J. J., & v Grünhagen, W. (2008). Inversion Impossible? Technical report, Bremen University of Applied Sciences and DLR Braunschweig, Germany.
3. Dauphin-Tanguy, G., Rahmani, A., & Sueur, C. (1999). Bond graph aided design of controlled systems. *Simulation Practice and Theory, 7*(5–6), 493–513.
4. van Dixhoorn, J. (1977). Simulation of bond graphs on minicomputers. *Journal of Dynamic Systems, Measurement and Control, 99*, 9–14.
5. Gandanegara, G. (2003). Méthodologie de conception systémique en Génie Electrique à l'aide de l'outil Bond Graph – Application à une chaîne de traction ferroviaire. Ph.D. thesis, UMR INP Toulouse.
6. Gawthrop, P. (1995). Bicausal bond graphs. In F. Cellier & J. Granda (Eds.), *ICBGM'95, International Conference on Bond Graph Modeling and Simulation* (Simulation Series, Vol. 27, No. 1, pp. 83–88). SCS Publishing.

7. Gawthrop, P. J. (1997). Control system configuration: Inversion and bicausal bond graphs. In J. J. Granda & G. Dauphin-Tanguy (Eds.), *1997 International Conference on Bond Graph Modeling, and Simulation (ICBGM'97)* (Simulation Series, Vol. 29, No. 1, pp. 97–102). SCS Publishing. ISBN: 1-56555-103-6.
8. Junco, S. (2001). Lyapunov second method and feedback stabilization directly on bond graphs. In J. J. Granda & G. Dauphin-Tanguy (Eds.), *Proceedings of 2001 International Conference on Bond Graph Modeling and Simulation (ICBGM'01)* (SCS Simulation Series, Vol. 33, No. 1, pp. 137–142). Phoenix: SCS.
9. Mason, S. J. (1956). Feedback theory – further properties of signal flow graphs. In *Proceedings of IRE 44* (pp. 920–926)
10. Moylan, P. J. (1977). Stable inversion of linear systems. *IEEE Transactioss on Automatic Control, 22*, 74–78.
11. Ngwompo, R., Scavarda, S., & Thomasset, D. (1996). Inversion of linear time-invariant SISO systems modelled by bond graph. *Journal of the Franklin Institute, 333*(B)(2), 157–174.
12. Ngwompo, R., Scavarda, S., & Thomasset, D. (1997). Structural invertibility and minimal inversion of multivariable linear systems – a bond graph approach. In J. Granda & G. Dauphin-Tanguy (Eds.), *Proceedings of International Conference on Bond Graph Modeling and Simulation (ICBGM '97)* (Simulation Series, Vol. 29, No. 1, pp. 109–114). Phoenix: SCS Publishing.
13. Ngwompo, R. F., Scavarda, S., & Thomasset, D. (2001). Physical model-based inversion in control systems design using bond graph representation part 1: Theory. *Proceedings of the IMechE Part I Journal of Systems and Control Engineering, 215*(2), 95–103.
14. Sain, M. K., & Massey, J. L. (1969). Invertibility of linear time-invariant dynamical systems. *IEEE Transactions Automatic Control, AC-14*(2), 141–149.
15. Seraji, H. (1989). Minimal inversion, command matching and disturbance decoupling in multivariable systems. *International Journal of Control, 49*(6), 2093–2121.
16. Silverman, L. M. (1969). Inversion of multivariable linear systems. *IEEE Transactions on Automatic Control, AC-14*(3), 270–276.
17. Sueur, C., & Dauphin-Tanguy, G. (1991). Bond-graph approach for structural analysis of MIMO linear systems. *Journal of the Franklin Institute, 328*(1), 55–70.
18. Tan, S., & Vandewalle, J. (1988). Inversion of singular systems. *IEEE Transactions on Circuits and Systems, 35*(5), 583–587.
19. Willsky, A. S. (1974). On the invertibility of linear systems. *IEEE Transactions on Automatic Control, 19*, 272–274.

Chapter 3
Fault Diagnosis

Fault diagnosis, i.e. detection, isolation, identification, and estimation of faults, is an indispensable task in order to ensure safety and reliability of mobile engineering systems as well as of industrial plants and is a prerequisite for fault tolerant control (FTC) and failure prognosis. Smart system components equipped with sensors and embedded systems can collect and preprocess data and assess its health state. Values of sampled data exceeding certain thresholds indicate a deviation from normal desired system behaviour. Due to a fault a system component shows some malfunction which affects the system behaviour. Smart system components can communication this information to other components of a cyber-physical system and to a remote supervision system. Once a fault in a system component has occurred, the system is no longer the one for which the control in use has been designed. Depending on the type of a fault and its severity, it may become necessary to adapt the controller parameters or even change the control law. If online estimation of a detected and isolated fault indicates that the fault's magnitude is slowly increasing with time, it is important to estimate how long a faulty system can be safely operated if necessary with acceptable limited functionality until the fault would cause a failure of the faulty component and could even entail a failure of the entire system.

These brief considerations show that fault diagnosis, fault tolerant control, and failure prognosis are tightly linked. This chapter, first of all, addresses fault diagnosis with a focus on bond graph model-based fault diagnosis.

3.1 Types of Faults

Faults may happen inside an engineering system to be operated, as well as in its sensors and actuators. Sensors may become faulty due to external environmental disturbances such as significant temperature changes, or humidity. As a result, their

W. Borutzky, *Bond Graph Modelling for Control, Fault Diagnosis and Failure Prognosis*, https://doi.org/10.1007/978-3-030-60967-2_3

offset or their gain may change, their readings may show drift so that they provide faulty information. This can be a serious problem with fatal consequences as some recent cases in commercial aviation have demonstrated. The problem can be reduced if redundant hardware sensors and virtual sensors are available and allow for a voting that can discard wrong information. If actuators become faulty, e.g. due to friction, material fatigue, or a failure of their power supply, then a desired dynamic system behaviour cannot be achieved anymore.

System faults may be parametric or structural. For instance, despite filters in some places of a hydraulic system, contamination of the fluid may occur and may result in a partial blockage of some valves. This blockage affects flows and pressures that are measured but can be mapped to a change of the discharge coefficient in the valve's constitutive equation. Short circuit, or open circuit faults in electronic circuits due to broken switching devices in electric circuits, a stuck valve in a hydraulic circuit, or a broken transmission in a mechanical system may be considered structural faults. They change a system's structure. Once a parametric or structural fault has been diagnosed, the system model used in model-based fault diagnosis can be changed accordingly and can be considered the model of a non-faulty system in order to be able to detect subsequent new faults.

Furthermore, system faults may be due to wear and tear during normal system operation, human operator errors, postponed maintenance, or a faulty design of a system component, or errors in the assembly of system components from different engineering domains. They may result in either suddenly occurring faults or faults of gradually increasing severity with possibly disastrous consequences.

Faults may occur almost instantly, their magnitude may jump to a new persistent value, a resistance, for instance, may drop to a very low value, or faults may be significantly progressive with time and cause an increasing deviation of a component's behaviour from that in normal operation mode, or may develop slowly with small magnitude in comparison to abrupt faults due to tear and wear or ageing. The onset of these slowly developing incipient faults may be difficult to detect as a feedback controller will try to beat down the slowing increasing error. A system component's behaviour may deviate from a desired behaviour but may still be within given bounds of acceptable behaviour. With progression of time, however, the magnitude of the incipient fault will reach a value that turns the component's behaviour into a malfunction that cannot be tolerated anymore as it can result in a failure.

Abrupt, progressive, and incipient faults are present at each time instant. In contrast, faults may also be intermittent. Abnormal dynamic behaviour can happen only for short time intervals. As the occurrence and the length of these time intervals are unpredictable, it is difficult to react properly. Even if the sampling of signals at discrete time instances and fault diagnosis are interleaved, intermittent faults may remain undetected when signals of the intermittent fault are present when the fault diagnosis module is not ready to receive them.

On the other hand, after an instance of an abnormal behaviour has been detected, the intermittent fault may vanish for an unpredictable period of time and will not be detectable at subsequent sampling times. What is the appropriate reaction to the noticed abnormal behaviour over a short time when it does not reappear in a certain time horizon?

Finally, there is commonly a distinction between additive and multiplicative faults. Assume that the dynamic behaviour of an engineering system can be sufficiently accurately described by a linear time invariant (LTI) state space model. Sensor and actuator fault are then considered additional external input signals and are taken into account as additive terms in the state space equations. They are denoted as *additive* faults.

Let **A**, **B**, **C**, **D**, **E**, **F**, **G**, **K** be constant coefficient matrices of appropriate dimensions and let $\boldsymbol{x}(t)$ denote the state vector, $\boldsymbol{u}(t)$ the vector of known inputs, $\boldsymbol{y}(t)$ the vector of measured outputs, $\boldsymbol{f}_a(t)$ actuator faults, $\boldsymbol{f}_s(t)$ sensor faults, and $\boldsymbol{d}(t)$ disturbances. The dynamic behaviour of a process subject to additive faults can then be described by the linear state space model

$$\dot{\boldsymbol{x}}(t) = \mathbf{A}\boldsymbol{x}(t) + \mathbf{B}\boldsymbol{u}(t) + \mathbf{E}\boldsymbol{d}(t) + \mathbf{K}\boldsymbol{f}_a(t)\,, \quad \boldsymbol{x}(0) = \boldsymbol{x}_0 \tag{3.1a}$$

$$\boldsymbol{y}(t) = \mathbf{C}\boldsymbol{x}(t) + \mathbf{D}\boldsymbol{u}(t) + \mathbf{F}\boldsymbol{d}(t) + \mathbf{G}\boldsymbol{f}_s(t) \tag{3.1b}$$

In contrast, contamination of the fluid in a hydraulic system may be considered as an example of an internal process fault. Because contamination may result in a change of the discharge coefficient of some valves, such process faults are taken into account as a change of the process parameters. They affect the actual parameters $\boldsymbol{\Theta}$ in the matrices of the state space equations and are called *multiplicative* faults.

$$\dot{\boldsymbol{x}}(t) = \mathbf{A}(\boldsymbol{\Theta})\boldsymbol{x}(t) + \mathbf{B}(\boldsymbol{\Theta})\boldsymbol{u}(t)\,, \quad \boldsymbol{x}(0) = \boldsymbol{x}_0 \tag{3.2a}$$

$$\boldsymbol{y}(t) = \mathbf{C}(\boldsymbol{\Theta})\boldsymbol{x}(t) + \mathbf{D}(\boldsymbol{\Theta})\boldsymbol{u}(t) \tag{3.2b}$$

3.2 Signal Preprocessing

There are many articles and books dedicated solely to signal processing. As signal preprocessing is essential for data-driven approaches to fault diagnosis, as bond graph model-based fault diagnosis also needs smoothed raw measurement data and, in general, needs time derivatives of measured signals, and since Chap. 4 proposes a combined bond graph model-based, data-driven approach to failure prognosis, this section briefly considers filtering of measured signals.

As measured signals are corrupted with random noise, the information they carry must be extracted by means of appropriate filters. There are various classes of filters. The type of filter in use and its parameters depend on the characteristics of the ground truth signal and on the way the signal is subsequently used by a fault diagnosis module. If the ground truth signal is sufficiently smooth, low pass filters of some order or moving average filters can be used to remove the superimposed noise.

If needed, the filtered signal can be differentiated with respect to time with sufficient accuracy. However, if the true signal shows discontinuities or has sharp edges like a clock signal, smoothing by means of median filters provides better results as edges are preserved. Median filters also remove outliers. An approximation of a signal with narrow jumps and sharp edges by means of a Fourier series shows that high frequency components are needed to keep the error small. Low pass filters, however, attenuate high frequency components. A moving average filter and low pass filters smooth narrow jumps, i.e. their height is reduced and their width is increased. That is, the original signal is distorted. Differentiation in discrete time of such signals by differentiating the filtered signal is not possible without some loss of accuracy. A good trade-off between smoothing and sufficiently accurate time derivatives of a signal obtained from measured data can be achieved by means of the Savitzky–Golay (SG) filter [58] introduced back in 1964 and widely used since then in various applications. This discrete time-domain low pass-filter can smooth given noisy data as well as provide time derivatives up to some order of the signal and is implemented in widely used software packages such as GNU Octave and Matlab®. It is fast, good in de-noising and signal reconstruction. The online computation of a filtered value can be obtained by just multiplying the first row of an offline precomputed matrix with a vector of noisy data values. A generalisation of the SG filter that may track more accurately abrupt deviations in time series has recently been proposed in [19].

In the following, the idea and the pros and cons of the Savitzky–Golay are briefly reviewed.

3.2.1 *Savitzky–Golay Filter*

Let $f(t)$ denote a signal corrupted by additive independent and identically distributed (iid) noise $n(t)$ with zero mean and variance σ^2, and $x_i = f(t_i) + n(t_i)$ the time series of the the noisy signal. Given a series of N equally spaced data values fed into the filter, its task is to reconstruct the underlying signal $f(t)$ at a point t_0. The output of the filter is denoted by $y(t_0)$.

Let w be a moving symmetric window with a length of $N = 2m + 1$ data around the current reconstruction point t_c.

$$w = \{x_{-m}, x_{-m+1}, \ldots, x_{-1}, x_0, x_1, \ldots, x_m\} \tag{3.3}$$

The SG filter fits a polynomial $P(t) = a_0 + a_1 t_i + \ldots + a_n t_i^n$ of low order n to the N observed equally spaced time series data in the moving symmetric window by minimising the mean square error in the current window.

$$\epsilon = \sum_{i=-m}^{m} (p(t_i) - x_i)^2 \tag{3.4}$$

The filter output is the value of the polynomial at the centre point $t_0 = 0$ of the moving window w.

$$y(0) = p(0) = a_0 \tag{3.5}$$

and replaces the noisy data value $x(t_c)$. The reconstruction of the centre point needs future values, which means some delay. In order to compute the replacement of the next noisy sample, the window is shifted by one time step Δt, and a new least-squares fit is performed which provides a replacement $y(t_0) = p(t_0) = a_0'$ for the noisy data value $x(t_c + \Delta t)$. The index $-m \leq i \leq m$ is used for the time points in each moving window. Computing the coefficients of the fitting polynomial in each window again is costly. However, Savitzky and Golay showed that the same filtering result can be obtained by a discrete convolution of samples in the windows with a fixed impulse response.

$$\begin{aligned} y_k &= h_k * x_k \\ &= \sum_{i=-m}^{m} h_i\, x_{k-i} = \sum_{i=-m}^{m} h_{k-i}\, x_i \end{aligned} \tag{3.6}$$

where $h_i = c_i$ is the impulse response of length N of the filter. That is, the filter output can be expressed as a weighted sum of the samples in the current window.

The filter coefficients can be determined in the following manner. A necessary condition for a minimal error ϵ_i is that the derivatives with respect to the unknown polynomial coefficients a_k vanish which yields a set of $n+1$ equations for the $n+1$ unknown a_k.

$$\sum_{i=-m}^{m} \sum_{k=0}^{n} a_k t_i^k t_i^j = \sum_{i=-m}^{m} x_i t_i^j \qquad j = 0, 1, \ldots, n \tag{3.7}$$

Let $\boldsymbol{x} = [x_{-m} \ldots x_{-1} x_0\ x_1 \ldots x_m]^T$, $\boldsymbol{a} = [a_0\ a_1 \ldots a_n]^T$ and

$$\mathbf{V}^T = \begin{bmatrix} t_{-m}^0 & t_{-m+1}^0 & \cdots & t_0^0 & t_1^0 & \cdots & t_m^0 \\ t_{-m}^1 & t_{-m+1}^1 & \cdots & t_0^1 & t_1^1 & \cdots & t_m^1 \\ \vdots & \vdots & \vdots & \vdots & \vdots & \cdots & \vdots \\ t_{-m}^{n-1} & t_{-m+1}^{n-1} & \cdots & t_0^{n-1} & t_1^{n-1} & \cdots & t_m^{n-1} \\ t_{-m}^n & t_{-m+1}^n & \cdots & t_0^n & t_1^n & \cdots & t_m^n \end{bmatrix} \tag{3.8}$$

be the transpose of the $N \times (n+1)$ Vandermonde matrix $\mathbf{V}$.

Equation 3.7 then reads in matrix form

$$(\mathbf{V}^T\mathbf{V})\boldsymbol{a} = \mathbf{V}^T\boldsymbol{x} \tag{3.9}$$

and can be solved for the unknown coefficients $\boldsymbol{a}$.

$$\boldsymbol{a} = \underbrace{(\mathbf{V}^T\mathbf{V})^{-1}\mathbf{V}^T}_{=:\,\mathbf{C}}\,\boldsymbol{x} \tag{3.10}$$

The important result is that the $(n+1) \times N$ convolution matrix $\mathbf{C}$ does not depend on the samples but only on the window length $m = (L-1)/2$ and the order n of the fitting polynomial. That is, the convolution coefficients are the same for all windows. As the output of the filter is a_0, only the first row of matrix $\mathbf{C}$ is needed. As a result,

$$y(0) = a_0 = \sum_{i=-m}^{m} c_{-i} x_i \tag{3.11}$$

The samples $x(t)$ in the current window are fitted by a polynomial and the data value $x(t_c)$ in the moving window's centre is replaced by the filter output $y(0)$. A value for the time derivative of the function underlying the samples at the centre point of the window can be obtained by differentiating the polynomial. In addition to a value of the underlying function $f(t)$, the SG filter can also provide the kth time derivative at $t_0 = 0$

$$y^{(k)}(0) = k!\, a_k \tag{3.12}$$

To that end, the kth row of the convolution matrix $\mathbf{C}$ must be computed. However, tables of convolution coefficients for various window lengths and orders of the fitting polynomial can be found in the literature [47, 58].

The design parameters of SG filters are the length of the moving window, i.e. the number $N = 2m + 1$ of data values and the order n of the fitting polynomial. There is no definite answer as how to choose proper values for these parameters. Given a fixed window length, numerical experiments have shown that higher order polynomials better fit narrow features of a waveform with regard to their height and width at the expense of less smoothing broader features [47]. The smoothing of a polynomial of given order is better for larger but not too large windows. Experiments with a sampled Gaussian and chosen fixed parameters reported in [44] indicate that the sampled error at the centre point first decreases with larger window lengths. However, beyond a window length $N = 25$ the error increases significantly. In [51], the authors use Chebyshev orthogonal polynomials and show that the optimal window length depends on the characteristics of the waveform under consideration, on the noise power, the number of samples, and the order of the fitting polynomial.

For illustration, the SG filter is applied to three test functions of different type corrupted with white Gaussian noise, i.e. to a smooth sinusoidal waveform, and to two non-smooth functions. To that end, the GNU Octave function `sgolay()` contained in the signal processing package `signal` has been used.

```
G = sgolay(K,F,m,ts)
```

where

$K = p$: order of the fitting polynomial (default = 3)
$F = n$: filter length (n odd and $n > p$)
m : mth derivative (default = 0)
ts : scaling factor

This function computes the filter coefficients for all Savitzky–Golay smoothing filters of order p. The result together with a vector **y** of sampled data is then passed to the Octave function `sgolayfilt()`, i.e.

```
yGm = sgolayfilt(y,G)
```

to get a smoothed waveform $yGm(t)$ of the mth derivative ($m = 0, 1$) of a function given by the sampled data in vector $\boldsymbol{y}$.

Figure 3.1 shows a sinusoidal waveform and its first derivative obtained by a SG filter with a noisy sinusoid as input.

The filter well reconstructs the underlying sinusoid and provides a first order derivative (yG1) of the filter signal (yG0) that is close to the analytical derivative (ydiff) of the original function y although 10% of noise has been added. The result has been obtained with a fitting polynomial of order $p = 4$ and a window length of $N = 95$ samples. The noisy signal has been sampled at time points $t_n = n \cdot 5e{-}2$, $0 \le n \le 251$.

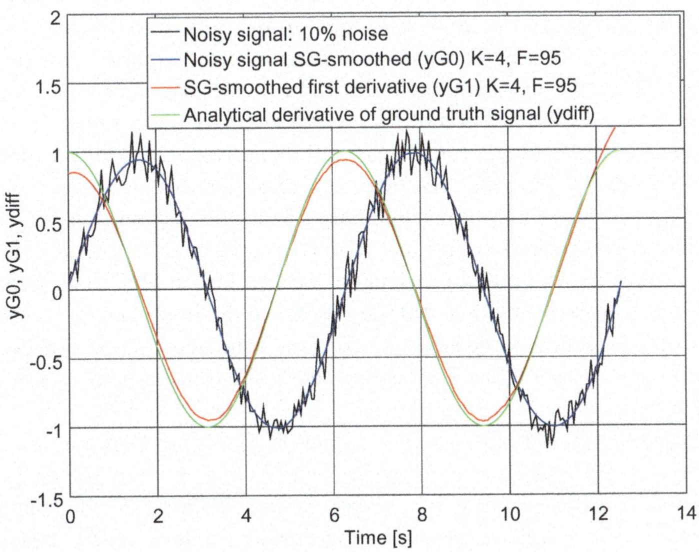

Fig. 3.1 Sinusoidal waveform and its first derivative obtained by a SG filter

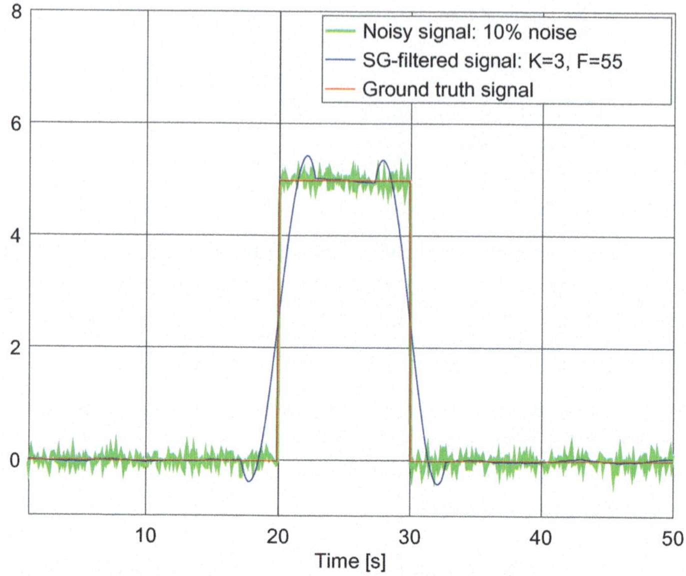

Fig. 3.2 SG filter applied to a rectangular pulse

The smoothed first derivative deviates from the analytical one at the beginning and at the end of the considered time interval. This is due to the fact that the SG filter needs m_s data values on both sides of a sample at current time instant t_c that is replaced by the output of the filter $y(0)$ equal to the polynomial value at the centre point of the current symmetric window. For the initial sample m_s left samples are not available. Equally, m_s samples to the right of the last sample are missing. In [29], Gorry presents an extension of the Savitzky–Golay convolution approach that enables a general order polynomial fit, and all its derivatives, at all positions.

Figure 3.2 shows that the SG filter replaces the discontinuities of a noisy rectangular pulse by smooth transients. Apart from the discontinuities, the filtered signal is close to the ground truth signal. The 10% noise level has been well removed by means of a third order polynomial and a window length of 55 data samples. The noisy signal has been sampled at 500 time points in the interval $0 \leq t \leq 50$ s.

The third test function is smooth. Its first time derivative, however, has a discontinuity at $t = 1$ s. Figure 3.3 indicates the limitations of what can be achieved with a SG filter.

The smoothed signal (yG0) is close to the original one. Given a noise level of 2%, the filter has difficulties to approximate sufficiently accurate the analytical first order derivative of the test function. A polynomial of order $K = 2$ and window length $F = 145$ can well approximate the constant values of the analytical first order derivative but not the discontinuity. If the order of the polynomial is increased to $K = 4$ to better fit the Discontinuity, then the approximation of the constant values of the derivative gets less smooth (Fig. 3.4).

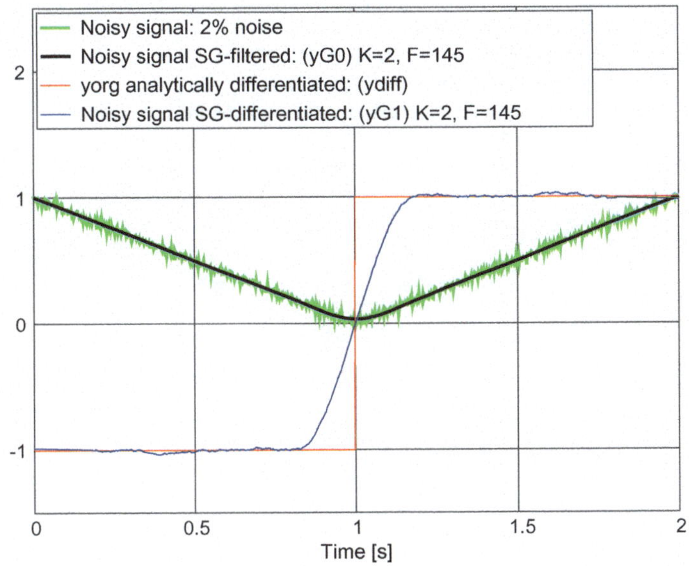

Fig. 3.3 SG filter applied to a continuous test function with a discontinuous first derivative

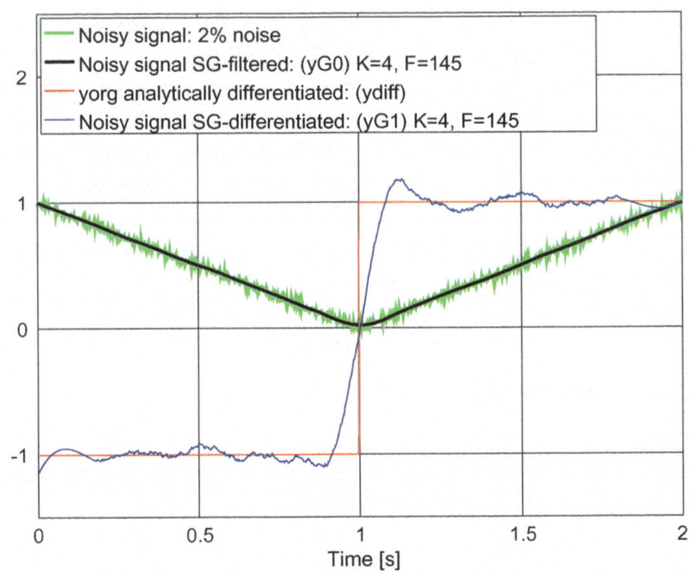

Fig. 3.4 SG filter applied with a higher order polynomial to a continuous test function with a discontinuous first derivative

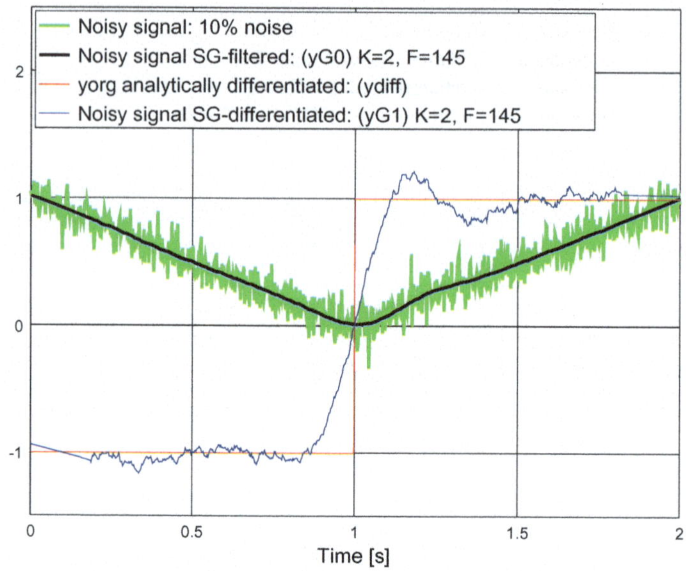

Fig. 3.5 SG filter applied in the case of a higher noise level

If the noise level is increased to 10% (Fig. 3.5), the approximation of the first order derivative gets less accurate in comparison to Fig. 3.3.

The application of the SG filter to the three test functions suggests that smoothing of continuous signals superimposed with noise of even higher level can be quite accurate. Discontinuities are replaced by a smooth polynomial approximation. The first order derivative of a continuously differentiable function such as a sinusoid can approximate with good accuracy. Approximating the discontinuous first order derivative of a continuous function is quite less accurate and becomes worse for higher noise levels.

3.2.2 *State Variable Filters*

Time derivatives of a *smoothed* signal may also be obtained by means of state variable filters (SVFs) given by a transfer function

$$G_f(s) = \frac{X_f(s)}{X(s)} = \frac{c_0}{c_n s^n + c_{n-1} s^{n-1} + \ldots + c_1 s + c_0} \tag{3.13}$$

where $x(t)$ denotes the input signal into the filter and $x_f(t)$ its output. The coefficients c_i may be determined in a Butterworth design. The cut-off frequency of a SVF should be higher than the highest system frequency of interest. The block

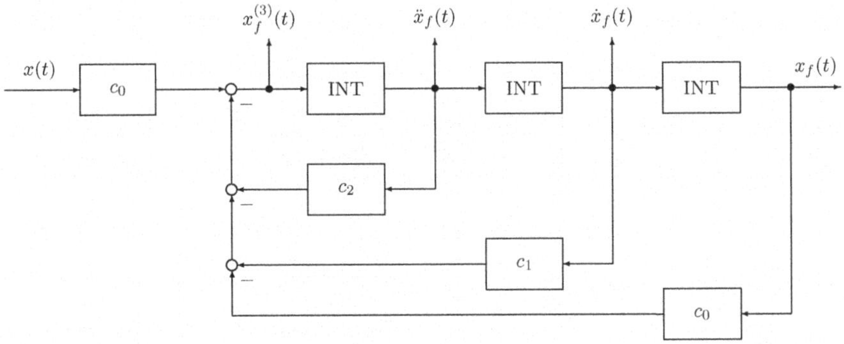

Fig. 3.6 Block diagram of a third order state variable filter

diagram of a third order filter (Fig. 3.6) indicates that the filtered derivatives $x_f^{(i)}$ of the input signal $x(t)$ are just the internal states of the filter.

A discrete SVF version has been given in [45].

3.3 Data-Driven Methods

Preprocessing of raw sensor data such as the filtering of noise and the removal of artefacts is a first step towards fault diagnosis. The filtered signals can then be used in data-driven as well as in model-based fault diagnosis. As fault diagnosis is a prerequisite to failure prognosis addressed in Chap. 4, and since combinations of data-driven and model-based approaches to fault diagnosis and failure prognosis aim at exploiting the strengths of both methods, some features of data-based methods shall be briefly revisited.

Given preprocessed data, a fundamental task that can be performed in time domain as well as in frequency domain is to extract features that should be uncorrelated, can be attributed a physical meaning, can be mathematically described, and may serve as fault indicators. Extracting features from preprocessed data results in a multidimensional feature space. Clusters of features enable to distinguish classes of faults.

Classification of features can be achieved by comparing time-varying feature vectors with known fault patterns stored in a library or with predetermined fault alarm conditions. Admissible fault limits such as the so-called Bollinger bands (moving average of a signal $\pm$ moving variance) can be computed concurrently to the time evolution of a feature. Other options for fault classification are neural networks (NNs) that are trained with historical data, support vector machines (SVMs) [63], or fault classification based on fuzzy-logic. Fuzzified features are input into an inference engine together with a rule base. The outputs are defuzzified to get the probability of a particular fault.

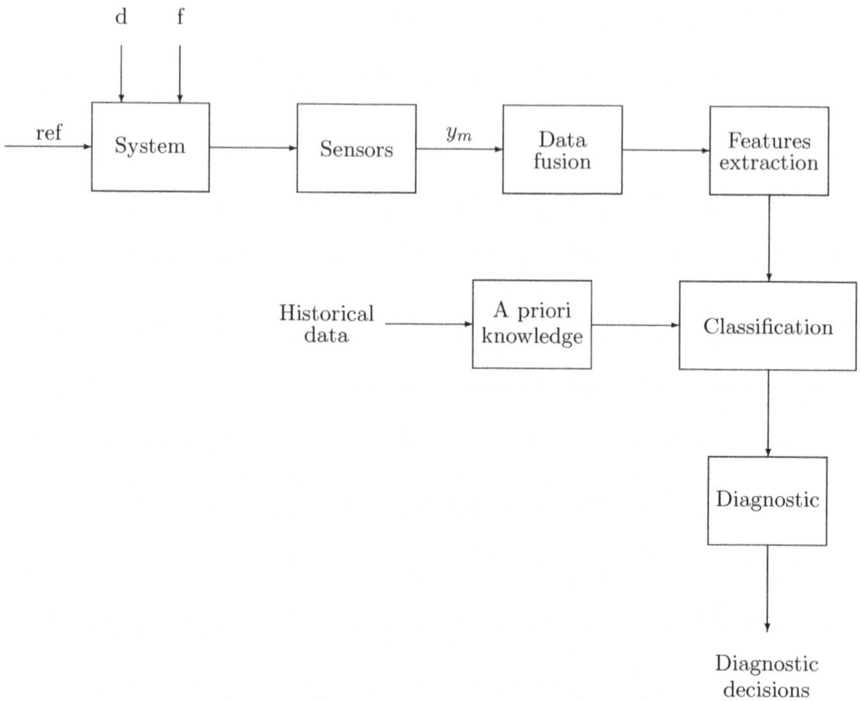

Fig. 3.7 Flowchart of data-based fault diagnosis

Figure 3.7 displays a flowchart of data-based fault diagnosis, in which d denotes disturbances and f faults.

3.4 Filters for Estimating the State of Health of a System

Beyond filtering noisy signals, filters such as Kalman filters (KFs) or particle filters (PFs) are commonly used to estimate the state of a system that cannot be directly measured. A sufficient approximation of the unknown current state vector of a system on the basis of known control inputs, noisy measurements, and process noise is important in data-driven fault diagnosis and for estimating the future states with regard to failure prognosis. An abnormal change of the system state is an indication of a fault that has happened. Over time, the fault may cause some degradation of the functionality of the faulty system component and may ultimately lead to a failure of the component or even of the entire system if no preventive counter measures are taken in due time. Therefore, the rationale as well as pros and cons of the standard Kalman filter and of the particle filter shall be briefly outlined in the following.

3.4.1 Discrete-Time Linear Kalman Filter

The discrete-time linear Kalman filter (KF) was introduced by Kalman back in 1960 [32] and has been widely used in various applications since then. Moreover, various extension have been developed such as the extended Kalman filter (EKF) [50], or unscented KF (UKF) [31, 62]. The KF filter algorithm is purely formulated in the time domain in contrast to many other filters that are formulated in the frequency domain. Given a time series of inaccurate and uncertain observations, the KF is an optimal estimator of the unknown state of a linear system in the sense that it minimises the estimated state error covariance when some presumed conditions are met.

The KF algorithm starts from two models, a *linear* state evolution equation and a *linear* observation equation.

$$\boldsymbol{x}_k = \mathbf{A}_{k-1}\boldsymbol{x}_{k-1} + \mathbf{B}_{k-1}\boldsymbol{u}_{k-1} + \boldsymbol{w}_{k-1} \tag{3.14a}$$

$$z_k = \mathbf{H}_k\boldsymbol{x}_k + \boldsymbol{v}_k \tag{3.14b}$$

where $\boldsymbol{x}_k \in \mathbb{R}^n$, $z_k \in \mathbb{R}^p$, and $\boldsymbol{u}_k \in \mathbb{R}^m$ are state, measurement, and input vectors, respectively, and $z_k \in \mathbb{R}^m$ the observation vector. In these models, the random vector $\boldsymbol{w}_k$ accounts for process noise and the random vector $\boldsymbol{v}_k$ for measurement noise. It is assumed that both vectors are uncorrelated *Gaussian*, zero mean random sequences with *known covariances* and are uncorrelated with the initial state $\boldsymbol{x}_0$, which is a random vector with known mean $\mu_0 = E[\boldsymbol{x}_0]$ and covariance $\mathbf{P}_0 = E[(\boldsymbol{x}_0 - \mu_0)(\boldsymbol{x}_0 - \mu_0)^T]$. That is,

$$E[\boldsymbol{w}_k] = 0 \quad E[\boldsymbol{v}_k] = 0 \tag{3.15}$$

$$E[\boldsymbol{w}_k\boldsymbol{w}_k{}^T] = \mathbf{Q}_k \quad E[\boldsymbol{v}_k\boldsymbol{v}_k{}^T] = \mathbf{R}_k \tag{3.16}$$

where $\mathbf{Q}_k$ denotes the $n \times n$ process noise covariance matrix and $\mathbf{R}_k$ the $m \times m$ measurement noise covariance matrix.

$$E[\boldsymbol{w}_k\boldsymbol{w}_j{}^T] = 0 \quad E[\boldsymbol{v}_k\boldsymbol{v}_j{}^T] = 0 \quad k \neq j \tag{3.17}$$

$$E[\boldsymbol{w}_k\boldsymbol{x}_0^T] = 0 \quad E[\boldsymbol{v}_k\boldsymbol{x}_0^T] = 0 \quad \forall\, k \tag{3.18}$$

Moreover, process noise and measurement noise are assumed to be uncorrelated.

$$E[\boldsymbol{w}_k\boldsymbol{v}_j^T] = 0 \quad \forall\, k, j \tag{3.19}$$

The Kalman filter algorithm consists of two stages.

Time Update

In a time update, or prediction step from time instant t_{k-1} to t_k, the estimated state $\hat{\boldsymbol{x}}_{k-1}$ is used to predict the state $\hat{\boldsymbol{x}}_k$ and the estimated state error covariance matrix $\mathbf{P}_{k-1}$ is also updated.

$$\hat{\boldsymbol{x}}_{k|k-1} = \mathbf{A}_{k-1}\hat{\boldsymbol{x}}_{k-1} + \mathbf{B}_{k-1}\boldsymbol{u}_{k-1} \tag{3.20}$$

$$\mathbf{P}_{k|k-1} = \mathbf{A}_{k-1}\mathbf{P}_{k-1}\mathbf{A}_{k-1}^T + \mathbf{Q}_{k-1} \tag{3.21}$$

where the index $k|k-1$ means evaluation at time instant t_k given the matrix is known at the previous time instant t_{k-1}. That is, the time update step yields a priori estimates of the state and of the state error covariance matrix for the next time step.

Measurement Update

The a priori estimates are corrected in a subsequent measurement update by scaling the difference between the actual measurement z_k and the predicted output $\mathbf{H}_k\hat{\boldsymbol{x}}_{k|k-1}$ also called the measurement innovation, or residual. The scaling factor is the $n \times m$ Kalman gain matrix

$$\mathbf{K}_k = \mathbf{P}_{k|k-1}\mathbf{H}_k^T \left(\mathbf{H}_k\mathbf{P}_{k|k-1}\mathbf{H}_k^T + \mathbf{R}_k\right)^{-1} \tag{3.22}$$

where $\mathbf{S}_k := \mathbf{H}_k\mathbf{P}_{k|k-1}\mathbf{H}_k^T + \mathbf{R}_k$ is the measurement prediction covariance matrix.

The Kalman gain matrix $\mathbf{K}_k$ minimises the a posteriori estimate of the error covariance matrix

$$\mathbf{P}_k = E[\boldsymbol{e} \cdot \boldsymbol{e}^T] \tag{3.23}$$

where $\boldsymbol{e} := \boldsymbol{x}_k - \hat{\boldsymbol{x}}_k$.

The a posteriori estimates of the state $\hat{\boldsymbol{x}}_k$ and the error covariance matrix $\mathbf{P}_k$ then read

$$\hat{\boldsymbol{x}}_k = \hat{\boldsymbol{x}}_{k|k-1} + \mathbf{K}_k \left(z_k - \mathbf{H}_k\hat{\boldsymbol{x}}_{k|k-1}\right) \tag{3.24}$$

$$\mathbf{P}_k = (\mathbf{I} - \mathbf{K}_k\mathbf{H}_k)\,\mathbf{P}_{k|k-1} \tag{3.25}$$

where $\mathbf{I}$ denotes the identity matrix.

Equation (3.24) indicates that the state estimate $\hat{\boldsymbol{x}}_k$ of the KF is the sum of the predicted state estimate $\hat{\boldsymbol{x}}_{k|k-1}$ plus the error weighted by the Kalman gain $\mathbf{K}_k$. The Kalman filtering algorithm is depicted as a flowchart in Fig. 3.8.

Due to its two stages, the recursive KF algorithm resembles a predictor-corrector algorithm. Estimating the state of a system by means of a standard Kalman filter may be considered a sort of feedback control. The filter provides a state estimate $\hat{\boldsymbol{x}}_k$ and receives as a feedback a new noisy measurement z_k. Figure 3.9 [60] displays

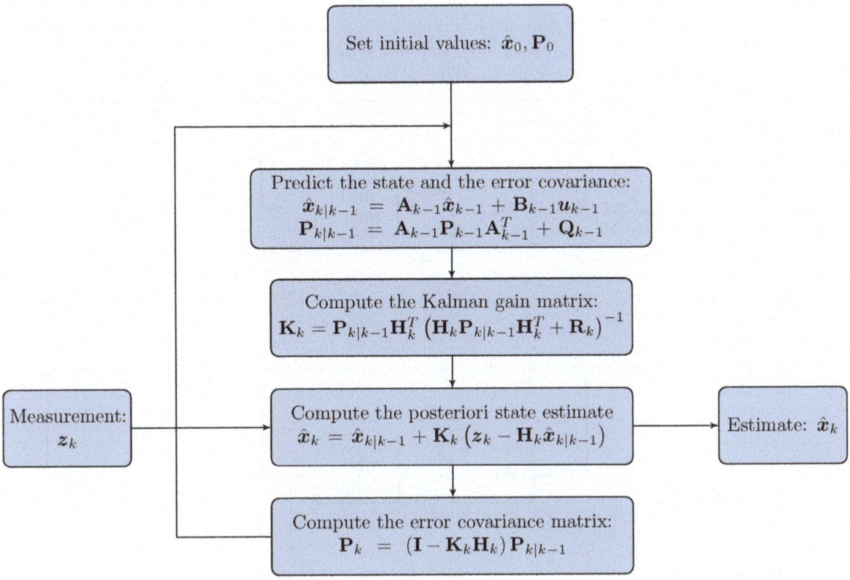

Fig. 3.8 Flowchart of the recursive Kalman filtering algorithm

a block diagram of the KF representing the model equations (3.14) and the filter equations (3.20) and (3.24).

Some Remarks

The covariance matrix $\mathbf{P}_0$ depends on the choice of the initial state $\boldsymbol{x}_0$ and affects the initial convergence of the filter. The covariance matrices $\mathbf{Q}_k$ and $\mathbf{R}_k$ significantly affect the estimation performance of the filter. If more process noise is assumed, then the state estimation is more uncertain, which affects the state error covariance matrix $\mathbf{P}_{k|k-1}$ (Eq. 3.21). According to (3.22) and (3.24) this results in a heavier correction of the a priori state estimate $\boldsymbol{x}_{k|k-1}$.

$$\lim_{\mathbf{R}_k \to \mathbf{0}} \mathbf{K}_k = \mathbf{H}_k^{-1} \tag{3.26}$$

In other words, a decrease of the measurement error covariance matrix $\mathbf{R}_k$ means that the actual measurement z_k becomes more correct and the predicted measurement $\mathbf{H}_k\hat{\boldsymbol{x}}_{k|k-1}$ less accurate.

On the other side, more measurement noise results in a less trusted measurement, which leads to less correction of the a priori state estimate.

$$\lim_{\mathbf{P}_{k|k-1} \to \mathbf{0}} \mathbf{K}_k = \mathbf{0} \tag{3.27}$$

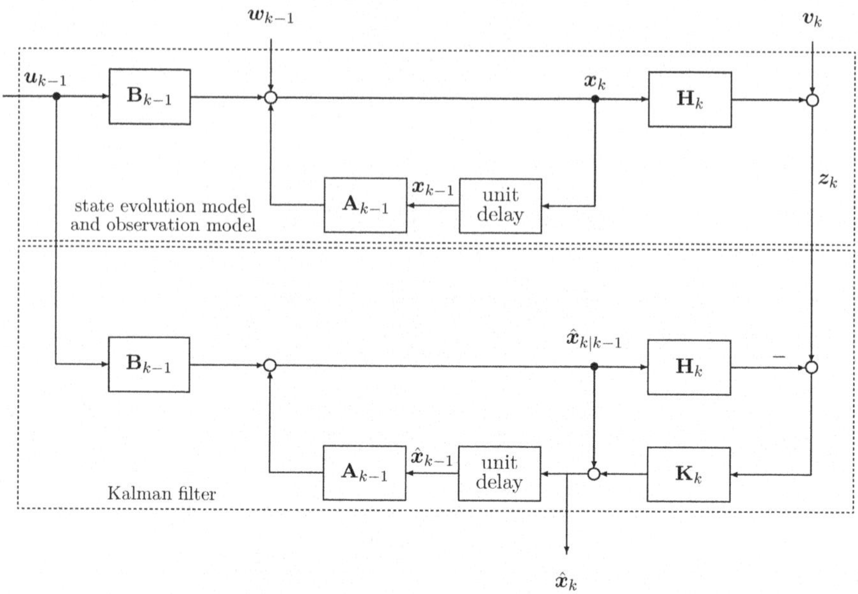

Fig. 3.9 Block diagram of the Kalman filter

A decrease of the a priori state error covariance matrix $\mathbf{P}_{k|k-1}$ means that the actual measurements are to be considered less trustworthy in contrast to the predicted measurements $\mathbf{H}_k\hat{\boldsymbol{x}}_{k|k-1}$. The measurement noise covariance matrix $\mathbf{R}_k$ may be determined. In contrast, the process states to be estimated cannot be directly observed. Therefore, the process noise covariance matrix $\mathbf{Q}_k$ has to be tuned to improve the filter performance.

For the state error prediction covariance matrix $\mathbf{P}_{k|k-1}$, a discrete matrix Riccati equation can be obtained from the Kalman filter equations. In the case of a LTI system, its solution converges to a constant symmetric positive definite steady state covariance matrix $\bar{\mathbf{P}}$ under some assumptions and yields a constant steady state Kalman gain matrix [50]. The KF then becomes time invariant.

$$\mathbf{0} = \mathbf{A}\bar{\mathbf{P}}\mathbf{A}^T - \mathbf{A}\bar{\mathbf{P}}\mathbf{H}^T[\mathbf{H}\bar{\mathbf{P}}\mathbf{H}^T + \mathbf{R}]^{-1}\mathbf{H}\bar{\mathbf{P}}\mathbf{A}^T - \bar{\mathbf{P}} \tag{3.28}$$

where $\bar{\mathbf{P}} = \lim_{k\to\infty} \mathbf{P}_{k|k-1}$

In the case of errors in the system modelling, the filter estimate may differ significantly from the ground truth although the state error covariance is low.

In conclusion, the standard recursive KF algorithm can be used in real-time. The KF only uses present measurements, the calculated state one time step back and the state error covariance matrix. No other past information need to be stored in a buffer. The algorithm is easily implemented. A weakness of the standard KF is that it is limited to linear systems and assumes additive noise that is Gaussian with zero mean. Particle filters briefly considered in Sect. 3.4.2 do not have these limitations.

Simultaneous Estimation of States and Parameters

For fault detection and control applications, it is important to constantly determine the state of health of a system. The standard Kalman filter enables to estimate the state. However, while some components of the state vector may change rapidly with time, system parameters may change slowly with time so that they cannot be directly determined by means of known inputs and measured outputs. The state estimation by means of a Kalman filter considered so far can be extended so that states and parameters can be estimated simultaneously. One approach called Joint EKF is to regard parameters as variables in the dynamical model, to augment the state vector with the vector of system parameters and to apply an EKF on the augmented model. Another option called the Dual EKF uses an EKF for state estimation and a second cooperating EKF for parameter estimation. In the following, only the joint EKF is briefly considered. Further information on the use of the KF for parameter estimation may be found in [18, 36, 38, 46].

Slow changes of a component parameter vector $\boldsymbol{\theta} \in \mathbb{R}^p$ over time may be modelled by some *fictitious* noise $\boldsymbol{r}_k$ assumed to be a Gaussian noise process with zero mean that is independent of the process noise $\boldsymbol{w}_k$ and the measurement noise $\boldsymbol{v}_k$.

$$\boldsymbol{\theta}_k = \boldsymbol{\theta}_{k-1} + \boldsymbol{r}_{k-1} \tag{3.29}$$

As the parameter vector is regarded a variable, the equation for the forecast of the state $\boldsymbol{x}$ is nonlinear in general.

$$\boldsymbol{x}_k = \mathbf{f}(\boldsymbol{x}_{k-1}, \boldsymbol{u}_{k-1}, \boldsymbol{\theta}_{k-1}) + \boldsymbol{w}_{k-1} \tag{3.30}$$

In the joint EKF approach, the augmented state space model then captures both the state evolution and the parameter evolution.

$$\begin{bmatrix} \boldsymbol{x}_k \\ \boldsymbol{\theta}_k \end{bmatrix} = \begin{bmatrix} \mathbf{f}(\boldsymbol{x}_{k-1}, \boldsymbol{u}_{k-1}, \boldsymbol{\theta}_{k-1}) \\ \boldsymbol{\theta}_{k-1} \end{bmatrix} + \begin{bmatrix} \boldsymbol{w}_{k-1} \\ \boldsymbol{r}_{k-1} \end{bmatrix} \tag{3.31}$$

$$z_k = \mathbf{g}(\boldsymbol{x}_{k-1}, \boldsymbol{u}_{k-1}, \boldsymbol{\theta}_{k-1}) + \boldsymbol{v}_{k-1} \tag{3.32}$$

As the augmented state space model is nonlinear, the EKF approach first linearises the state space at each time instant around the current state estimate $\hat{\boldsymbol{x}}_k$ and then proceeds with the standard KF algorithm, which starts with a guess for $E[\boldsymbol{x}_0]$, $E[\boldsymbol{\theta}_0]$ and an initialisation of the error covariance matrix for the augmented state vector.

The EKF allows for a nonlinear state evolution model and a nonlinear observation model. Their functions are required to be differentiable. However, the EKF is still limited to Gaussian noise processes and unlike the standard KF it is not an optimal estimator in general. Moreover, the EKF is sensitive to the initial estimate of the state and tends to underestimate the true error covariance matrix.

3.4.2 Particle Filters

The particle filter (PF) was introduced by Gordon et. al. back in 1993 [28]. There are various flavours of PFs. In the following, only the principle of the basic particle filter (PF) is outlined. For details and more general particle filters, it is referred to the literature [3, 42, 52, 53].

A particle filter is a recursive Bayesian filter that enables to compute the posterior probability density function (pdf) of the state of a dynamic system on the basis of all currently available information. The state evolution model and the observation model do not need to be linear. Process noise and measurement noise do not need to be Gaussian but may have arbitrary distributions. That is, the state space model may be of the form

$$\boldsymbol{x}_k = \mathbf{f}_{k-1}(\boldsymbol{x}_{k-1}, \boldsymbol{w}_{k-1}) \tag{3.33}$$

$$z_k = \mathbf{h}_k(\boldsymbol{x}_k, \boldsymbol{v}_k) \tag{3.34}$$

where $\boldsymbol{x}_k$ is the system state to be estimated, $\mathbf{f}_{k-1}$ a known, possibly nonlinear state transition function, $\boldsymbol{w}_{k-1}$ a sequence of process noise with a *known* probability density function (pdf), z_k the vector of measurements provided by the system sensors and a *feature extraction* module related to the state $\boldsymbol{x}_k$ and the measurement noise $\boldsymbol{v}_k$ with a *known* pdf by means of a known observation function $\mathbf{h}_k$. Process noise and measurement noise are assumed to be mutually independent.

From a probabilistic point of view, (3.33) describes a first order Markov process and (3.33) is equivalent to the *transition density* $p(\boldsymbol{x}_k|\boldsymbol{x}_{k-1})$. Equation 3.34 is equivalent to the conditional pdf $p(z_k|\boldsymbol{x}_k)$, which is the likelihood that a state $\boldsymbol{x}_k$ gives rise to an observation z_k. In order to estimate the state, i.e. to compute the posterior pdf $p(\boldsymbol{x}_k|z_{1:k})$ at time instant t_k, the filter needs to know the prior pdf of the initial state $p(\boldsymbol{x}_0)$ at t_0, where $z_{1:k} = \{z_k|k = 1, 2, \ldots, k\}$ denotes the sequence of all available measurements up to and including time instant t_k.

The operation of a recursive PF consists of a prediction state and an update stage. In the prediction step, the PF propagates an available posterior pdf $p(\boldsymbol{x}_{k-1}|z_{1:k-1})$ forward to the next time instant t_k. The Chapman–Kolmogorov equation and (3.33) yield the prior pdf

$$\begin{aligned} p(\boldsymbol{x}_k|z_{1:k-1}) &= \int p(\boldsymbol{x}_k, \boldsymbol{x}_{k-1}|z_{1:k-1}) d\boldsymbol{x}_{k-1} \\ &= \int p(\boldsymbol{x}_k|\boldsymbol{x}_{k-1}, z_{1:k-1}) p(\boldsymbol{x}_{k-1}|z_{1:k-1}) d\boldsymbol{x}_{k-1} \\ &= \int p(\boldsymbol{x}_k|\boldsymbol{x}_{k-1}) p(\boldsymbol{x}_{k-1}|z_{1:k-1}) d\boldsymbol{x}_{k-1} \end{aligned} \tag{3.35}$$

The prior pdf $p(\boldsymbol{x}_k|z_{1:k-1})$ can be updated by means of Bayes rule and the new measurement z_k to give the new posterior pdf $p(\boldsymbol{x}_k|z_{1:k})$ of $\boldsymbol{x}_k$ at time instant t_k.

$$p(\boldsymbol{x}_k|z_{1:k}) = \frac{p(z_k|\boldsymbol{x}_k)p(\boldsymbol{x}_k|z_{1:k-1})}{p(z_k|z_{1:k-1})} \tag{3.36}$$

The normalisation $\int p(\boldsymbol{x}_k|z_{1:k})d\boldsymbol{x}_k = 1$ yields for the denominator in (3.36)

$$p(z_k|z_{1:k-1}) = \int p(z_k|\boldsymbol{x}_k)p(\boldsymbol{x}_k|z_{1:k-1})d\boldsymbol{x}_k \tag{3.37}$$

Equations 3.35 and 3.36 together with the initial condition $p(\boldsymbol{x}_0|z_{1:0}) := p(\boldsymbol{x}_0)$, i.e. no measurements are received yet, are the constitutive equations of a recursive Bayesian filter (Fig. 3.10).

However, the input $p(\boldsymbol{x}_{k-1}|z_{1:k-1})$ into the prediction stage is usually not available in closed form and the integral in (3.35) for the prior pdf $p(\boldsymbol{x}_k|z_{1:k-1})$ is difficult to determine analytically. An analytic computation of the posterior pdf $p(\boldsymbol{x}_k|z_{1:k})$ is only possible in few special cases. For a linear state space model with Gaussian process noise and Gaussian measurement noise being mutually independent, the equations lead to those of the standard Kalman filter.

The key idea of particle filters is to approximate the posterior pdf $p(\boldsymbol{x}_k|z_{1:k})$ of $\boldsymbol{x}_k$ by a weighted sum of N_s random samples, also called particles associated with normalised weights w_k^i ($\sum_i^{N_s} w_k^i = 1$).

$$p(\boldsymbol{x}_k|z_{1:k}) \approx \sum_{i=1}^{N_s} w_k^i \delta(\boldsymbol{x}_k - \boldsymbol{x}_k^i) \tag{3.38}$$

where δ denotes the Dirac impulse function. For large N_s the approximation is close to the true pdf.

Assume that a set $\{\boldsymbol{x}_{k-1}^{i\star}\}_{i=1}^{N_s}$ of N_s random samples have been drawn from the posterior pdf $p(\boldsymbol{x}_{k-1}|z_{1:k-1})$. In the prediction phase, the state evolution equation is used together with samples $\boldsymbol{w}_{k-1}^i$ drawn from the known pdf of the process noise to generate a set of prior samples $\{\boldsymbol{x}_k^i\}_{i=1}^{N_s}$.

$$\boldsymbol{x}_k^i = \mathbf{f}_{k-1}(\boldsymbol{x}_{k-1}^*, \boldsymbol{w}_{k-1}^i) \tag{3.39}$$

As a result, a set of samples is produced from the prior pdf $p(\boldsymbol{x}_k|z_{1:k-1})$.

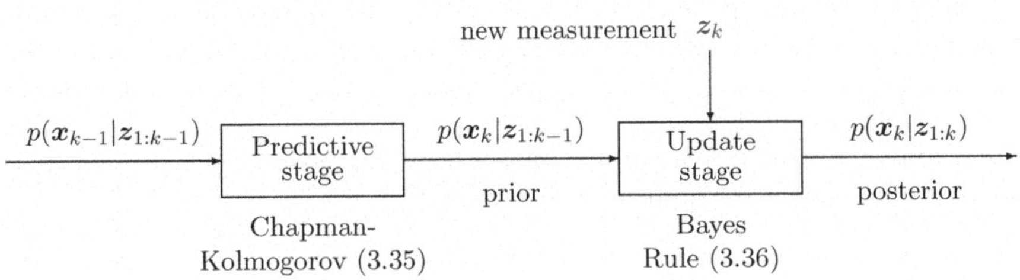

Fig. 3.10 Scheme of a recursive Bayesian filter

In the update phase, a *resampling* with replacement of prior particles is performed to generate a new set of particles. The newly available measurement z_k is used to calculate a normalised *importance weight* for each sample.

$$\tilde{w}_k^i = p(z_k|\boldsymbol{x}_k^i) \tag{3.40}$$

$$w_k^i = \frac{\tilde{w}_k^i}{\sum_{j=1}^{N_s} \tilde{w}_k^j} \tag{3.41}$$

From the set of prior samples $\{\boldsymbol{x}_k^i\}_{i=1}^{N_s}$ obtained by means of the state evolution model a member $\boldsymbol{x}_k^{i\star}$ is chosen with a probability Pr equal to its weight.

$$Pr\{\boldsymbol{x}_k^{i\star} = \boldsymbol{x}_k^j\} = w_k^j \qquad \forall\, i, j \tag{3.42}$$

This choice is repeated N_s times. That is, the cardinality of the new set $\{\boldsymbol{x}_k^{i\star}\}_{i=1}^{N_s}$ equals one of the set of prior samples. However, samples with high weight may be chosen multiple times at the expense of samples with low weight. The result is a density of particles in areas under the posterior pdf curve with high probability. High values of the measurement likelihood $p(z_k|\boldsymbol{x}_k^i)$ evaluated at the prior samples $\boldsymbol{x}_k^i$ in the numerator of the normalised weights indicate state values that are likely. Low values suggest that the state value associated with the latest measurement is unlikely. The resulting new samples $\boldsymbol{x}_k^{i\star}$ are then the ones of the posterior pdf $p(\boldsymbol{x}_k|z_{1:k})$ to be computed.

Note that this resampling with replacement is performed whenever a new measurement z_k is available. Each sample 'survives' in proportion to its weight. Particles with a high weight are replicated. If particles are not resampled and are used again in the state evolution model, then this affects the transition probability for the next time step and may lead to unlikely states, a propagation of samples with low weights and to an increase of particles with low probability. This phenomenon is sometimes called 'particle depletion'. As a result, areas under the posterior pdf curve with high probability are no longer represented by a sufficient number of particles.

In the literature, the outlined algorithm, which starts from a pdf of the initial state $\boldsymbol{x}_0$, is known as Sampling Importance Resampling (SIR) filter or bootstrap filter [28] or Monte Carlo Filter [37]. The particle filter approach to online state estimation has become popular since it has not got the limitations of the standard KF and the extended KF and due to the simplicity of the basic algorithm, in which only the number of samples is a tuning parameter. Figure 3.11 shows a flowchart of the particle filter algorithm.

Moreover, recently, a dual particle filter scheme has been proposed that enables a simultaneous state and parameter estimation [20].

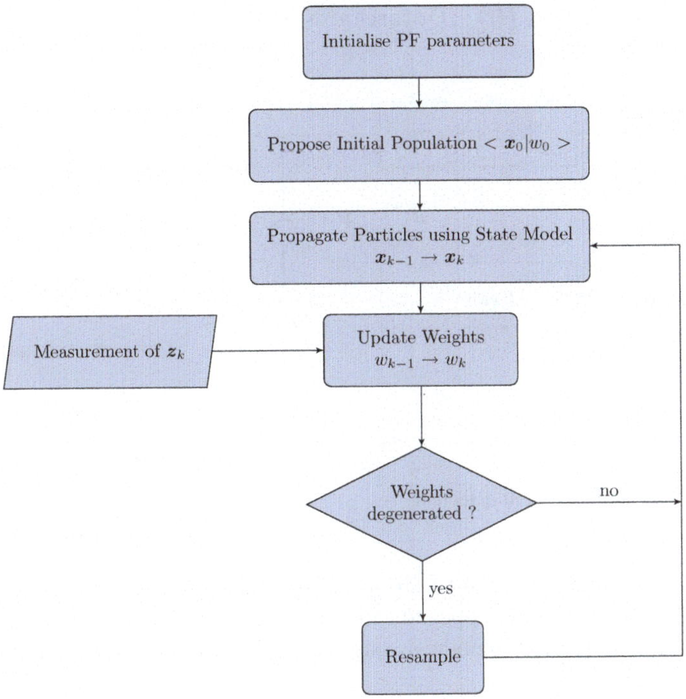

Fig. 3.11 Flowchart of the particle filter algorithm

3.5 Bond Graph Model-Based Fault Detection and Isolation

Various model-based fault detection and isolation (FDI) techniques reported in the literature build on either a comparison of measurements obtained from a real physical system and outputs of a model, i.e. on output residuals, or on constraints between measurements and known control signals called *Analytical Redundancy Relations* (ARRs). The result of a numerical evaluation of an ARR is denoted as ARR residual. Model-based approaches to fault diagnosis often use residuals as features. The following sections show how Bond Graph methodology can support both approaches. Bond Graph methodology can be used to create the model that both approaches need.

3.5.1 Observer-Based Fault Detection

If process and measurement noise are significant and have to be taken into account by a dynamic system model, then Kalman or particle filters considered in the previous section may be used for fault detection. The continuous comparison of measurement delivered by sensors with outputs computed by the filter algorithm

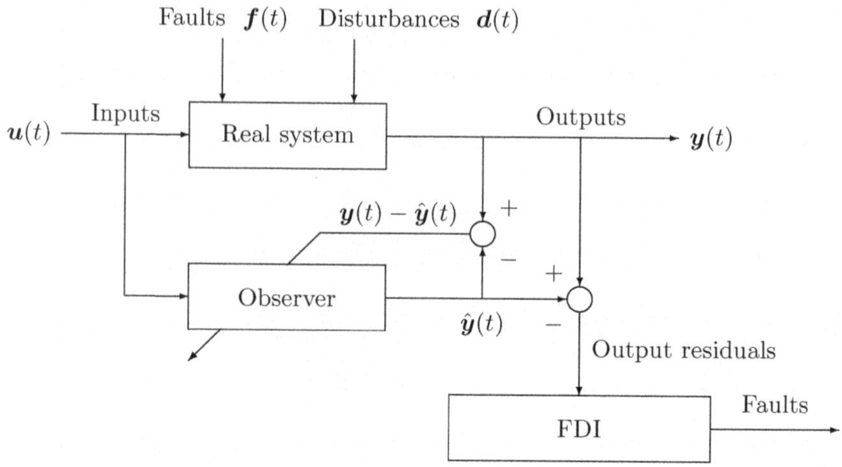

Fig. 3.12 Observer-based fault detection

give rise to the time evolution of an output residual that may indicate as of some time instant that a fault has happened. A bank of filters with each of them accounting for a different single fault hypothesis may be used to isolate a detected fault [55].

Unknown System Inputs

If measurement noise has been appropriately filtered, an observer may be used for fault detection and a bank of observers for fault isolation (Fig. 3.12).

Suppose that the system or process under consideration may be presented by a linear time invariant model.

$$\dot{\boldsymbol{x}}(t) = \mathbf{A}\boldsymbol{x}(t) + \mathbf{B}\boldsymbol{u}(t) + \mathbf{E}\boldsymbol{d}(t) \tag{3.43a}$$

$$\boldsymbol{y}(t) = \mathbf{C}\boldsymbol{x}(t) \tag{3.43b}$$

where $\boldsymbol{x}(t) \in \mathbb{R}^n$ denotes the state vector, $\boldsymbol{y}(t) \in \mathbb{R}^m$ the vector of outputs, $\boldsymbol{u}(t) \in \mathbb{R}^r$ the known inputs, $\boldsymbol{d}(t) \in \mathbb{R}^q$ the vector of disturbances, or model uncertainties such as nonlinearities, parametric uncertainties, or noise, and $\mathbf{E}$ is a known matrix assumed to be of full column rank.

Due to the unknown input $\boldsymbol{d}(t)$, the design of a standard Luenberger observer without disturbance decoupling is inappropriate, since the latter one uses all inputs for an estimation of the state. For dynamic models of the form (3.43), commonly an Unknown Input Observer (UIO) is used [16], which is designed such that the state estimation error $\boldsymbol{e}(t) := \boldsymbol{x}(t) - \hat{\boldsymbol{x}}(t)$ approaches zero asymptotically *regardless* of the presence of an unknown input $\boldsymbol{d}(t)$. The dynamic equations of a full order UIO are

$$\dot{\boldsymbol{z}}(t) = \mathbf{F}\boldsymbol{z}(t) + \mathbf{TB}\boldsymbol{u}(t) + \mathbf{K}\boldsymbol{y}(t) \tag{3.44a}$$

Fig. 3.13 Block diagram of a full order UIO

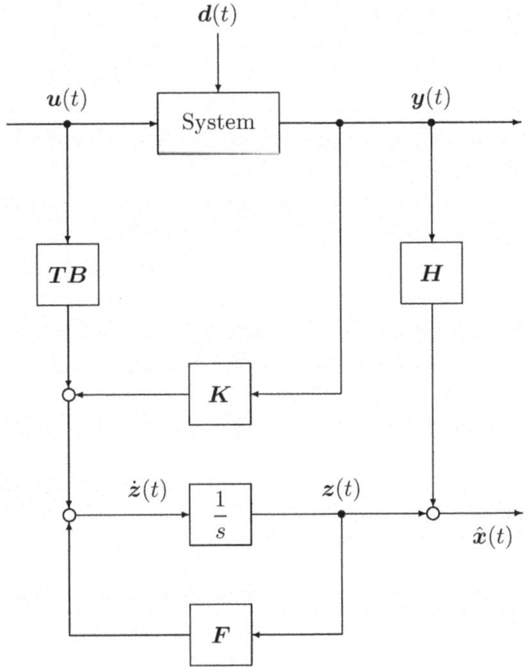

$$\hat{\boldsymbol{x}}(t) = z(t) + \mathbf{H}\boldsymbol{y}(t) \tag{3.44b}$$

$$\mathbf{K} = \mathbf{K}_1 + \mathbf{K}_2 \tag{3.44c}$$

where $z(t) \in \mathbb{R}^n$ is the state of the full order observer, $\hat{\boldsymbol{x}}(t) \in \mathbb{R}^n$ the estimate of the state $\boldsymbol{x}(t)$, and $\mathbf{F}, \mathbf{T}, \mathbf{K}, \mathbf{H}$ are matrices to be designed such that the unknown input $\boldsymbol{d}(t)$ is *decoupled*. Figure 3.13 [16] displays a block diagram of a full order UIO.

From (3.43) and (3.44) one obtains for the state estimation error:

$$\begin{aligned}\dot{e} &= (\mathbf{A} - \mathbf{HCA} - \mathbf{K}_1\mathbf{C})e + [\mathbf{F} - (\mathbf{A} - \mathbf{HCA} - \mathbf{K}_1\mathbf{C})]z \\ &\quad + [\mathbf{K}_2 - (\mathbf{A} - \mathbf{HCA} - \mathbf{K}_1\mathbf{C})\mathbf{H}]y \\ &\quad + [\mathbf{T} - (\mathbf{I} - \mathbf{HC})]\mathbf{B}u + (\mathbf{HC} - \mathbf{I})\mathbf{E}d\end{aligned} \tag{3.45}$$

As can be seen from (3.45), the state error estimation can be decoupled from the state of the observer and its inputs if the following conditions hold:

$$\mathbf{0} = \mathbf{F} - (\mathbf{A} - \mathbf{HCA} - \mathbf{K}_1\mathbf{C}) \tag{3.46}$$

$$\mathbf{0} = \mathbf{K}_2 - \mathbf{FH} \tag{3.47}$$

$$\mathbf{0} = \mathbf{T} - (\mathbf{I} - \mathbf{HC}) \tag{3.48}$$

$$\mathbf{0} = (\mathbf{HC} - \mathbf{I})\mathbf{E} \tag{3.49}$$

The equation for the state error estimation then reduces to

$$\dot{e} = \mathbf{F}e \tag{3.50}$$

If all eigenvalues of $\mathbf{F}$ are stable, then $\boldsymbol{e} \rightarrow \mathbf{0}$ for $t \rightarrow \infty$ regardless of the presence of unknown disturbance inputs.

Equation 3.49 is solvable if and only if $rank(\mathbf{CE}) = rank(\mathbf{E})$. In that case

$$\mathbf{H} = \mathbf{E}(\mathbf{CE})^+ \tag{3.51}$$

is a solution of (3.49), where $(\mathbf{CE})^+$ is a left inverse of $(\mathbf{CE})$.[1]

If in addition, the pair $(\mathbf{C}, \mathbf{A}_1)$ is detectable,[2] then the UIO (3.44) exists, where $\mathbf{A}_1 := \mathbf{A} - \mathbf{ECE}^+\mathbf{CA}$ [16].

If the UIO exists, the unknown disturbance $\boldsymbol{d}(t)$ can be estimated.

$$\dot{\hat{y}} = \mathbf{C}\dot{\hat{x}} = \mathbf{CA}\hat{x} + \mathbf{CB}\hat{u} + \mathbf{CE}d \tag{3.52}$$

Solving (3.52) for $\boldsymbol{d}$ yields

$$\hat{d} = (\mathbf{CE})^+[\dot{\hat{y}} - \mathbf{CA}\hat{x} - \mathbf{CB}\hat{u}] \tag{3.53}$$

Luenberger Observer

If there are no unknown disturbances and known disturbances tend to zero for $t \rightarrow \infty$, then a standard Luenberger observer may be used for fault detection. Now assume that process noise, measurement noise, and disturbances can be disregarded. The UIO then reduces to a standard Luenberger observer. With $\mathbf{T} = \mathbf{I}$ and $\mathbf{H} = \mathbf{0}$ the equations of an UIO read

$$\dot{z} = \mathbf{F}z + \mathbf{B}u + \mathbf{K}y \tag{3.54}$$

$$\hat{x} = z \tag{3.55}$$

$$\mathbf{K} = \mathbf{K}_1 + \mathbf{K}_2 \tag{3.56}$$

$$\mathbf{F} = \mathbf{A} - \mathbf{K}_1\mathbf{C} \tag{3.57}$$

$$\mathbf{K}_2 = \mathbf{0} \tag{3.58}$$

or

$$\begin{aligned}\frac{d}{dt}\hat{x} &= (\mathbf{A} - \mathbf{KC})\hat{x} + \mathbf{B}u + \mathbf{K}y \\ &= \mathbf{A}\hat{x} + \mathbf{B}u + \mathbf{K}(y - \hat{y})\end{aligned} \tag{3.59}$$

[1] $(\mathbf{CE})^+ = [(\mathbf{CE})^T(\mathbf{CE})]^{-1}(\mathbf{CE})^T$.

[2] A pair $(\mathbf{C}, \mathbf{A})$ is *detectable* when all unobservable modes for this pair are stable [15].

Combining a Luenberger State Observer with a Disturbance Estimator

When a system is subject to external disturbances, an estimation of its states by means of a standard Luenberger observer may result in a steady state error depending on the type of the disturbances.

Consider the LTI system given by Eq. 3.43 and reproduced as (3.60)

$$\dot{\boldsymbol{x}}(t) = \mathbf{A}\boldsymbol{x}(t) + \mathbf{B}\boldsymbol{u}(t) + \mathbf{E}\boldsymbol{d}(t) \tag{3.60a}$$

$$\boldsymbol{y}(t) = \mathbf{C}\boldsymbol{x}(t) \tag{3.60b}$$

and assume that the disturbance $\boldsymbol{d}(t)$ on the plant is a step with a height $\bar{\boldsymbol{d}}$ which becomes effective as of some time instant t_1. The state estimation error $\boldsymbol{e}(t)$ then depends on the disturbance.

$$\dot{\boldsymbol{e}}(t) = (\mathbf{A} - \mathbf{LC})\boldsymbol{e}(t) + \mathbf{E}\boldsymbol{d}(t) \tag{3.61}$$

In the case that the state estimation matrix $\mathbf{A}_o = \mathbf{A} - \mathbf{LC}$ is stable, the steady state output error is

$$\boldsymbol{y}(\infty) - \hat{\boldsymbol{y}}(\infty) = \mathbf{C}\boldsymbol{e}(\infty) \;=\; -\mathbf{C}(\mathbf{A} - \mathbf{LC})^{-1}\mathbf{E}\bar{\boldsymbol{d}} \neq \mathbf{0} \tag{3.62}$$

To account for the steady state error in the case of an external step disturbance as of some time instant t_1, recently it has been proposed to use a Luenberger state observer

$$\dot{\hat{\boldsymbol{x}}}(t) = \mathbf{A}\hat{\boldsymbol{x}}(t) + \mathbf{L}(\boldsymbol{y}(t) - \hat{\boldsymbol{y}}(t)) + \mathbf{B}\boldsymbol{u}(t) + \mathbf{E}\hat{\boldsymbol{d}}(t) \tag{3.63}$$

which in addition to the command signal $\boldsymbol{u}(t)$ and the output error $\boldsymbol{y}(t) - \hat{\boldsymbol{y}}(t)$ uses an estimate $\hat{\boldsymbol{d}}(t)$ of the disturbance and to combine the state observer with a disturbance estimator [2]. The state estimation error $\boldsymbol{e}(t) = \hat{\boldsymbol{x}}(t) - \boldsymbol{x}(t)$ then depends on the disturbance error $\tilde{\boldsymbol{d}}(t) := \boldsymbol{d}(t) - \hat{\boldsymbol{d}}(t)$.

$$\dot{\boldsymbol{e}}(t) = \underbrace{(\mathbf{A} - \mathbf{LC})}_{\mathbf{A}_o}\boldsymbol{e}(t) - \mathbf{E}\tilde{\boldsymbol{d}}(t) \tag{3.64}$$

The observer gain matrix $\mathbf{L}$ is chosen such that all eigenvalues of the state estimation matrix $\mathbf{A}_o$ have a strict negative real part, which ensures that the state estimation error tends to zero in the case of no disturbances. To get an expression for the disturbance error $\tilde{\boldsymbol{d}}(t)$, let the disturbance estimate be

$$\hat{\boldsymbol{d}}(t) = z + q\boldsymbol{y}(t) \tag{3.65}$$

and z an internal state.

$$\dot{z}(t) = -q\mathbf{C}(\mathbf{A}\hat{\boldsymbol{d}}(t)\hat{\boldsymbol{x}}(t) + \mathbf{B}\boldsymbol{u}(t) + \mathbf{E}\hat{\boldsymbol{d}}(t)) \tag{3.66}$$

This yields a matrix equation for the state estimation error and the disturbance estimation error.

$$\frac{d}{dt}\begin{bmatrix} \boldsymbol{e} \\ \tilde{\boldsymbol{d}} \end{bmatrix} = \underbrace{\begin{bmatrix} \mathbf{A}_o & -\mathbf{E} \\ q\mathbf{CA} & -q\mathbf{CA} \end{bmatrix}}_{\mathbf{A}_{ed}} \begin{bmatrix} \boldsymbol{e} \\ \tilde{\boldsymbol{d}} \end{bmatrix} \tag{3.67}$$

The gain parameter q is chosen such that the eigenvalues of $\mathbf{A}_{ed}$ have a strict negative real part. In that case, the state estimation error and the disturbance error both simultaneously converge asymptotically to zero. That is, the states of a system can be accurately estimated despite an external step disturbance and can be used for the detection of parametric faults, or updates of the disturbance estimate can be used in a state feedback controller as in [2].

Actuator Faults, Sensor Faults, and Parametric Plant Faults

Actuator or sensor faults, or parametric plant faults affect the output residual $\boldsymbol{r} := \boldsymbol{y}(t) - \hat{\boldsymbol{y}}(t)$. A significant change in $\boldsymbol{r}$ then indicates a fault. The dynamic system model reads

$$\dot{\boldsymbol{x}} = \mathbf{A}\boldsymbol{x} + \mathbf{B}\boldsymbol{u} + \mathbf{K}_a \boldsymbol{f}_a \tag{3.68}$$

$$\boldsymbol{y} = \mathbf{C}\boldsymbol{x} + \mathbf{G}_s \boldsymbol{f}_s \tag{3.69}$$

where $\boldsymbol{f}_a$ denotes a vector of actuator faults, $\boldsymbol{f}_s$ accounts for sensor faults, and $\mathbf{K}_a$, $\mathbf{G}_s$ are matrices of appropriate dimensions. Then one obtains for the state estimation error $\boldsymbol{e}_x$ and the output residual $\boldsymbol{r}$

$$\dot{\boldsymbol{e}}_x = \mathbf{F}\boldsymbol{e}_x + \mathbf{K}_a \boldsymbol{f}_a + \mathbf{G}_s \boldsymbol{f}_s \tag{3.70}$$

$$\boldsymbol{r} = \mathbf{C}\boldsymbol{e}_x + \mathbf{G}_s \boldsymbol{f}_s \tag{3.71}$$

As can be seen, if there are actuator or sensor faults, then the state estimation error and thus the output residual does not vanish. Likewise, in the case of a parametric plant fault the output residual also does not approach zero for $t \to \infty$. Assume that the parametric fault only affects the system matrix $\mathbf{A}$ and that $\mathbf{A}$ can be decomposed into a matrix $\mathbf{A}_n$ with coefficients being functions of the nominal parameters and a matrix $\Delta\mathbf{A}$ accounting for the parametric faults. Then one obtains for the state estimation error:

$$\begin{aligned} \dot{\boldsymbol{e}}_x &= \dot{\boldsymbol{x}} - \dot{\hat{\boldsymbol{x}}} \\ &= (\mathbf{A}_n + \Delta\mathbf{A})\boldsymbol{x} + \mathbf{B}\boldsymbol{u} - (\mathbf{F}\hat{\boldsymbol{x}} + \mathbf{B}\boldsymbol{u} + \mathbf{K}\boldsymbol{y}) \\ &= (\mathbf{A}_n - \mathbf{KC})\boldsymbol{x} - \mathbf{F}\hat{\boldsymbol{x}} + (\Delta\mathbf{A})\boldsymbol{x} \\ &= \mathbf{F}\boldsymbol{e}_x + (\Delta\mathbf{A})\boldsymbol{x} \end{aligned} \tag{3.72}$$

The Laplace transform of the output error $\boldsymbol{r}$ reads

$$\boldsymbol{r}(s) = \underbrace{(s\mathbf{I} - \mathbf{F})^{-1}}_{\mathbf{G}_{rf}(s)} \underbrace{(\Delta\mathbf{A})\boldsymbol{x}}_{\boldsymbol{f}(s)} \tag{3.73}$$

Let $\boldsymbol{f} = (\Delta\mathbf{A})\boldsymbol{x}$ be the vector of parametric faults. Then $\mathbf{G}_{rf}(s)$ is the transfer function matrix between the fault vector and the output error $\boldsymbol{r}$. If none of the entries of the matrix $\mathbf{G}_{rf}(s)$ vanishes, any non-zero fault affects the observer output error.

Decoupling Disturbances from the Observer Output Error

In the general case, where in addition to actuator, sensor, and parametric plant faults possible disturbances are taken into account, the output error is a weighted sum of the vector of faults $\boldsymbol{f}(s)$ and the vector of disturbances $\boldsymbol{d}(s)$

$$\boldsymbol{r}(s) = \mathbf{H}_f(s)\boldsymbol{f}(s) + \mathbf{H}_d(s)\boldsymbol{d}(s) \tag{3.74}$$

with transfer function matrices $\mathbf{H}_f(s)$, $\mathbf{H}_d(s)$.

In order to decouple the output error $\boldsymbol{r}$ from the disturbances $\boldsymbol{d}$ to enable a disturbance decoupled fault detection, a transformation matrix $\mathbf{T}(s)$ may be designed such that $\hat{\boldsymbol{r}}(s) := \mathbf{T}(s)\boldsymbol{r}(s)$ and $\mathbf{T}(s)\boldsymbol{d}(s) = \mathbf{0}$. If a fault happens, the transformed output error $\hat{\boldsymbol{r}}$ becomes different from zero, and the fault is detected. Moreover, if the matrix $\mathbf{T}(s)\mathbf{H}_f(s)$ is diagonal, a detected fault can also be isolated. This, however, would require quite a number of sensors, which may be costly and technically not feasible in all required locations. Therefore, instead of this perfect decoupling, one aims at maximising the dependency of the output residual from the faults and to attenuate its dependency from disturbances.

Determination of the Gain Matrix Entries of an Observer Directly from a BG

In the case of a standard Luenberger observer only the feedback gain matrix $\mathbf{L}$ needs to be determined (Fig. 3.14).

The classical approach is to compare the coefficients of the characteristic polynomial $p_O(s) = p_{A-LC}(s)$ of the observer matrix $\mathbf{A} - \mathbf{L}C$ with the coefficients of a characteristic polynomial $p_d(s)$ given by desired poles, which are specified such that the decrease of the observation error is faster than the transient behaviour of the observed plant.

Mathematical Example

Consider the following simple LTI system.

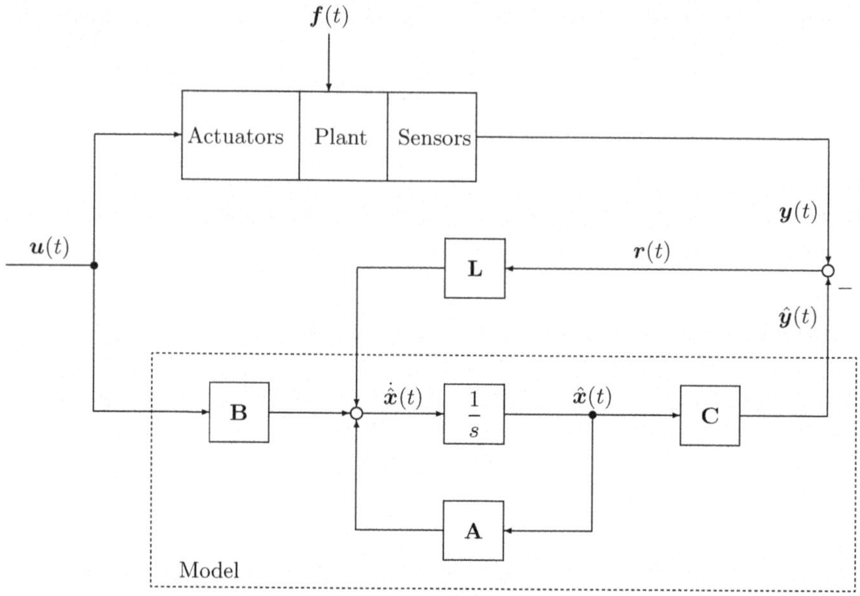

Fig. 3.14 Observer-based residual generation

$$\dot{\boldsymbol{x}}(t) = \underbrace{\begin{bmatrix} -1 & 0 \\ 1 & -1 \end{bmatrix}}_{\mathbf{A}} \boldsymbol{x}(t) + \underbrace{\begin{bmatrix} \frac{1}{2} \\ 1 \end{bmatrix}}_{\mathbf{B}} u(t) \tag{3.75a}$$

$$y(t) = \underbrace{\begin{bmatrix} 0 & 2 \end{bmatrix}}_{\mathbf{C}} \boldsymbol{x}(t) \tag{3.75b}$$

As

$$rank \mathscr{O} = rank \begin{bmatrix} \mathbf{A} \\ \mathbf{C} \end{bmatrix} = rank \begin{bmatrix} -1 & 0 \\ 1 & -1 \\ \hline 0 & 2 \end{bmatrix} = 2 \tag{3.76}$$

the system is completely state observable.

Let $\mathbf{L} = [L_1 \, L_2]^T$ be the observer gain matrix. The generic observer state estimation matrix then reads

$$\mathbf{A} - \mathbf{LC} = \begin{bmatrix} -1 & -2L_1 \\ 1 & -1 - 2L_2 \end{bmatrix} \tag{3.77}$$

The characteristic polynomial of the observer with unknown coefficients L_1 and L_2 is

$$det(s\mathbf{I} - (\mathbf{A} - \mathbf{LC})) = s^2 + 2(1 + L_2)s + 2L_2 + 2L_1 + 1 \tag{3.78}$$

The eigenvalues of the system matrix $\mathbf{A}$ are $-1, -1$. Let the desired eigenvalues of the observer be $-4, -4$. Then the observer characteristic polynomial reads

$$det(s\mathbf{I} - (\mathbf{A} - \mathbf{LC})) = (s + 4)^2 = s^2 + 8s + 16 \tag{3.79}$$

A comparison of the polynomial coefficients yields $L_1 = 9/2$ and $L_2 = 3$. As a result, the observer equation is

$$\begin{aligned} \dot{z}(t) &= \mathbf{A}z(t) + \mathbf{B}u(t) + \mathbf{L}(y(t) - \mathbf{C}z(t)) \\ &= (\mathbf{A} - \mathbf{LC})z(t) + \mathbf{B}u(t) + \mathbf{L}y(t) \\ &= \begin{bmatrix} -1 & -9 \\ 1 & -7 \end{bmatrix} z(t) + \begin{bmatrix} \frac{1}{2} \\ 1 \end{bmatrix} u(t) + \begin{bmatrix} \frac{9}{2} \\ 3 \end{bmatrix} y(t) \end{aligned} \tag{3.80}$$

As can be checked, the poles of the observer are, in fact, $-4, -4$.

The matrices of a LTI system can be automatically derived from a causal BG by means of a BG preprocessor software such as CAMP-G [30]. The command `place` from the Octave Control package then enables to compute the observer gain matrix for desired observer poles.

$$\mathbf{L} = \texttt{place}(\mathbf{A}', \mathbf{C}', [-4\ , -4]) \tag{3.81}$$

An alternative approach may be to apply Mason's loop rule directly on the causal BG (Sect. 2.2.2). The characteristic polynomial of the system matrix of a multivariable system is the denominator of each transfer function of the transfer function matrix and transfer functions can be directly obtained from a causal BG by applying Mason's rule. That is, the entries of the Luenberger gain matrix can be graphically determined by identifying causal paths on the observer BG. This shall be illustrated by means of a simple example.

Illustrative Example

Consider the passive network depicted in Fig. 3.15.

It is straightforward to convert the network in Fig. 3.15 into a causal BG (cf. the upper part of Fig. 3.16). A check of the result reveals that the network is structurally completely state observable with the single voltage sensor across the capacitor C : C_2. Moreover, the network is also numerically completely state observable with the single sensor of voltage V_2. The state space model reads

$$\begin{bmatrix} \dot{V}_1 \\ \dot{V}_2 \end{bmatrix} = \underbrace{\begin{bmatrix} -(R_1C_1)^{-1} & (R_1C_1)^{-1} \\ (R_1C_2)^{-1} & (RC_2)^{-1} \end{bmatrix}}_{\mathbf{A}} \underbrace{\begin{bmatrix} V_1 \\ V_2 \end{bmatrix}}_{\boldsymbol{x}} + \underbrace{\begin{bmatrix} C_1^{-1} & 0 \\ 0 & C_2^{-1} \end{bmatrix}}_{\mathbf{B}} \begin{bmatrix} F \\ E \end{bmatrix} \tag{3.82}$$

$$y = \underbrace{\begin{bmatrix} 0 \; 1 \end{bmatrix}}_{\mathbf{C}} \boldsymbol{x} \tag{3.83}$$

where $R^{-1} := R_1^{-1} + R_2^{-1}$. The Kalman observability matrix $\mathscr{O}$ has rank two.

$$\mathscr{O} = \begin{bmatrix} \mathbf{C} \\ \mathbf{CA} \end{bmatrix} = \begin{bmatrix} 0 & 1 \\ (R_1C_2)^{-1} & -(RC_2)^{-1} \end{bmatrix} \tag{3.84}$$

As a consequence, the observer gain matrix to be determined is minimal, i.e. $\mathbf{L} = [L_1 \; L_2]$.

Figure 3.16 shows a BG of the network together with a BG of the observer. The output error $\boldsymbol{r} = [r_1 \, r_2]$ weighted by the observer gain matrix $\mathbf{L}$ is added to the outputs of the storage elements via modulated flow sources according to the dynamic equation of the Luenberger observer (3.59) displayed by the block diagram in Fig. 3.14.

An inspection of causal paths in the BG of the observer reveals the following causal paths

$$\begin{aligned} p_1: &\quad \mathrm{C}:C_1 \overset{2}{\rightarrow} 0_1 \overset{4}{\rightarrow} 1_1 \overset{5}{\rightarrow} \mathrm{R}:R_1 \\ p_2: &\quad \mathrm{C}:C_2 \overset{7}{\rightarrow} 0_2 \overset{9}{\rightarrow} 1_2 \overset{10}{\rightarrow} \mathrm{R}:R_2 \\ p_3: &\quad \mathrm{C}:C_2 \overset{7}{\rightarrow} 0_2 \overset{6}{\rightarrow} 1_1 \overset{5}{\rightarrow} \mathrm{R}:R_1 \end{aligned}$$

and due to the feedback of the output $\hat{y}_2 = \hat{V}_2$ the following signal loops

$$l_1: \quad V_2 \rightarrow L_1 \rightarrow \mathrm{MSf} \overset{3}{\rightarrow} 0_1 \overset{2}{\rightarrow} C:C_1 \overset{2}{\rightarrow} 0_1 \overset{4}{\rightarrow} 1_1 \overset{5}{\rightarrow} \mathrm{R}:R_1$$

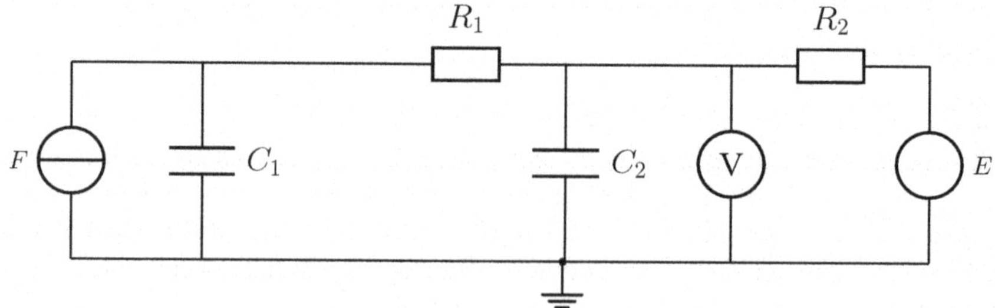

Fig. 3.15 Simple passive network

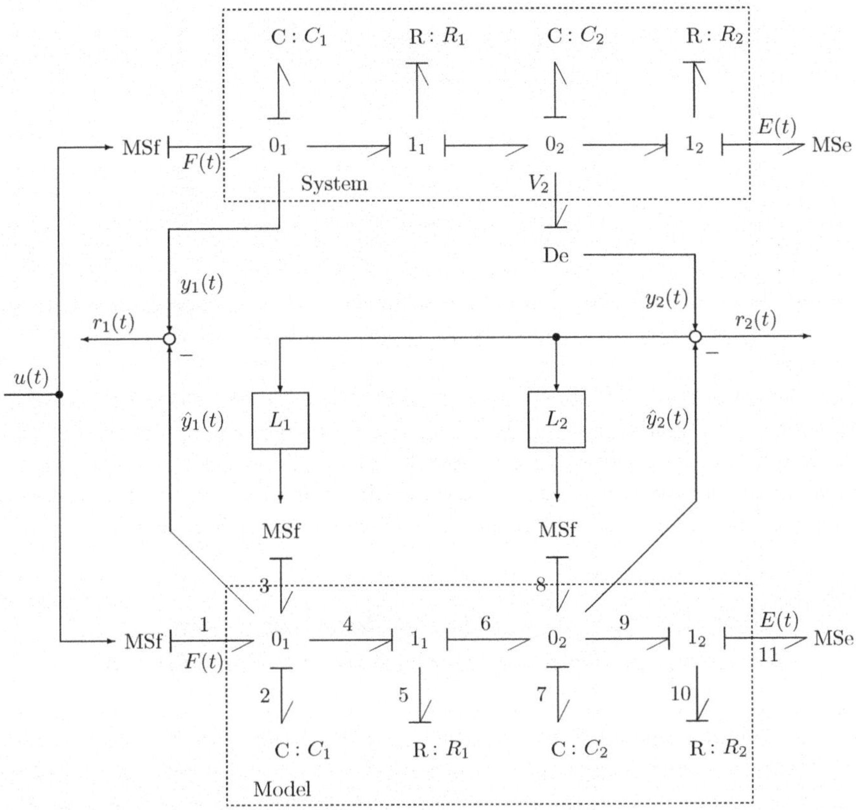

Fig. 3.16 BG of the network in Fig. 3.15 together with a BG of a Luenberger observer

$$\xrightarrow{5} 1_1 \xrightarrow{6} 0_2 \xrightarrow{7} \mathrm{C} : C_2 \rightarrow V_2$$

$$l_2 : \quad V_2 \rightarrow L_2 \rightarrow \mathrm{MSf} \xrightarrow{8} 0_2 \xrightarrow{7} \mathrm{C} : C_2 \rightarrow V_2$$

According to Mason's loop rule (2.17), the graph determinant Δ in this case reads

$$\Delta = 1 + \frac{1}{R_1 s C_1} + \frac{1}{R_2 s C_2} + \frac{1}{R_1 s C_2} + \frac{L_2}{s C_2} + \frac{L_1}{R_1 s C_1 s C_2} + \frac{1}{R_1 s C_1} \frac{1}{R_2 s C_2} + \frac{1}{R_1 s C_1} \frac{L_2}{s C_2} \tag{3.85}$$

Multiplication by s^2 yields the characteristic polynomial of the observer $p_O(s) = p_{A-LC}(s)$ with unknowns L_1, L_2.

$$p_O(s) = s^2 + \underbrace{\left(\frac{1}{R_1C_1} + \frac{1}{R_2C_2} + \frac{1}{R_1C_2} + \frac{L_2}{C_2}\right)}_{a_1} s$$

$$+ \underbrace{\frac{L_1}{R_1C_1C_2} + \frac{1}{R_1C_1}\frac{1}{R_2C_2} + \frac{1}{R_1C_1}\frac{L_2}{C_2}}_{a_0} \tag{3.86}$$

A comparison of the coefficients a_1, a_0 with those of an observer characteristic polynomial $p_d(s)$ given by desired poles p_1, p_2, yields the entries of the observer gain matrix.

In [48], Rahmani has shown that a bond graph-based determination of the coefficients of a characteristic polynomial can be facilitated by means of a table with families of causal paths. Nevertheless, beyond medium size bond graph models software support will be needed for the symbolic computation of the coefficients of a characteristic polynomial and the entries of the observer gain matrix.

3.5.2 Fault Detection and Isolation Based on Analytical Redundancy Relations Derived from a Bond Graph

Another common approach to model-based FDI besides the use of observers is to build constraints between known control signals and measurements called *Analytical Redundancy Relations* (ARRs). As long as a monitored system is in normal operation, an evaluation of ARRs yields residuals that are close to zero. Due to parameter uncertainties, residuals can be within small boundaries around zero. If the time evolution of a residual crosses, however, a fault threshold, then this event indicates that a parametric fault has happened in some system component and needs to be isolated. Fault threshold may be constant values or adaptive with the time evolution of measured signals. The task is to make sure that true faults are detected and false alarms are avoided.

Derivation of ARRs from a Diagnostic Bond Graph

In a bond graph framework, ARRs may be set up for fault detection and isolation by attaching detectors to certain junctions of a bond graph model according to the sensors in use and to sum the efforts or flows, respectively, at these junctions. Sensors deliver measurements. That is, they contribute a known input into their ARR that has been properly filtered before it is used in the ARR. Therefore, their detector causalities are inverted when possible. Detectors in a BG that cannot be inverted indicate redundant hardware sensors. The number of ARRs that can be used for fault detection and isolation equals the number of sensors [57].

Assuming that there are no unknown disturbances into the system, the unknown variables in an ARR can be expressed by known control signals and known

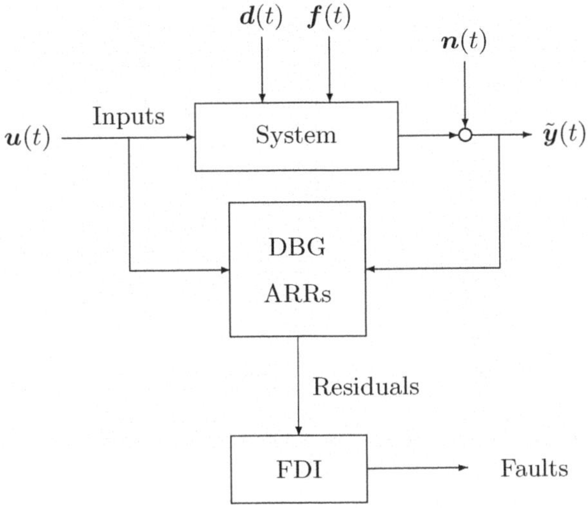

Fig. 3.17 Fault detection based on a DBG

measurements. Relations determining the unknowns can be obtained by following causal paths in a BG. If these relations can be solved analytically for the unknowns, they can be replaced in the ARRs and ARRs in closed symbolic form can be obtained. This may not be possible due to algebraic loops in the BG and nonlinear constitute element equations so that a set of equations must be numerically solved to evaluate an ARR.

When measurements are continuously taken from a real process, the initial state of storage elements is unknown. Therefore, storage elements in the bond graph model should be in preferred derivative causality when possible. As a result, not only known inputs and measurements constitute an ARR but also their derivatives if a state variable in an ARR is to be replaced. If a storage element must take integral causality to avoid a causal conflict, then the resulting ARR is differentiated with respect to time to get rid of the initial condition which does not affect the ARR's structural dependency from components parameters. That is, resulting ARRs are *dynamic* constraints in general. Therefore, it is important to use a filter that does not only smooth measurement noise but can also provide sufficient accurate time derivatives of smoothed measurements such as the Savitzky–Golay filter considered in Sect. 3.2. A BG with detectors in inverted causality and storage elements in derivative causality has been termed by Samantaray a *diagnostic bond graph* (DBG) [57]. A DBG model is a residual generator in online model-based FDI as depicted in Fig. 3.17 where $\boldsymbol{d}(t)$ denote disturbances, $\boldsymbol{f}(t)$ plant faults, and $\boldsymbol{n}(t)$ measurement noise. The disturbances and measurement noise affect the outputs $\tilde{\boldsymbol{y}}$ of the physical system and have an effect on the generated residuals. Measurement noise can be reduced by appropriate filtering before the sensor readings are used as input signals into the DBG.

In a FDI software module, residuals obtained by an online evaluation of offline generated ARRs are used to detect and to isolate faults. As known inputs and

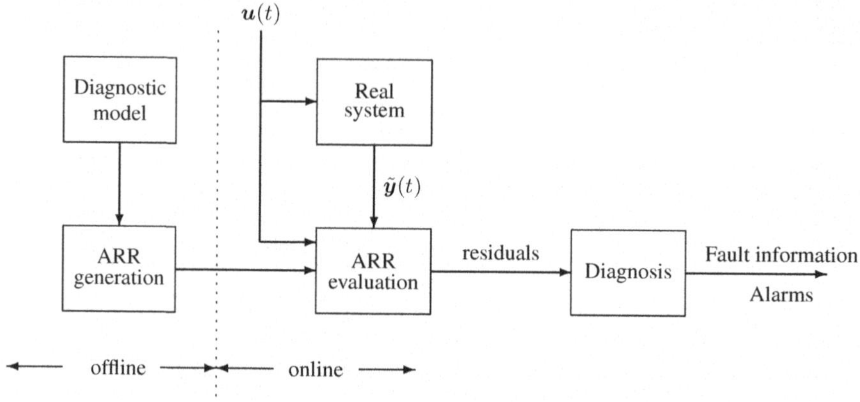

Fig. 3.18 Offline generation of ARRs and their online evaluation

measurements from the real system are inputs into the ARRs, their evaluation must take place in real-time. The ARRs, however, can be generated offline by deducing them from a diagnostic bond graph model as indicated in Fig. 3.18 [10].

In systems with various modes of operation represented by a hybrid model with discrete and time continuous states, some parts may be switched off temporarily so that they do not contribute to the system's dynamic behaviour in some modes of operation. As a result, fault detection is mode dependent.

In summary, ARR based fault detection may be formulated mathematically as follows. Let $\boldsymbol{\Theta}_n$ be the vector of all parameters with nominal value, $\boldsymbol{u}(t)$ the vector of known input signals, $\boldsymbol{y}(t)$ the vector of filtered measurement signals and let $\boldsymbol{b}(t)$ denote the vector of all discrete switch state values at time instant t. For as long as a system is in one mode of operation, no discrete switch states change and $\boldsymbol{b}(t)$ is constant. Furthermore, let the kth ARR be expressed by means of a possibly nonlinear real-valued function f_k and let ε_k^j be the error bound for the absolute values of residuals $r_k(t)$ in system mode j given by a physically feasible combination of discrete switch states. Then

$$r_k(t) = f_k(\boldsymbol{u}(t), \dot{\boldsymbol{u}}(t), \ldots, \boldsymbol{u}^{(m)}(t), \boldsymbol{y}(t), \dot{\boldsymbol{y}}(t), \ldots, \boldsymbol{y}^{(n)}(t), \boldsymbol{\Theta}_n, \boldsymbol{b}(t)) \tag{3.87}$$

and

$$|r_k(t)| < \varepsilon_k^j \quad \forall\, k \quad \forall\, t > 0 \tag{3.88}$$

indicates a non-faulty system. However, in case of a single fault or multiple simultaneous faults, the absolute values of some residuals exceed system mode dependent fault thresholds $thr_i^j(t)$.

$$\exists\, i \;\; \exists\, D_i^j \subset \mathbb{R}^+ \quad \text{such that} \quad |r_i(t)| \geq thr_i^j(t) \quad \forall\, t \in D_i^j \tag{3.89}$$

In an online FDI software module each residual $r_k(t)$ is checked against its threshold $thr_k^j(t)$ and the result is entered into a *coherence vector* $\boldsymbol{C} = [c_1(t), \ldots, c_n(t)]$ where n is the number of residuals and

$$c_k(t) = \begin{cases} 0 & \text{if} \quad |r_k(t)| < \varepsilon_k^j, \quad i = 1, \ldots, n \\ 1 & \text{otherwise} \end{cases} \tag{3.90}$$

If in some system mode j, vector $\boldsymbol{C} \neq \boldsymbol{0}$, then a fault has happened that needs to be isolated and an alarm is raised.

Note that ARR residuals do not depend on the state vector, which is considered unknown. Derivative causality at the storage elements decouples the state vector from ARRs.

The ARRs in (3.87) use nominal parameters. However, in real systems actual parameter values may be different from their nominal values. They may be even slowly time-varying. That is, parameter values are uncertain. As a result, modelling uncertainties and parameter uncertainties affect the numerical evaluation of ARRs and may lead to non-zero residuals although no fault has happened. In order to detect faults unambiguously and to avoid false alarms, ARRs should be significantly sensitive to true faults and only slightly sensitive to parameter variations. To support model-based FDI robust with regard parameter uncertainties for continuous time models, two approaches have been reported in the literature. One of them is based on uncertain bond graphs in linear transformation form (LFT) [21, 23, 39], the other one on incremental bond graphs [10, 11]. The two approaches enable to define adaptive ARR residual bounds. As long as the time evolution of an ARR residual stays inside these bounds, no fault alarm is raised. FDI robust with regard to parameter uncertainties is addressed in Sect. 3.6.

Nonlinearities in a model may prevent a formulation of ARRs in explicit symbolic closed form. In practice, this is not really a problem. As long as implicit equations do have a solution, they can be solved numerically by iteration for each time instant. This, however, means that instead of the evaluation of a set of ARRs in closed form, a larger set of equations must be solved numerically. The higher computational effort may matter as the generation of ARR residuals must be performed online.

Samantaray has implemented the offline derivation of ARRs from a diagnostic bond graph in a module of the modelling and simulation software package Symbols 2000 [40, 43].

Structural Fault Signature Matrix

When a fault has been detected, it is assumed that it can be mapped onto a parameter of a system component, a sensor, or of an actuator. Commonly, not all components of the parameter vector $\boldsymbol{\Theta}_n$ are present in each ARR. The information which component contributes to which ARR can be represented in a so-called *Structural Fault Signature Matrix* (FSM) $\mathbf{S} = (S_{ij})$, $i = 1, \ldots, p$, $j = 1, \ldots, N$, where p is the number of component parameters Θ_i and N is the number of ARR residuals

r_j equal to the number of sensors. If parameter Θ_i affects residual r_j in all system modes, then $S_{ij} = 1$. If Θ_i does not contribute at all to r_j, then $S_{ij} = 0$. If the contribution of parameter Θ_i to ARR_j depends on a function $f_{ij}()$ of the discrete switch states and if the value of $f_{ij}()$ is $b_{ij} \in \{0, 1\}$, then $S_{ij} = b_{ij}$. The entries in a row of the FSM constitute a so-called *component fault signature*. If the latter one is unique, the fault can be isolated. The columns of a FSM indicate the fault signatures of the ARRs of the residuals. A matrix entry S_{ij} that is non-zero for some system modes means that in these system modes, residual r_j is *structurally* sensitive to faults in the i-th component. Structurally independent ARRs, i.e. ARRs that cannot be algebraically constructed from other ARRs, have a unique fault signature. If there is a subset of ARRs in which each ARR depends only on a single component parameter, then the subset is called *structured*. The submatrix of **S** is diagonal, the fault signatures are unique and potential faults in this subset of components can be isolated. In general, the number of sensors, N, is less than the number of component parameters, $\boldsymbol{p}$, so that a FSM is not square. As a result, some component parameters may have the same component fault signature (in some modes) so that faults in these parameters cannot be isolated just by structural inspection of the FSM.

It is common to augment a FSM by two additional columns. If a parameter of the ith component contributes to an ARR_k in all modes, then the parameter may cause the residual $r_k(t)$ to exceed its threshold in case the parameter becomes faulty. That is, the fault can be detected. This is expressed by an entry equal to one in a detectability column with the heading D_b. For a component fault signature that depends on switch states, the detectability of the fault is given by the logical OR of the switch states. If in the same system mode, a parameter of the jth component, $j \neq i$, is also present in the fault signature of ARR_k, it cannot be decided to which component a residual $r_k(t)$ outside its admissible boundaries is due. That is, the parametric fault cannot be isolated. This is indicated by an entry equal to zero in the isolatability column with the heading I_b. If the fault signature of a detectable fault is unique in all system modes, it can also be isolated, which is denoted by an entry equal to one in the isolatability column. If parameters of different components may be the cause for an ARR residual outside its thresholds, then the detected fault can be isolated by estimation of the candidate parameters.

Example

For illustration, consider the simple switched RLC network depicted in Fig. 3.19.

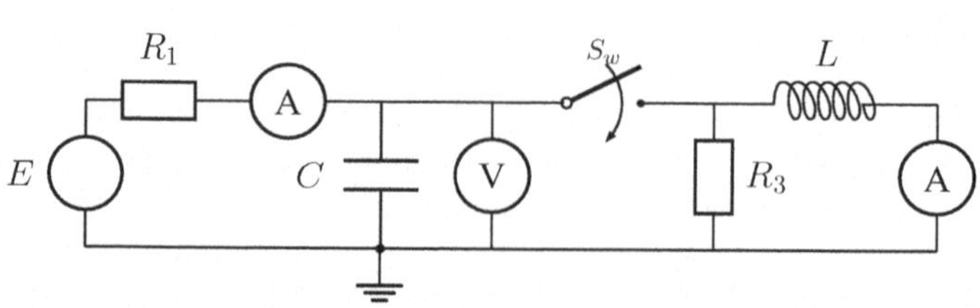

Fig. 3.19 Switched RLC network with three sensors

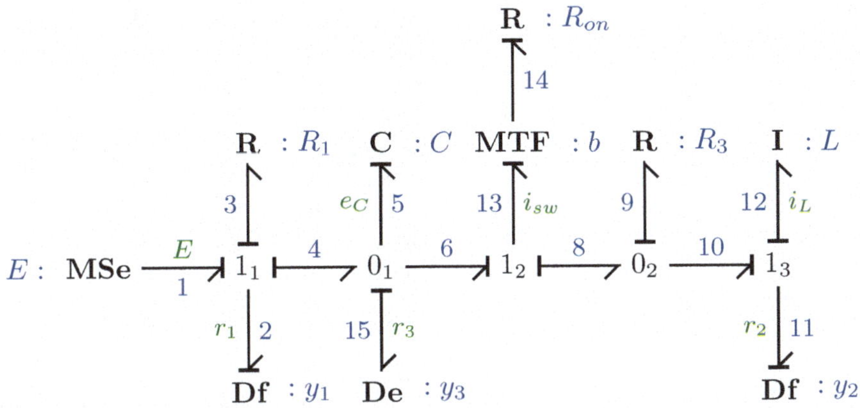

Fig. 3.20 Diagnostic BG of the network in Fig. 3.19

Figure 3.20 shows a diagnostic BG of the network in Fig. 3.19. The switch has been modelled as a linear resistor R : R_{on} that is switched on when the switched is closed ($b = 1$). The advantage of this switch model is that it has fixed conductance causality at its port independent of the switch state. As can be seen, all detectors can take inverted causality and all storage elements are in preferred derivative causality.

ARRs in explicit closed symbolic form are obtained by summing efforts at junctions 1_1 and 1_3 and flows at junction 0_1 according to the weak constitutive law of the junctions to which a detector is attached. Depending on the switch state, the parallel connection of resistor R : R_3 and the inductor I : L are either connected or disconnected to the rest of the network. This is reflected by the discrete state b in the ARRs derived from the DBG. A part of ARR$_2$ and ARR$_3$ is switched off and on.

$$S_w : \quad i_{sw} = \frac{b}{R_{\text{on}}}[y_3 - R_3(i_{\text{sw}} - y_2)] \tag{3.91}$$

$$= \frac{b}{R_{\text{on}} + bR_3}(y_3 + R_3 y_2) \tag{3.92}$$

$$1_1 : \quad r_1 = E - R_1 y_1 - y_3 \tag{3.93}$$

$$0_1 : \quad r_3 = y_1 - C\dot{y}_3 - i_{\text{sw}} \tag{3.94}$$

$$1_3 : \quad r_2 = R_3(i_{\text{sw}} - y_2) - L\dot{y}_2 \tag{3.95}$$

The structure of the ARRs is captured by the FSM in Table 3.1.

The FSM in Table 3.1 indicates that apart from the switch, all potential component faults can be detected. Clearly, a faulty resistance of the switch can only be detected when the switch is on. Note that given the three sensors only a

Table 3.1 Structural fault signature matrix of the DBG in Fig. 3.19

Element	ARR_1	ARR_2	ARR_3	D_b	I_b
MSe : E	1	0	0	1	0
R : R_1	1	0	0	1	0
C : C	0	0	1	1	1
Sw : R_{on}	0	b	b	b	0
R : R_3	0	1	b	1	0
I : L	0	1	0	1	0
Df : y_1	1	0	1	1	0
Df : y_3	1	0	1	1	0
Df : y_2	0	1	0	1	0

capacitor fault could be isolated independent of the switch state. In case the voltage source MSe : E can be considered faultless, its row can be removed from the FSM. As a result, the fault signature of R : R_1 is unique and R : R_1 can also be isolated. Moreover, in case the sensors are faultless, their rows can be removed from the FSM as well.

In the case the switch is closed and its ON resistance can be neglected, the switch is turned into an ideal one. As a consequence, resistance R : R_3 can be isolated. This result can be obtained if the non-ideal switch model is replaced by an ideal one with the *implicit* constitutive equation

$$\text{Sw}: \quad (1-b)\, i_{sw} + b\, u_{sw} = 0 \tag{3.96}$$

as depicted in the DBG of Fig. 3.21.

From the DBG in Fig. 3.21 one obtains

$$y_3 - u_{sw} = R_3\,(i_{sw} - y_2) \tag{3.97}$$

Solving (3.97) for u_{sw} and substituting the result into the implicit switch equation yields an equation for i_{sw} that holds for both switch states.

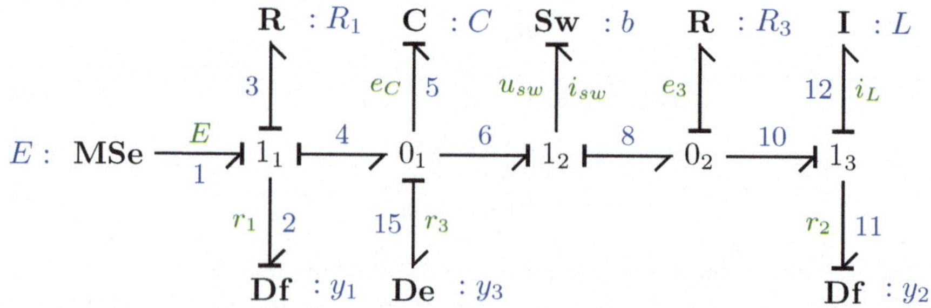

Fig. 3.21 Diagnostic BG of the network in Fig. 3.19 with an ideal switch Sw : b

$$i_{sw} = -\frac{1-b}{b-(1-b)R_3}(y_3 + R_3 y_2) \tag{3.98}$$

The FSM then takes the form displayed in Table 3.2.

Table 3.2 Structural fault signature matrix of the DBG in Fig. 3.21

Element	ARR$_1$	ARR$_2$	ARR$_3$	D$_b$	I$_b$
MSe : E	1	0	0	1	0
R : R_1	1	0	0	1	0
C : C	0	0	1	1	1
R : R_3	0	1	b	1	b
I : L	0	1	0	1	0

Construction of a Structural FSM from a Diagnostic BG

The number of ARRs is equal to the number of sensors. In order to see which component parameters contribute to which ARR, it is not necessary to derive equations from a DBG and to eliminate unknown variables. It is sufficient to identify causal paths from model inputs (including inputs from detectors) to sensors. Elements that are traversed on these causal paths contribute to the ARR of a residual related to a sensor element. An output of a source or an element that is followed directly or indirectly by switches on the causal path to a sensor element provides an entry in the FSM equal to the product of the switch states.

ARR$_1$: Causal paths to detector Df : y_1:
MSe $\rightarrow E \rightarrow r_1$
Df : $y_1 \rightarrow f_3 \rightarrow$ $\boxed{\mathrm{R} : R_1}$ $\rightarrow e_3 \rightarrow r_1$

ARR$_3$: Causal paths to detector De : y_3:
De : $y_3 \rightarrow e_5 \rightarrow$ $\boxed{\mathrm{C} : C}$ $\rightarrow f_5 \rightarrow r_3$
ARR$_2$: Causal paths to detector Df : y_2:
Df : $y_2 \rightarrow f_{12} \rightarrow$ $\boxed{\mathrm{I} : L}$ $\rightarrow e_{12} \rightarrow r_2$
Df : $y_2 \rightarrow f_{10} \rightarrow f_9$ $\boxed{\mathrm{R} : R_3}$ $\rightarrow e_9 \rightarrow e_{10} \rightarrow r_2$
Df : $y_2 \rightarrow f_{10} \rightarrow f_9 \rightarrow \mathrm{R} : R_3 \rightarrow e_9 \rightarrow e_8 \rightarrow e_{13} \rightarrow$ $\boxed{\mathrm{MTF} : b}$ $\rightarrow e_{14} \rightarrow$ $\boxed{\mathrm{R} : R_{on}}$ $\rightarrow f_{14} \rightarrow f_{13} \rightarrow f_8 \rightarrow f_9 \rightarrow \mathrm{R} : R_3 \rightarrow e_9 \rightarrow e_{10} \rightarrow r_2$

From the identified causal paths, the FSM in Table 3.1 can be constructed.

Construction of a Structural FSM from a Diagnostic BG with Bicausal Paths

A structural FSM can also be set up from a bicausal BG by identifying bicausal paths from detectors to passive elements and sources. Elements on a bicausal path from a detector to an element at the end of the bicausal causal path contribute to the ARR associated with the weak law of the junction to which the bicausal detector bond is connected. Figure 3.22 displays a bicausal BG of the RLC network in Fig. 3.19 with some bicausal paths highlighted by additional signal arrows.

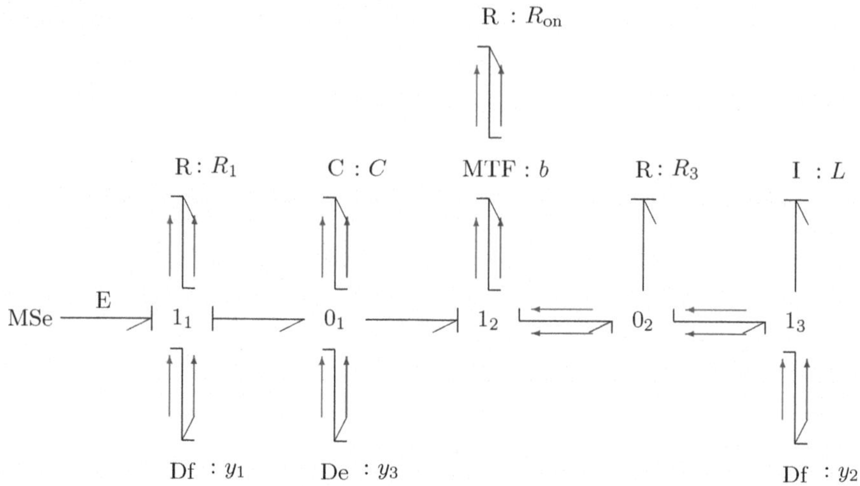

Fig. 3.22 Bicausal diagnostic BG of the network in Fig. 3.19 with three disjoint bicausal paths

In the bicausal BG of Fig. 3.22 the following bicausal paths are highlighted:

$$
\begin{aligned}
\text{ARR}_1: &\quad \text{Df}: y_1 \rightarrow \boxed{\text{R}: R_1}\\
\text{ARR}_3: &\quad \text{Df}: y_3 \rightarrow \boxed{\text{C}: C}\\
\text{ARR}_2: &\quad \text{Df}: y_2 \rightarrow \boxed{\text{MTF}: b} \rightarrow \boxed{\text{R}: R_{\text{on}}}
\end{aligned}
$$

Further bicausal paths not shown in the bicausal BG of Fig. 3.22 can be identified. For instance, starting the assignment of bicausality at detector Df : y_2 another two bicausal paths can be found.

$$
\begin{aligned}
\text{ARR}_2: &\quad \text{Df}: y_2 \rightarrow \boxed{\text{I}: L}\\
\text{ARR}_2: &\quad \text{Df}: y_2 \rightarrow \boxed{\text{R}: R_3}
\end{aligned}
$$

Again, the boxed elements are the ones that contribute to the ARR associated with the weak law of the junction to which a sensor is connected.

3.5.3 Avoiding Differentiation of Measurements

Storage elements in a DBG are assigned derivative causality in order to get rid of their initial conditions that are unknown in online measurement and fault detection. However, even with appropriate filtering of sensor readings, numerical differentiation of measurements becomes increasingly less accurate for higher order derivatives which causes problems for the detection, the isolation, and the identification of faults. In order to avoid the differentiation of measurements,

Benmoussa introduced a procedure with three steps for linear systems that leads to time integrals of ARR residuals [4].

1. ARRs linear with regard to known control inputs and known system outputs are Laplace transformed.
2. The Laplace transformed ARRs are differentiated with respect to the complex variable s as often as necessary to get rid of all initial conditions.
3. All resulting ARRs are multiplied by s^{-n} (n-times integration) and transformed back into time domain, where n denotes the highest order of differentiation in the initial ARRs.

For illustration, these steps shall be performed on a linear ARR that contains first order derivatives at most for the sake of simplicity.

$$r_i(t) = \sum_{j=1}^{n_s} p_{ij}^0(\Theta_n) y_j(t) + \sum_{j=1}^{n_s} p_{ij}^1(\Theta_n) \dot{y}_j(t) + \sum_{k=1}^{m} q_{ij}^0(\Theta_n) u_k(t) + \sum_{k=1}^{m} q_{ij}^0(\Theta_n) \dot{u}_k(t) \tag{3.99}$$

where n_s is the number of sensors, m the number of control input signals. The ARR takes a simpler form when written in vector form.

$$r_i(t) = \boldsymbol{p}_i^0 \boldsymbol{y}(t) + \boldsymbol{p}_i^1 \dot{\boldsymbol{y}}(t) + \boldsymbol{q}_i^0 \boldsymbol{u}(t) + \boldsymbol{q}_i^1 \dot{\boldsymbol{u}}(t) \tag{3.100}$$

where $\boldsymbol{p}_i^0$, $\boldsymbol{p}_1^0$, $\boldsymbol{q}_i^0$, and $\boldsymbol{q}_i^1$ are row vectors with coefficients that depend on the vector $\boldsymbol{\Theta}_n$ of parameters.

Laplace transform yields

$$R_i(s) = \boldsymbol{p}_i^0 \boldsymbol{Y}(s) + \boldsymbol{p}_i^1 (sY(s) - y(0)) + \boldsymbol{q}_i^0 \boldsymbol{U}(s) + \boldsymbol{q}_i^1 (sU(s) - u(0)) \tag{3.101}$$

Let a prime (') denote the differentiation with respect to s.

$$R_i' = \boldsymbol{p}_i^0 \boldsymbol{Y}' + \boldsymbol{p}_i^1 (Y + sY') + + \boldsymbol{q}_i^0 \boldsymbol{U}' + \boldsymbol{q}_i^1 (U + sU') \tag{3.102}$$

Multiplication with $1/s$ gives

$$\frac{1}{s} R_i' = \frac{1}{s} \boldsymbol{p}_i^0 \boldsymbol{Y}' + \frac{1}{s} \boldsymbol{p}_i^1 Y + \boldsymbol{p}_i^1 Y' + \frac{1}{s} \boldsymbol{q}_i^0 \boldsymbol{U}' + \boldsymbol{q}_i^1 U + \boldsymbol{q}_i^1 U' \tag{3.103}$$

The differentiation d^n/ds corresponds to a multiplication of the factor $(-1)^n t^n$ and s^{-n} to an n-fold integration. Accordingly, transformation back into the time domain yields

$$-\int_0^t \tau r_i(\tau)d\tau = \int_0^t \boldsymbol{p}_i^0(-1)\tau \boldsymbol{y} d\tau + \int_0^t \boldsymbol{p}_i^1 \boldsymbol{y} d\tau + \boldsymbol{p}_i^1(-1)t\boldsymbol{y}$$
$$+ \int_0^t \boldsymbol{q}_i^0(-1)\tau \boldsymbol{u} d\tau + \boldsymbol{q}_i^1(-1)t\boldsymbol{u} + \int_0^t \boldsymbol{q}_i^1 \boldsymbol{u} d\tau \quad (3.104)$$

Using this integral form of the ARRs derived from a DBG does not only avoid numerical differentiation but also reduces measurement noise as the integration performs an averaging of the measurements. Moreover, for the time evolution $r_i(t)$ of residuals with strictly positive or negative values, the integral over the product $t \cdot r_i(t)$ deviates more distinctly from zero than the time evolution of $r_i(t)$ which is of benefit for an early detection of incipient faults.

3.5.4 Parametric Fault Isolation and Fault Estimation

Depending on the number of sensors and their locations all possible parametric faults may be detectable but only few or even none of them can be isolated by just an inspection of a structural FSM because parameters of different components may have the same fault signature so that it cannot be decided which one of two potentially faulty parameters has given rise to an ARR residual outside its threshold.

Parameter Sensitivity Matrix of ARRs

The unstructured part of a FSM could be reduced by using additional physical sensors if they can be placed in the real system and if costs permit. In case this is not feasible, then let $f_k()$ be the real-valued function of the kth ARR. One option to overcome the problem is to replace the entry S_{ik} in the structural FSM by the sensitivity of $f_k()$ with respect to the parameter Θ_i, i.e. by the partial derivative $\partial f_k / \partial \Theta_i$ termed *residual sensitivity function* and to replace the sensitivity by zero in case its value is small. As a result, the number of unique component fault signatures in a given FSM may be increased so that more component faults can be isolated.

Parameter sensitivities of ARR residuals can be obtained from an *incremental bond graph* (incBG) [9, 10]. The latter bond graph can be systematically developed from an initial bond graph with nominal parameters by replacing elements with parameters to be estimated by their incremental component model. Inputs into an incBG are variations of the parameters to be estimated multiplied by a power variable of the initial BG. Outputs may be parameter variations of ARR residuals. They are a weighted sum of the parameter variations and the weighting factors are just the residual sensitivity functions. Alternatively, sensitivities of ARR residuals with respect to a component parameter can also be obtained from *sensitivity bond graphs* [25], *sensitivity pseudo bond graphs* [8, 10, 14], or from *diagnostic sensitivity bond graphs* [54].

Parameter Estimation by Least Squares ARR Residuals Minimisation

If there is an *unstructured* part of a FSM in which parameters of different components have the same fault signature, then the values of the parameters involved can be determined by estimation. A comparison with their nominal values isolates the faulty parameters. Advantages are that multiple simultaneous faults can be isolated and that their size can be determined. However, even if only a subset of all system parameters is to be estimated, the computation of their magnitude by minimisation of a real-valued functional can be costly which may be an issue in online FDI.

Assume that ARRs can be derived in closed symbolic form from a DBG. Let $\boldsymbol{\Theta}$ denote the subvector of parameters and $\boldsymbol{r}$ the subvector of the ARRs in the *unstructured* part of a FSM. If nominal parameter values are used in the ARRs, their residuals significantly differ from zero. In contrast, if their true faulty values were known, the ARR residuals would be close to zero. The unknown faulty parameter values can be estimated by using the ARR residuals considered as functions of the parameter vector $\boldsymbol{\Theta}$ in a quadratic cost function $f()$ and to vary $\boldsymbol{\Theta}$ so that the value of the cost function becomes minimal.

$$f(\boldsymbol{\Theta}) = \frac{1}{2} \sum_{j=k-q}^{k} \boldsymbol{r}^T(t_j, \boldsymbol{\Theta})\boldsymbol{r}(t_j, \boldsymbol{\Theta}) \tag{3.105}$$

where q denotes the length of a moving observation window.

In general, the residual vector is a *nonlinear* function of the parameter vector $\boldsymbol{\Theta}$. That is, a well known nonlinear least squares problem (NLSP) is to be solved, for which various optimisation algorithms such as the Levenberg–Marquardt algorithm [49], the quasi-Newton method, or the Gauss-Newton (GN) algorithm [7] may be used that need the above residual sensitivities with respect to parameters. If the numerical computation of the gradient of an objective function shall be avoided, and if accuracy requirements are not too high, a direct method such as the Nelder-Mead simplex algorithm [41] may be used. Figure 3.23 displays the fault isolation approach based on ARR residuals generated by a DBG. Details may be found in [10].

Estimation of a Single Faulty Parameter on a Bicausal BG

The estimation of an element parameter based on a bicausal path between the element and a detector presented in Sect. 2.5 can be used in particular for the estimation of faulty parameters. For illustration, consider the bicausal BG in Fig. 3.22. If the values of faulty parameters in the three ARRs associated with the junctions to which a detector is connected, were known, then their residuals would vanish.

The bicausal path from detector Df : y_1 to resistor R : R_1 delivers both effort and flow into the resistor so that its parameter can be determined given that the voltage source MSe : E is faultless.

$$R_1 = \frac{E - y_3}{y_1} \tag{3.106}$$

Following the bicausal path from detector Df : y_3 to the capacitor C : C yields

$$C = \frac{y_1 - i_{sw}}{\dot{y}_3} \tag{3.107}$$

The sum of flows at junction 0_2 gives for the switch current:

$$i_{sw} = \frac{1}{R_3} L \dot{y}_2 + y_2 \tag{3.108}$$

Finally, by following the bicausal path from Df : y_2 to the ON resistor of the switch, one obtains

$$R_{on} = \frac{b(y_3 - L\dot{y}_2)}{i_{sw}} \tag{3.109}$$

Note that although the three bicausal paths highlighted in the BG of Fig. 3.22 are disjoint, the parameters of their terminating elements can only be estimated if parameters E and R_3 can be assumed to be non-faulty. Two of the three bicausal

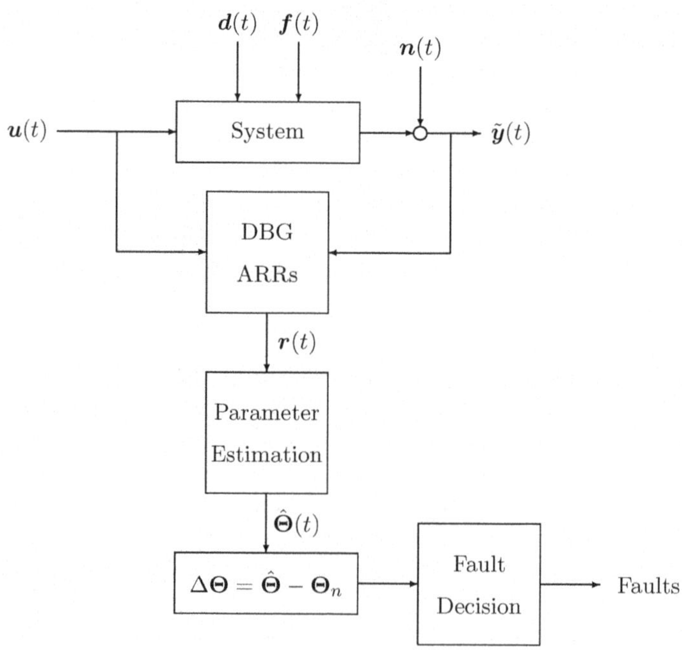

Fig. 3.23 Fault isolation by least squares ARR residuals minimisation

paths share variables with other bicausal paths not displayed in the BG of Fig. 3.22. The bicausal path Df : $y_1 \rightarrow$ R : R_1 shares y_1 with the bicausal path Df : $y_1 \rightarrow$ MSe : E, and Df : $y_2 \rightarrow$ R : R_{on} has variables with the bicausal path Df : $y_2 \rightarrow$ R : R_3 in common. That is, an element parameter can be estimated by following a bicausal path from a detector to that element if the path is disjoint to all other bicausal paths such as the one Df : $y_2 \rightarrow$ C : C or if a single fault hypothesis can be adopted.

In the case that the sensors deliver output signals from the healthy system into the DBG, the estimation would give the nominal parameter values. Equations (3.106)–(3.109) can also be obtained from the DBG in Fig. 3.20 with $r_1 = r_3 = r_2 = 0$.

3.6 Robustness with Regard to Parameter Uncertainties

Analytical redundancy relations are not only constraints between known inputs and noisy measurements but also depend on system parameters. Parameter values, however, are often uncertain. In addition to an available nominal parameter value, it may only be known that the actual value is within a known interval. As ARR residuals serve as fault indicators, it is important that ARRs are robust with regard to parameter uncertainties in order to minimise misdetections and false alarms. To that end, a DBG as a residual generator should account for parameter uncertainties. A natural approach is to split passive elements in a DBG that are linear with regard to their parameter into one part depending on nominal parameters and another part that includes the parameter uncertainties. The decomposition of elements can be performed in different ways.

Consider the constitutive equations of a linear 1-port resistor and a linear 1-port C storage element with absolute parameter uncertainties ΔR, ΔC, respectively.

$$
\begin{aligned}
R: \quad e_R &= (R_n + \Delta R) f_R \\
&= R_n f_R + (\Delta R) f_R
\end{aligned}
\tag{3.110}
$$

$$
\begin{aligned}
C: \quad f_C &= (C_n + \Delta C) \dot{e}_C \\
&= C_n \dot{e}_C + (\Delta C) \dot{e}_C
\end{aligned}
\tag{3.111}
$$

The parameter uncertainties ΔR and ΔC can be taken into account by means of an additional resistor, an additional C storage element, respectively, as depicted in Fig. 3.24, or by means of modulated sinks as displayed in Fig. 3.25.
Transformer and gyrator elements are decomposed accordingly.

3.6.1 Uncertain BGs

A BG with elements decomposed as in Fig. 3.24 is called an uncertain BG [33]. From an uncertain BG the *canonical form of an uncertain state space model* can be derived.

$$\dot{x} = [\mathbf{A}_n + \Delta\mathbf{A}]\,x + [\mathbf{B}_n + \Delta\mathbf{B}]u \tag{3.112}$$

$$y = [\mathbf{C}_n + \Delta\mathbf{C}]\,x + [\mathbf{D}_n + \Delta\mathbf{D}]u \tag{3.113}$$

The constant coefficient matrices $\mathbf{A}_n$, $\mathbf{B}_n$, $\mathbf{C}_n$, $\mathbf{D}_n$ are of appropriate dimensions. Their entries depend on the components of the vector of nominal parameters $\boldsymbol{\Theta}_n$. The entries in the incremental matrices $\Delta\mathbf{A}$, $\Delta\mathbf{B}$, $\Delta\mathbf{C}$, $\Delta\mathbf{D}$, in general, are nonlinear functions of the parameter variations $\Delta\boldsymbol{\Theta}$ and the nominal parameters. $\Delta\mathbf{A} := \mathbf{A}(\boldsymbol{\Theta}) - \mathbf{A}(\boldsymbol{\Theta}_n)$, $\boldsymbol{\Theta} := \boldsymbol{\Theta}_n + \Delta\boldsymbol{\Theta}$. The matrices $\Delta\mathbf{B}$, $\Delta\mathbf{C}$, $\Delta\mathbf{D}$ are defined likewise.

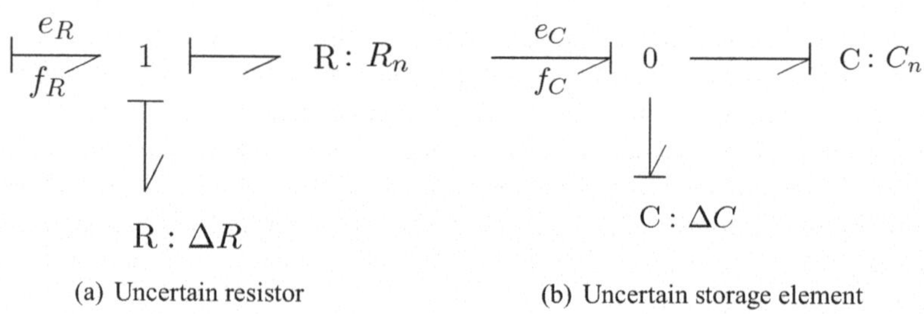

Fig. 3.24 Decomposition of a 1-port resistor and a 1-port capacitor with uncertain parameter. (**a**) Uncertain resistor. (**b**) Uncertain storage element

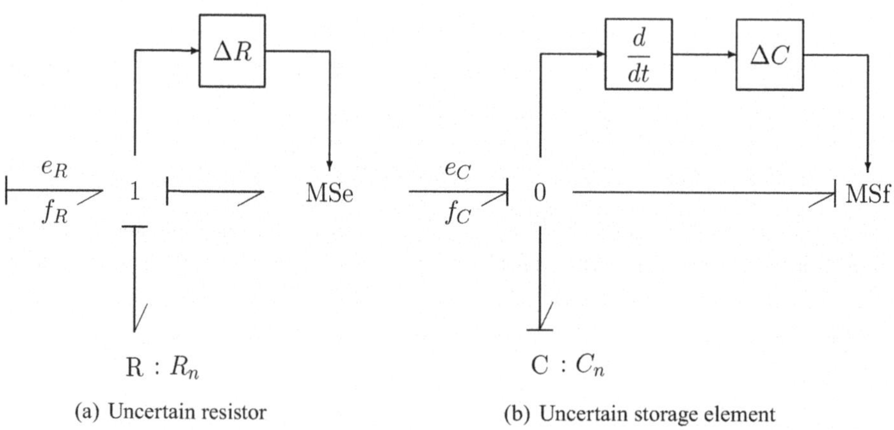

Fig. 3.25 BG representation of the parameter uncertainty by means of a modulated sink. (**a**) Uncertain resistor. (**b**) Uncertain storage element

All multiplicative parametric faults may also be collected into unknown vectors $\boldsymbol{F}_1(t)$, $\boldsymbol{F}_2(t)$.

$$\dot{\boldsymbol{x}} = \mathbf{A}_n \boldsymbol{x} + \mathbf{B}_n \boldsymbol{u} + \underbrace{(\Delta\mathbf{A})\boldsymbol{x} + (\Delta\mathbf{B})\boldsymbol{u}}_{\boldsymbol{F}_1(t)} \tag{3.114}$$

$$\boldsymbol{y} = \mathbf{C}_n \boldsymbol{x} + \mathbf{D}_n \boldsymbol{u} + \underbrace{(\Delta\mathbf{C})\boldsymbol{x} + (\Delta\mathbf{D})\boldsymbol{u}}_{\boldsymbol{F}_2(t)} \tag{3.115}$$

In a BG, the components of the vectors $\boldsymbol{F}_1(t)$, $\boldsymbol{F}_2(t)$ accounting for parametric faults may be represented by modulated sources as they depend on the states $\boldsymbol{x}$ and the system inputs $\boldsymbol{u}$.

3.6.2 BGs in Linear Fractional Transformation Form

Another approach is to decompose a passive element in such a way that the relative uncertainty $\delta := \Delta\theta/\theta$ of its parameter θ is separated as displayed for a resistor in Fig. 3.26.

The BG model of an uncertain R element in Fig. 3.26 may be obtained by reformulating its constitutive equation and by introducing auxiliary variables.

$$e_R = (R_n + \Delta R) f_R \;=\; R_n f_R - \underbrace{\left(-\frac{\Delta R}{R_n}\right)}_{:= -\delta_R} e_{R_n} \tag{3.116}$$

$$z_R = e_{R_n} \tag{3.117}$$

$$w_R = -\delta_R z_R \tag{3.118}$$

A capacitor with uncertain parameter may be decomposed in a similar way as depicted in Fig. 3.27.

BGs in which passive elements, transformers, and gyrators with uncertain parameters are decomposed accordingly are called bond graphs in linear fractional

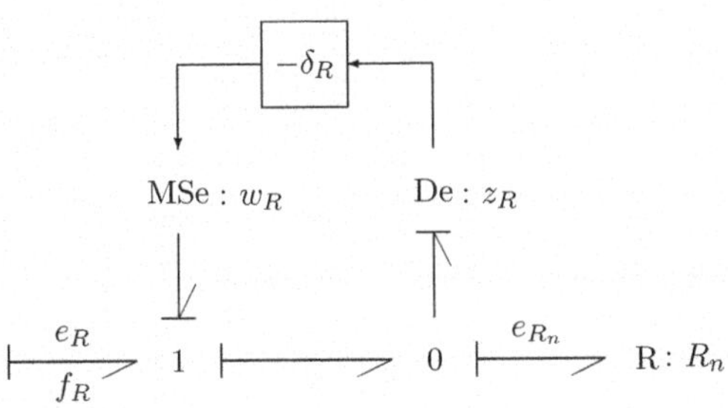

Fig. 3.26 BG model of a resistor with a separation of the relative parameter uncertainty δ_R

transformation form (LFT-BG) in the literature [34]. From LFT-BGs the equations for the *standard interconnection model* in Fig. 3.28 can be derived.

The model in Fig. 3.28 is also known as an *internal feedback loop model*. The interconnection matrix $\mathbf{M}(s)$, $s \in \mathbb{C}$ depends on nominal parameter values only, while $\boldsymbol{\Delta}$ is a diagonal matrix (δ_{ij}) with $|\delta_{ii}| < 1$. According to Fig. 3.28, the following LTI state space model with matrices of appropriate dimensions can be set up.

$$\dot{\mathbf{x}} = \mathbf{A}_n \boldsymbol{x} + \underbrace{[\mathbf{B}_{1_n} \mathbf{B}_{2_n}]}_{\hat{\mathbf{B}}_n} \begin{bmatrix} \boldsymbol{w} \\ \mathbf{u} \end{bmatrix} \tag{3.119a}$$

$$\begin{bmatrix} z \\ y \end{bmatrix} = \underbrace{\begin{bmatrix} \mathbf{C}_{1_n} \\ \mathbf{C}_{2_n} \end{bmatrix}}_{\hat{\mathbf{C}}_n} \boldsymbol{x} + \underbrace{\begin{bmatrix} \mathbf{D}_{11_n} & \mathbf{D}_{12_n} \\ \mathbf{D}_{21_n} & \mathbf{D}_{22_n} \end{bmatrix}}_{\hat{\mathbf{D}}_n} \begin{bmatrix} \boldsymbol{w} \\ \boldsymbol{u} \end{bmatrix} \tag{3.119b}$$

$$\mathbf{w} = \boldsymbol{\Delta} \cdot z \; . \tag{3.119c}$$

With the above introduced matrices, the matrix $\mathbf{M}(s)$ in Fig. 3.28 reads

$$\mathbf{M}(s) = \hat{\mathbf{C}}_n (s\mathbf{I} - \mathbf{A}_n)^{-1} \hat{\mathbf{B}}_n + \hat{\mathbf{D}}_n \; . \tag{3.120}$$

Fig. 3.27 BG model of a capacitor in derivative causality with a separation of the relative parameter uncertainty δ_C

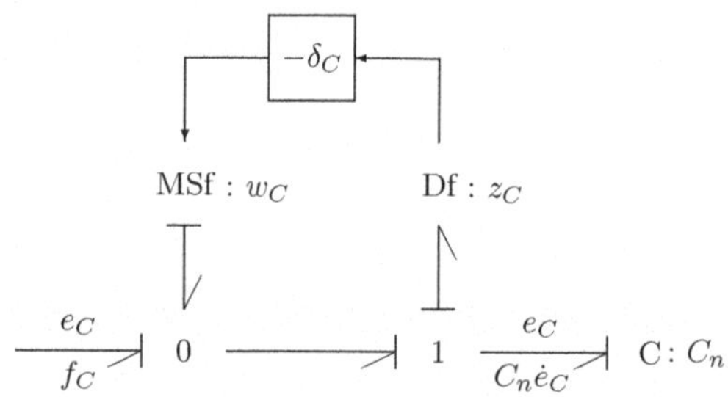

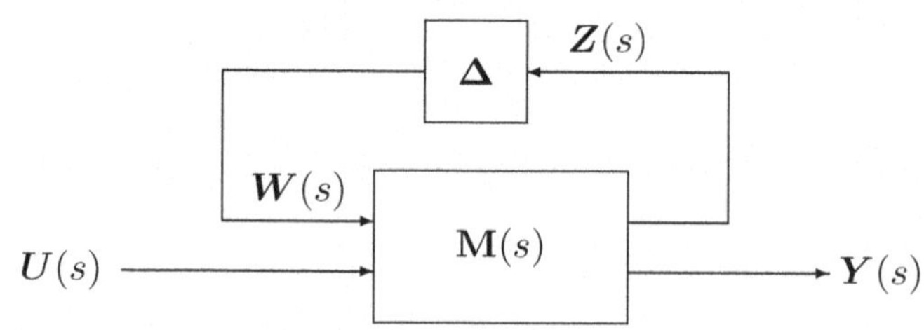

Fig. 3.28 Standard interconnection model

The standard interconnection form is used, for instance, by the μ-analysis and synthesis method.

Given that detectors are in inverted causality and storage elements are assigned derivative causality, ARRs can be derived from the two BG representations that account for parameter uncertainties $\Delta\boldsymbol{\Theta}$.

3.6.3 Incremental BGs and Adaptive Fault Thresholds

Another BG representation accounting for parameter uncertainties are *incremental BGs* introduced by the author [9]. Like LFT-BGs, incremental BGs enable to set up ARRs, and adaptive thresholds for their residuals that can be used to avoid false alarms. That is, as long as residual variations are bounded by these fault thresholds, no fault is reported to a fault diagnosis module.

The basic idea of incremental BGs is that a deviation $\Delta\theta$ of a 1-port element parameter θ from its nominal value causes variations of its power port variables. Their actual value at a time instant t differs by an increment from the value in the case of a parameter with nominal value. In this case, power variables and parameters are assigned a subscript n. For illustration, consider a linear 1-port resistor with the nominal resistance R_n and an uncertainty ΔR.

$$e_{Rn}(t) \;=\; R_n\, f_{Rn}(t) \tag{3.121}$$

The constitutive relation becomes

$$(e_{Rn} + \Delta e_R)(t) \;=\; (R_n + \Delta R)\,(f_{Rn} + \Delta f_R)(t) \tag{3.122}$$

Substituting Equation 3.121 into Equation 3.122 and neglecting the higher-order term $(\Delta R)\,(\Delta f_R)$ results in a linear relation between the incremental power variables Δe_R and Δf_R.

$$\begin{aligned}(\Delta e_R)(t) &= R_n\,(\Delta f_R)(t) + (\Delta R)\, f_{Rn}(t)\\ &= R_n\,(\Delta f_R)(t) + \underbrace{\frac{\Delta R}{R_n}}_{\delta_R} e_{Rn}(t)\end{aligned} \tag{3.123}$$

Equation (3.123) could also be obtained by taking the total differential of the product $R \cdot f_R(t)$.

$$de_R = \underbrace{\left.\frac{\partial e_R}{\partial f_R}\right|_{R_n}}_{R_n} df_R + \underbrace{\left.\frac{\partial e_R}{\partial R}\right|_{R_n}}_{f_{R_n}(t)} dR \tag{3.124}$$

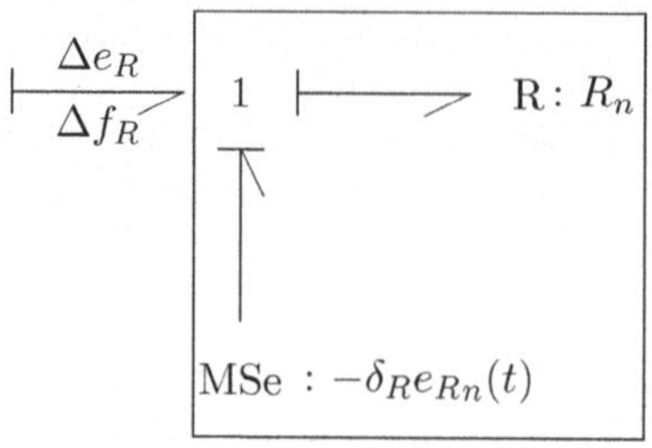

(a) Incremental BG of an uncertain resistor in resistive causality

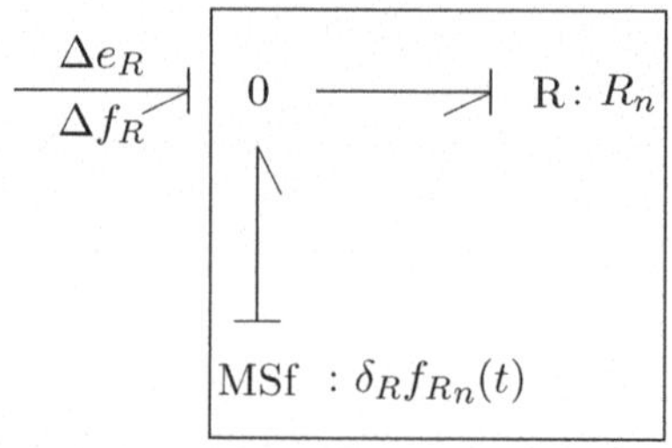

(b) Incremental BG of an uncertain resistor in conductance causality

Fig. 3.29 Incremental BG models of a linear 1-port resistor with an uncertain parameter. (**a**) Incremental BG of an uncertain resistor in resistive causality. (**b**) Incremental BG of an uncertain resistor in conductance causality

Fig. 3.30 Incremental BG model of a linear 1-port capacitor in derivative causality with an uncertain parameter

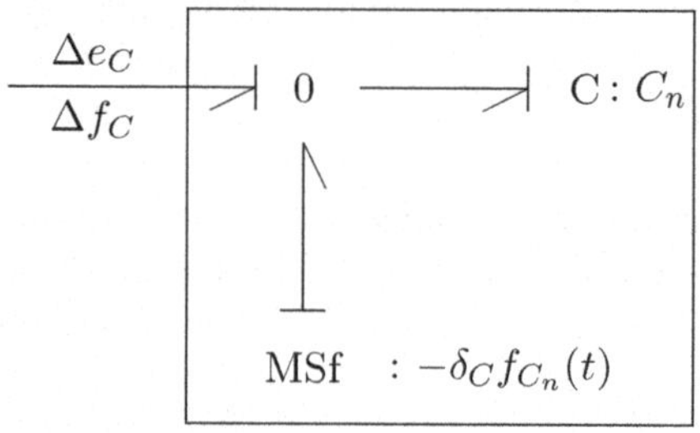

Equation 3.123 can be represented by one of the bond graphs depicted in Fig. 3.29 depending on the causality at the power port of the resistor.

Figure 3.30 shows an incremental BG of a C element with an uncertain parameter.

Incremental BG models for transformers and gyrators can be developed in a similar manner [9]. It turns out that incremental BG models of BG elements with nominal parameters differ from the latter ones only in two ways.

- Bonds carry the increments of power variables.
- The output of a sink modulated by a nominal power variable is added to the junction to which the element with a nominal parameter is attached.

This suggests to introduce an incremental BG (incBG) for the *increments of power variables* due to parameter variations which receives inputs from a BG with nominal parameter values as indicated in Fig. 3.31 with fragments in the incremental BG and in the nominal BG.

Incremental BGs have the following features:

- Inputs into an incBG are normalised parameter variations $\delta_{\Theta_i} = \Delta\Theta_i/\Theta_i$ multiplied by a power variable from the nominal BG.

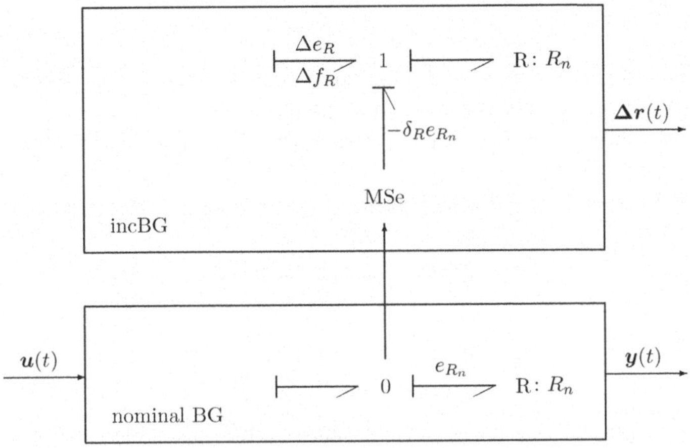

Fig. 3.31 Separation into an incremental BG and a nominal BG

- Outputs of an incBG with regard to FDI are variations $\Delta r_j(t)$ of ARR residuals $r_j(t)$ which, in the case of a LTI system, can be expressed as a weighted sum of the inputs.
- The coefficients of the normalised parameter variations may include transformer moduli $b_i(t) \in \{0, 1\}$. That is, variations of ARR residuals may be system mode dependent.
- The variation $\Delta r_j(t)$ of an ARR residual $r_j(t)$ due to parameter uncertainties is bounded by the sum of the absolute values of all its terms which yields an adaptive threshold. Accounting for these thresholds computed online for each time instant makes ARR residuals as fault indicators insensitive to parameter uncertainties and avoids false alarms.
- An incBG is obtained from a nominal BG by replacing just elements with uncertain parameters by their incremental model which introduces additional modulated sources.
- For a software that can generate equations from a BG, these additional modulated sources are the only difference between an incBG and its nominal BG. That is, existing software can be used to generate the equations for the variations of ARR residuals.

Remark 3.1 Adaptive thresholds obtained from variations of ARR residuals depend on the time derivatives of variables like ARRs derived from the nominal BG. That is, measurements from a real system fed into an incremental DBG are differentiated in discrete time. As a result, ARR thresholds depend on measurement uncertainties, noise, and the sampling time step.

A pure analytical alternative to incremental BGs would be to take the total differential of ARRs derived from the nominal DBG.

Illustrative Example

The determination of adaptive fault thresholds from an incremental BG is illustrated by means of the simple passive network depicted in Fig. 3.32.

Figure 3.33 shows a DBG of the network in Fig. 3.32. The upper index n indicates nominal parameter values.

From the DBG in Fig. 3.33 the following two nominal ARRs can be derived.

$$r_1 = F - C_1^n \dot{V}_1 - \frac{1}{R_1^n}(V_1 - V_2) \tag{3.125}$$

$$r_2 = \frac{1}{R_1^n}(V_1 - V_2) - C_2^n \dot{V}_2 - \frac{1}{R_2^n}(V_2 - E) \tag{3.126}$$

Figure 3.34 displays the associated incremental DBG. The superscript $(*)$ attached to the flow detectors Df* of the residual variations indicates that these detectors are virtual detectors of computed quantities.

From the incremental DBG in Fig. 3.34, the following variations of ARR residuals can be derived:

$$\begin{aligned} \Delta r_1 &= -\delta_{C_1} f_{C_1}^n - \delta_{R_1} f_{R_1}^n \\ &= -\delta_{C_1} C_1^n \dot{V}_1 + \delta_{R_1} \frac{1}{R_1^n}(V_1 - V_2) \end{aligned} \tag{3.127}$$

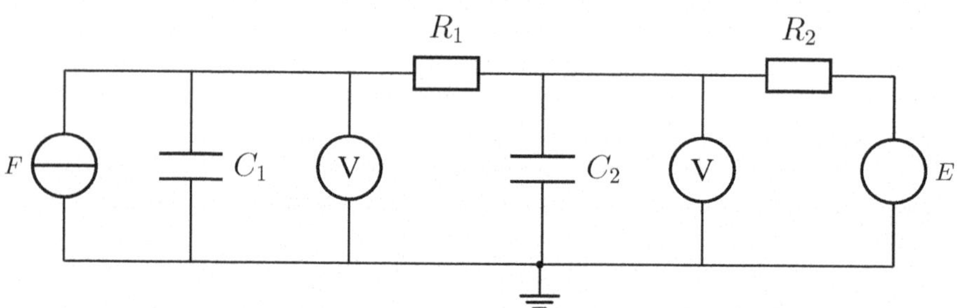

Fig. 3.32 Simple passive network with two voltage sensors

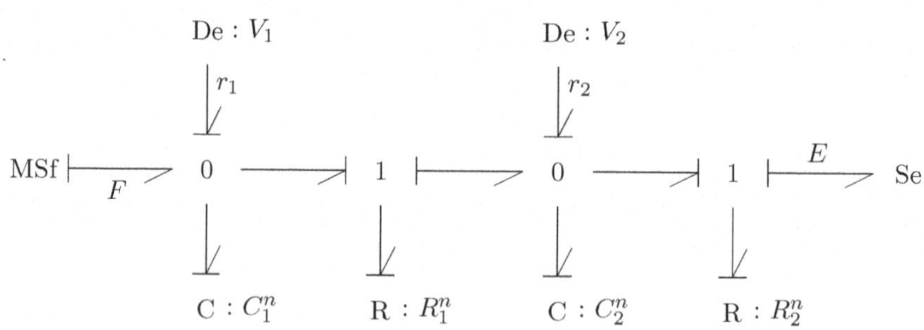

Fig. 3.33 DBG of the network in Fig. 3.32

$$\begin{aligned}\Delta r_2 &= -\delta_{R_1} f^n_{R_1} - \delta_{C_2} f^n_{C_2} + \delta_{R_2} f^n_{R_2} \\ &= -\delta_{R_1} \frac{1}{R^n_1}(V_1 - V_2) - \delta_{C_2} C^n_2 \dot{V}_2 + \delta_{R_2} \frac{1}{R^n_2}(V_2 - E)\end{aligned} \tag{3.128}$$

Application of the triangular inequality yields adaptive fault thresholds $thr_i(t), i = 1, 2$.

$$|\Delta r_1| \leq \delta_{C_1} C^n_1 |\dot{V}_1| + \frac{\delta_{R_1}}{R^n_1} |V_1 - V_2| =: thr_1(t) \tag{3.129}$$

$$|\Delta r_2| \leq \frac{\delta_{R_1}}{R^n_1} |V_1 - V_2| + \delta_{C_2} C^n_2 |\dot{V}_2| + \frac{\delta_{R_2}}{R^n_2} |V_2 - E| =: thr_2(t) \tag{3.130}$$

3.7 Measurement Uncertainties, Sensor Faults, and Actuator Faults

Measured signals are inputs into a DBG besides known control inputs. Even if they are filtered before they are used in the generation of ARR residuals, there is still some measurement uncertainty that may affect the evaluation of ARRs. This section shows how measurement uncertainties can be accounted for in a DBG and defines thresholds that can be used to make sure that measurement uncertainties do not trigger false alarms. In addition to the robustness with regard to

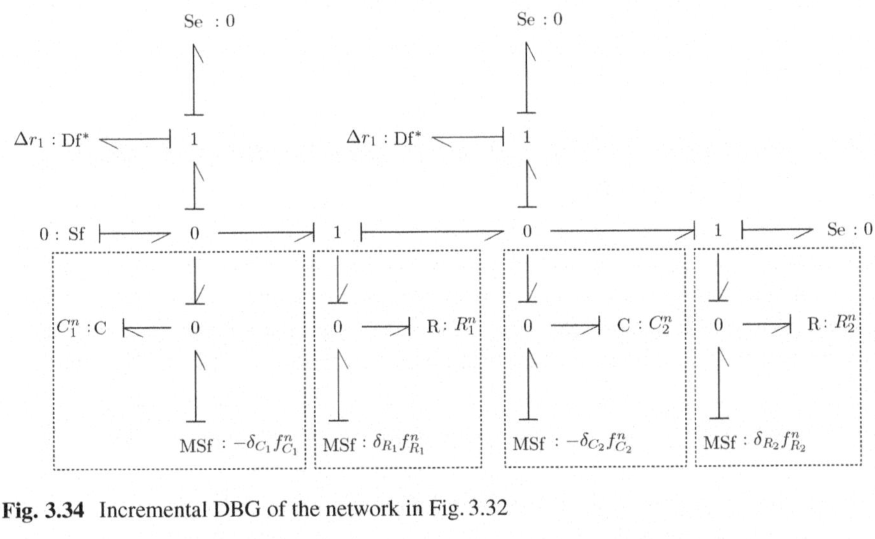

Fig. 3.34 Incremental DBG of the network in Fig. 3.32

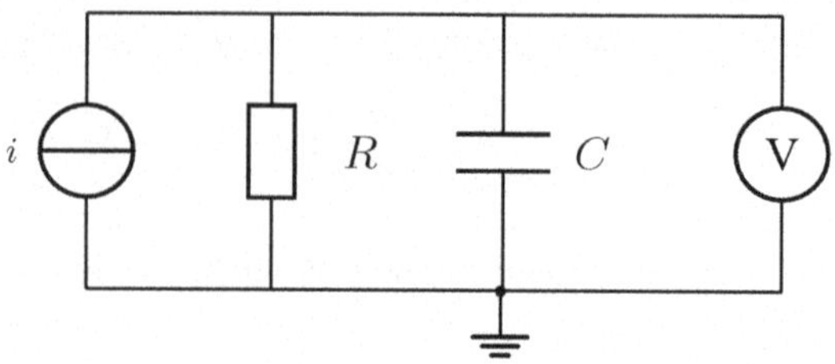

Fig. 3.35 Simple circuit with a faulty voltage sensor

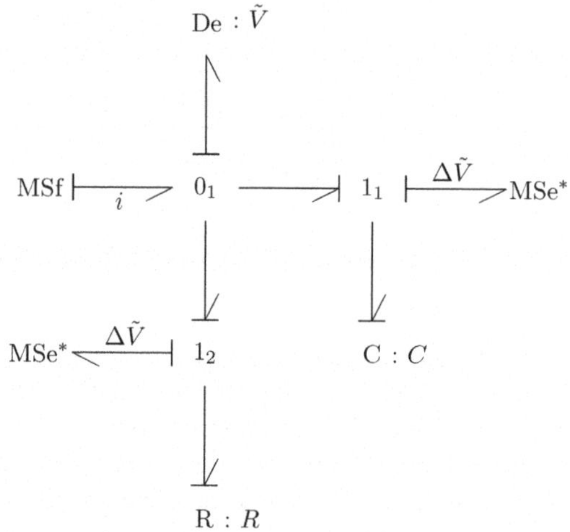

Fig. 3.36 DBG of the circuit in Fig. 3.35 accounting for a faulty voltage sensor

parameters considered in the previous section, ARR residuals shall be insensitive to measurement uncertainties that may be due to noise, to some bias, drift, incorrect calibration, or a parametric sensor fault.

3.7.1 *Accounting for Measurement Uncertainties and Sensor Faults in a BG*

It is assumed that sensor faults are additive and are bounded. These assumptions suggest to model unknown sensor faults by means of modulated sources or sinks and to insert them into each bond that is connected to a junction with a potentially faulty sensor attached [61]. The procedure is explained by means of two small examples.

DBG of a Simple Circuit Accounting for a Faulty Voltage Sensor

Consider the simple circuit in Fig. 3.35 and its DBG in Fig. 3.36.

Input variables into the DBG in Fig. 3.36 are the known current i and the measured voltage $\tilde{V}$ provided by faulty sensor indicated by an effort detector in inverted causality. Its measurement uncertainty $\Delta\tilde{V}$ is taken into account by two virtual modulated effort sinks. Their superscript (∗) indicates that these sinks do not represent real elements in the system. If the voltage $\tilde{V}$ including some measurement uncertainty ΔV, i.e. $\tilde{V} = V' + \Delta V$, is an input into the resistor and the capacitor, then the residual $r = r' + \Delta r$ due to the measurement uncertainty is different from zero. If there is no deviation from the true voltage V', then $r = r' = 0$.

$$0 \neq r = i - C\dot{\tilde{V}} - \frac{1}{R}\tilde{V} \tag{3.131}$$

However, if the faulty voltage $\tilde{V}$ subtracted by the measurement uncertainty $\Delta\tilde{V}$, i.e. the true non-faulty voltage V' is input into the elements, then the sum in (3.131) must be equal to zero.

$$0 = i - C(\dot{\tilde{V}} - \Delta\dot{\tilde{V}}) - \frac{1}{R}(\tilde{V} - \Delta\tilde{V}) \tag{3.132}$$

Reformulation of (3.132) yields

$$-r = C\,\Delta\dot{\tilde{V}} + \frac{1}{R}\Delta\tilde{V} \tag{3.133}$$

Let $|\Delta\tilde{V}(t)| \leq b$. If the derivative $\Delta\dot{\tilde{V}}$ is approximated by the difference quotient, then residual r due the measurement uncertainty is also bounded.

$$|\Delta r| \leq C\,|\Delta\dot{\tilde{V}}(t)| + \frac{1}{R}|\Delta\tilde{V}| \;\leq\; C\frac{2b}{\Delta t} + \frac{1}{R}b \tag{3.134}$$

where Δt denotes the measurement sampling time step.

Moreover, given that $\Delta\tilde{V}(0) = 0$, the magnitude of the measurement uncertainty can be estimated. From (3.133) one obtains

$$\Delta\dot{\tilde{V}} + \frac{1}{RC}\Delta\tilde{V} \;=\; -\frac{1}{C}r \tag{3.135}$$

The same result can be obtained from a DBG with a bicausal path from the effort detector De : $\tilde{V}$ to one of the two modulated effort sinks representing the measurement uncertainty $\Delta\tilde{V}$ as indicated in Fig. 3.37.

DBG with Two Faulty Effort Sensors

Figure 3.38 shows a DBG of the circuit in Fig. 3.32, which assumes that the two sensors provide voltages with measurement uncertainties $\Delta\tilde{V}_1$ and $\Delta\tilde{V}_2$.

If the measurement uncertainties $\Delta\tilde{V}_1$ and $\Delta\tilde{V}_2$ of the two sensors are not subtracted from the faulty measured voltages $\tilde{V}_1$, $\tilde{V}_2$, i.e. if the modulated effort sinks in the DBG of Fig. 3.38 are disregarded, then the following two residuals different from zero are obtained.

$$0 \neq r_1 = r_1' + \Delta r_1 = i - C_1 \dot{\tilde{V}}_1 - \frac{1}{R_2}(\tilde{V}_1 - \tilde{V}_2) \tag{3.136a}$$

$$0 \neq r_2 = r_2' + \Delta r_2 = i - C_1 \dot{\tilde{V}}_1 - C_2 \dot{\tilde{V}}_2 - \frac{1}{R_3}(\tilde{V}_2 - E) \tag{3.136b}$$

If the measurement uncertainties are subtracted, the residuals are equal to zero and the constraints read

$$r_1 = r_1' = 0 = i - C_1(\dot{\tilde{V}}_1 - \Delta \dot{V}_1) - \frac{1}{R_2}[(\tilde{V}_1 - \Delta \tilde{V}_1 - (\tilde{V}_2 - \Delta \tilde{V}_2)] \tag{3.137a}$$

$$r_2 = r_2' = 0 = i - C_1(\dot{\tilde{V}}_1 - \Delta \dot{V}_1)C_2(\dot{V}_2 - \Delta \dot{V}_2) - \\ - \frac{1}{R_3}[(\tilde{V}_2 - \Delta \tilde{V}_2) - E] \tag{3.137b}$$

Fig. 3.37 DBG of the circuit in Fig. 3.35 with a bicausal path from the voltage sensor to a modulated sink for the measurement uncertainty

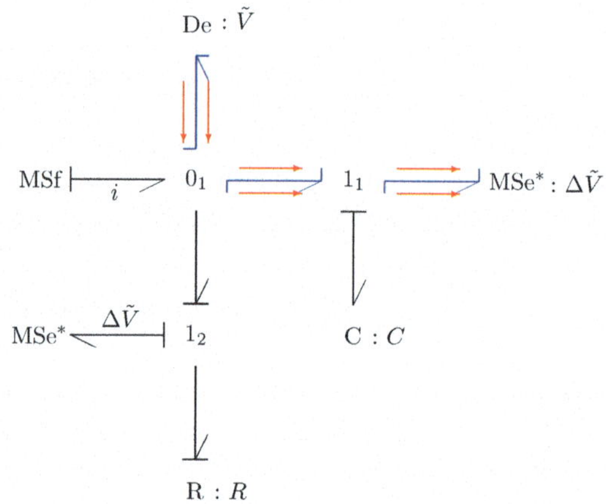

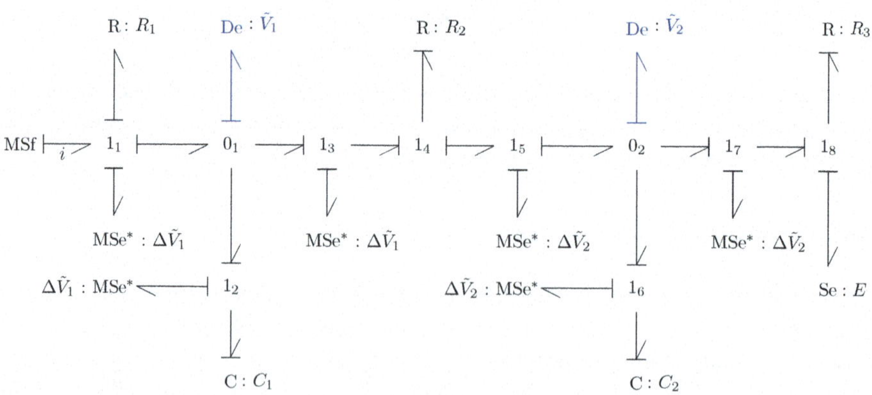

Fig. 3.38 DBG of the circuit in Fig. 3.32 accounting for two faulty voltage sensors

Substituting the residuals (3.136) into the constraints (3.137) yields

$$-\Delta r_1 = C_1 \Delta \dot{V}_1 + \frac{1}{R_2}(\Delta \tilde{V}_1 - \Delta \tilde{V}_2) \tag{3.138a}$$

$$-\Delta r_2 = C_1 \Delta \dot{V}_1 + C_2 \Delta \dot{V}_2 + \frac{1}{R_3} \Delta \tilde{V}_2 \tag{3.138b}$$

The result (3.138) can be verified by considering the two residuals as functions of the two voltages and their time derivatives and by taking the total differential.

In case the measurement uncertainties $\Delta \dot{V}_1$, $\Delta \dot{V}_2$ are bounded, the residuals (3.138) are also bounded.

3.7.2 Representing Actuator Faults in a BG

Once an actuator has become faulty, it is no longer able to deliver the actuating variable $\boldsymbol{u}$ that the controller demands but a deteriorated plant input $\tilde{\boldsymbol{u}}$. Consequently, sensors get an unavailable deteriorated plant output $\tilde{\boldsymbol{y}}$ different from the one in the case of fault free actuators which they themselves superimpose with measurement uncertainties. In Fig. 3.39, the resulting measurements are denoted by the vector $\tilde{\boldsymbol{y}}_m$.

If the dynamic behaviour of a plant can be described by a LTI model, then actuator faults can be taken into account by a state equation in the form

$$\begin{aligned} \dot{\boldsymbol{x}} &= \mathbf{A}\boldsymbol{x} + \mathbf{B}\,(\boldsymbol{u}^n + \boldsymbol{u}^f) \\ &= \mathbf{A}\boldsymbol{x} + \mathbf{B}\boldsymbol{u}^n + \mathbf{B}\,\boldsymbol{u}^f \end{aligned} \tag{3.139}$$

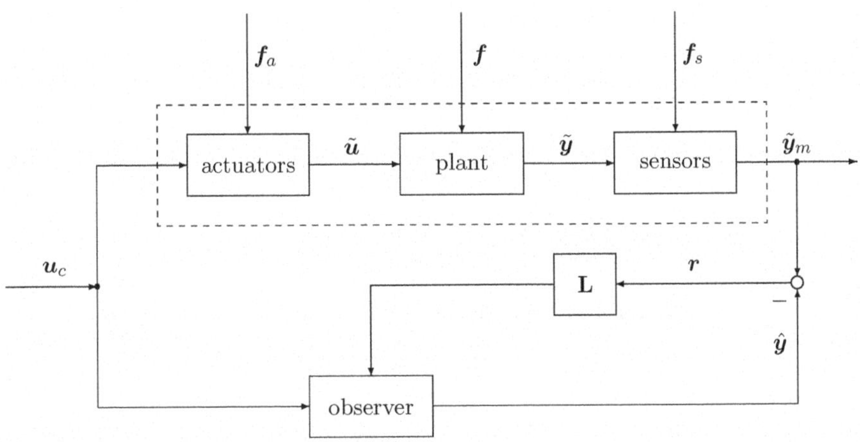

Fig. 3.39 Observer-based fault detection

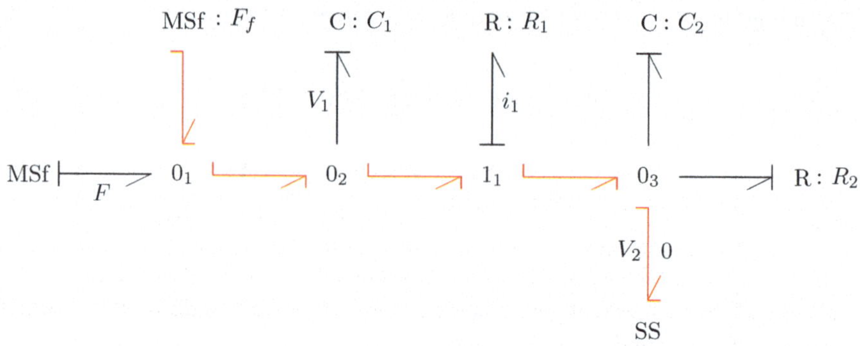

Fig. 3.40 Bicausal BG of a circuit with a faulty flow source

where $\boldsymbol{u}^n$ denotes the vector of system inputs in the case that all actuators are fault free, and the vector $\boldsymbol{u}^f$ collects all actuator faults. Suppose that the ith actuator fails, then $u_i^f = -u_i^n$. Actuator faults are additive.

If actuator dynamics can be neglected, actuators such hydraulic pumps or electrical sources are often represented by a modulated source. In that case an actuator can be accounted for in a BG by just adding a source to the junction the modulated source for the fault free actuator is connected to. The DBG in Fig. 3.40 may model either a hydrostatic two tank system with a displacement pump or an electrical analogon with a current source.

In the DBG of Fig. 3.40, a modulated flow source MSf : F_f attached to junction 0_1 accounts for an actuator fault F_f. Its magnitude can be estimated by following the bicausal path from the flow source MSf : F_f to the effort sensor De : V_2 in inverted causality replaced by a source sensor element SS.

$$F_f = -F + (C_1 \dot{V}_1 + i_1) \tag{3.140}$$

$$= -F + C_1(R_1 \frac{d}{dt} i_1 + \dot{V}_2) - (C_2 \dot{V}_2 + \frac{1}{R_2} V_2) \tag{3.141}$$

$$= -F + C_1[R_1 \frac{d}{dt}(C_2 \dot{V}_2 + \frac{1}{R_2} V_2) + \dot{V}_2] + (C_2 \dot{V}_2 + \frac{1}{R_2} V_2) \tag{3.142}$$

$$= -F + C_1[R_1(C_2 \ddot{V}_2 + \frac{1}{R_2} \dot{V}_2) + \dot{V}_2] + C_2 \dot{V}_2 + \frac{1}{R_2} V_2 \tag{3.143}$$

As a result, the actuator fault, F_f, depends on the measurement V_2, and its time derivatives.

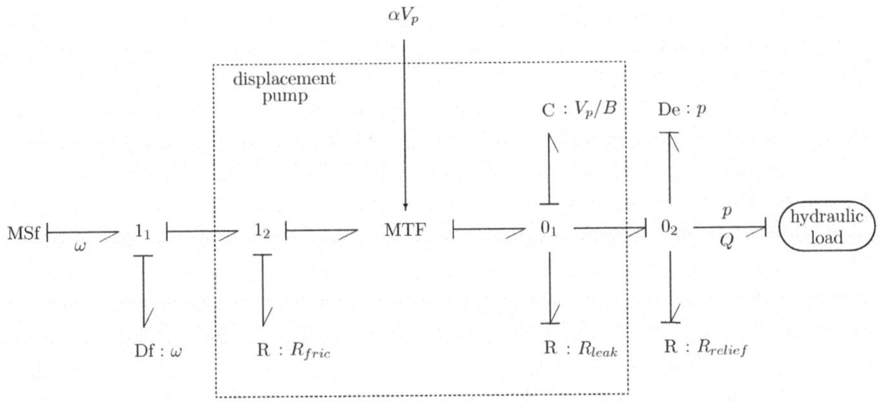

Fig. 3.41 Bond Graph model of a hydraulic variable displacement pump

If the dynamics of an actuator are of relevance for the dynamic behaviour of the system and if design details are known, models of various complexity can be developed for an actuator and stored in a library. In that case, the actuator is a subsystem of the overall system with its own internal states and parameters that may be subject to external disturbances, model, and parameter uncertainties. For instance, a model of a hydraulic variable displacement pressure regulated pump which provides a nearly constant pressure over a range of flow demand may reflect that the pump is an energy transducer and may account for the conversion of mechanical energy into hydraulic energy, for internal mechanical friction, for internal hydraulic leakage, and for fluid compressibility at its outlet as depicted in Fig. 3.41 [13, 59], where αV_p denotes the displacement of the pump according to the swash plate angle α, V_p/B the outlet fluid capacity, and R_{relief} the resistance of the relief valve. The motor driving the pump at the angular speed ω is simply represented by a flow source, which could be replaced by a detailed model as well that would help to detect component faults.

3.8 Sensor Placement on Diagnostic Bond Graphs and Fault Isolation

Some of the sensors attached to an engineering system are used to monitor process variables for control and performance assessment, e.g. position, speed, temperature, pressure, flow rates, etc. while others are directly related to fault modes identified as candidates for diagnosis, e.g. strain gauges, ultrasonic sensors, or accelerometers. With regard to fault isolation a question is how many sensors are to be placed in which locations in order to isolate a maximum of potentially faulty system components. Various approaches to the sensor placement problem based on bipartite graphs [24], on digraphs [1], or on bond graphs [6, 12, 17, 22, 35, 56] have been reported in the literature.

Given an engineering system with a set of sensors, one option is to develop a DBG model from which a structural FSM can be obtained by following causal paths. The number of columns in a FSM linked to an ARR residual equals the number of hardware sensors. In general, their number is less than the number of component parameters. That is, the matrix is not square, and only potentially faulty component parameters with a unique fault signature can be isolated by structural inspection of a FSM. In the case of non-unique component fault signatures, fault isolation may be achieved if the coherence vector matches a component fault signature that is unique in a subspace of the unstructured part of a FSM. Furthermore, the number of unique parameter fault signatures may be increased if binary entries in a FSM are replaced by parameter sensitivities of ARR residuals and if sensitivities with a small value are substituted by zero.

An alternative to a structural check of a FSM that is computationally more costly is an evaluation of multiple observer models with a single candidate of a parametric fault or a numerical estimation of all those parameters that may be faulty. In the case a bank of observer models with a single fault candidate is used, the model that matches the unknown component fault yields an observer output close to the measurements obtained from the faulty system and by this way isolates the fault [27].

Another computationally costly method is to perform a parameter estimation for components with the same fault signature. As a result, those parameters are identified that deviate from their nominal values and are thus faulty [10, 55].

3.8.1 Graphical Approach to Sensor Placement and Fault Isolation

In [12], the author proposes a graphical approach to the structural isolation of parametric component faults that aims at avoiding the limitations of an inspection of a structural FSM and the computational costs of the above numerical methods and to achieve a maximum number of isolated parametric component faults. It is assumed that sensors and actuators are faultless and that the dynamic behaviour of the plant can be described by a LTI state space model. Moreover, it is assumed that an abnormal system behaviour may be attributed to a deviation of some component parameters from their nominal values and that all element parameters could be potentially faulty.

The idea of the graphical approach is to start from a DBG of a system with a given set of sensors, to consider all causal paths from a detector to a source, or to a passive element, from a source to a passive element, and causal paths between passive elements. If there is a causal path from one of the given detectors to an element that is disjoint from other causal paths, then possibly faulty element parameter can be isolated. If there are paths from a detector to an element that are not disjoint from other causal paths, then the graphical procedure attaches detectors to those junctions of the DBG that are crossed by the highest number of these causal paths so that some causal paths vanish and further disjoint causal paths from detectors to possibly faulty elements are obtained.

If there are causal paths from different detectors to a potentially faulty element, the set of these non-disjoint causal paths must be unique. Potentially faulty elements with a unique causal path from a detector to the element can be isolated as can be verified by a FSM.

If causal paths from a detector to several elements remain and all of them are attached to one and the same junction, then this means that the parameters of these elements cannot be isolated. An additional junction with a detector attached must be inserted. For instance, a flow sensor is not enough to isolate the parameters of electrical elements connected in series. By this way, finally all elements could be isolated. The placement of a detector on the DBG of a system, however, must take into account whether a sensor can be added to the real system in a desired place. If this is not the case, virtual sensors, or an observer may be used.

In the following, the problem of fault isolation and sensor placement is explained and illustrated by means of two examples, a small RLC circuit and a DC motor drive.

Example: RLC Circuit

The circuit schematic in Fig. 3.42 with two flow sensors assumed to be non-faulty is rule-based converted into the BG in Fig. 3.43 with preferred integral causality.

From the BG in preferred integral causality with two flow detectors the following state space model can be derived.

$$\begin{bmatrix} \dot{e}_5 \\ \dot{f}_{12} \end{bmatrix} = \underbrace{\begin{bmatrix} -\dfrac{R}{CR_1R_{12}} & -\dfrac{R_3}{CR_{12}} \\ \dfrac{R_3}{LR_{23}} & -\dfrac{R_2R_3}{LR_{23}} \end{bmatrix}}_{\mathbf{A}} \begin{bmatrix} e_5 \\ f_{12} \end{bmatrix} + \underbrace{\begin{bmatrix} \dfrac{1}{R_1} \\ 0 \end{bmatrix}}_{\mathbf{B}} E \tag{3.144}$$

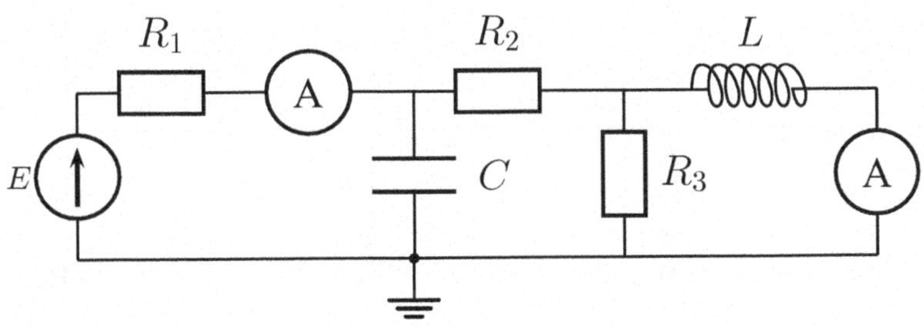

Fig. 3.42 RLC circuit schematic

$$\begin{bmatrix} y_1 \\ y_2 \end{bmatrix} = \underbrace{\begin{bmatrix} \frac{1}{R_1} & 0 \\ 0 & 1 \end{bmatrix}}_{\mathbf{C}} \begin{bmatrix} e_5 \\ f_{12} \end{bmatrix} + \underbrace{\begin{bmatrix} \frac{1}{R_1} \\ 0 \end{bmatrix}}_{\mathbf{D}} E \tag{3.145}$$

where $R := R_1 + R_2 + R_3$, $R_{12} := R_1 + R_2$, and $R_{23} := R_2 + R_3$.

Given matrices **A**, **C**, it can be proven that the system with the two flow sensors is numerically completely state observable. The system is even fully state observable with flow sensor Df : y_2. From the BG in Fig. 3.43, the following equations determining the states can be derived.

$$i_L = y_2 \tag{3.146}$$

$$(R_2 + R_3)CR_1\dot{f}_3 + Rf_3 = E + (R_2 + R_3)C\dot{E} + R_3y_2 \tag{3.147}$$

$$e_C = E - R_1f_3 \tag{3.148}$$

Given the input E and measurement y_2, (3.147) may be considered the equation of a virtual sensor that provides the output y_1.

In the sequel, it is shown that all element parameters considered potentially faulty can be detected, given the two flow sensors, but not all of them can be isolated. Figure 3.44 displays a diagnostic BG of the RLC circuit with flow detectors Df : y_1, Df : y_2 in inverted causality and storage elements in derivative causality.

From the DBG with both detectors in inverted causality and storage elements in preferred derivative causality (Fig. 3.44), two ARRs are obtained by summing efforts at the two junctions 1_1 and 1_3 with detectors attached and eliminating unknowns by following causal paths.

$$\text{ARR}_1: \quad r_1 = E + (R_2 + R_3)C\dot{E} - Ry_1 - (R_2 + R_3)CR_1\dot{y}_1 + R_3y_2 \tag{3.149}$$

$$\text{ARR}_2: \quad r_2 = R_3(y_1 - y_2) - R_3C\dot{E} + R_3CR_1\dot{y}_1 - L\dot{y}_2 \tag{3.150}$$

where $R = R_1 + R_2 + R_3$.

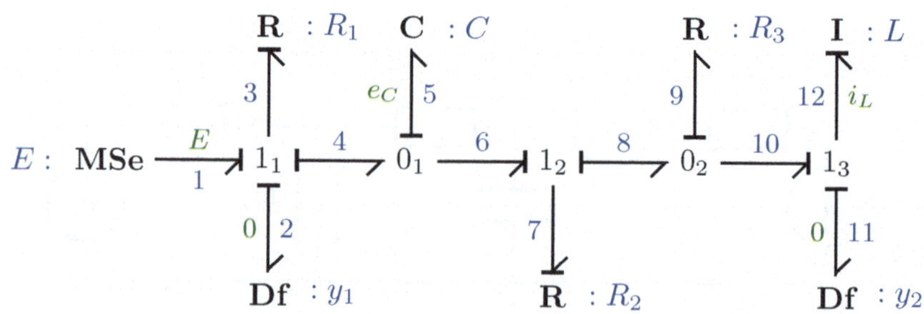

Fig. 3.43 BG in preferred integral causality of the RLC circuit in Fig. 3.42

Table 3.3 Structural fault signature matrix of the circuit in Fig. 3.42 with two current sensors

Element	r_1	r_2	D_b	I_b
Se : E	1	1	1	0
C : C	1	1	1	0
I : L	0	1	1	①
R : R_1	1	1	1	0
R : R_2	1	0	1	①
R : R_3	1	1	1	0

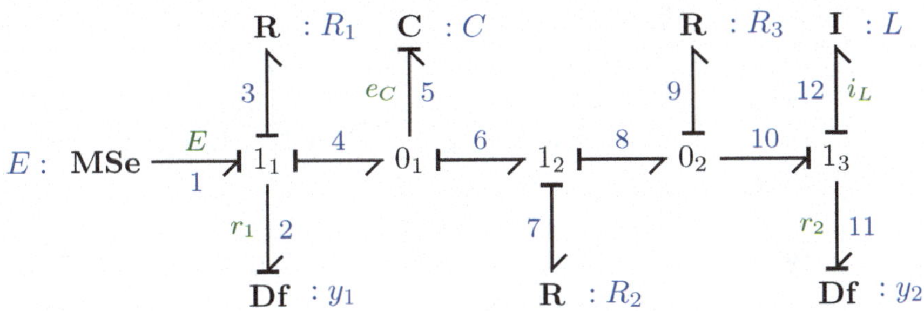

Fig. 3.44 Diagnostic BG of the RLC circuit in Fig. 3.42 with two flow sensors Df : y_1, Df : y_2

Accordingly, the structural fault signature matrix in Table 3.3 displays which element parameters affect which residual. As the entry '1' highlighted in blue in the last column indicates, only potentially faulty element parameters R_2 and L can structurally isolated.

Increasing the Number of Isolatable Parametric Faults

The question is how many sensors in which locations are to be added to isolate a maximum number of potentially faulty element parameters. An inspection of the DBG in Fig. 3.44 yields the causal paths p_i listed in Table 3.4.

Parameters E, R_1, R_2, R_3, C belong to paths through junction 1_1 to which flow detector Df : y_1 is attached. That is, they affect ARR_1. Likewise, parameters E, R_1, R_3, C, L belong to paths through junction 1_3 with flow detector Df : y_2, which means that they contribute to the residual of ARR_2. The only two paths that start from different detectors and are disjoint are p_3 from Df : y_1 to R : R_2 and p_7 from Df : $y2$ to I : L. Hence, parameters R_2 and L can be isolated, which is in accordance with the structural FSM in Table 3.3.

All causal paths except p_3 and p_7 have bonds in common along the power line from junction 1_1 to junction 1_3. In order to increase the number of direct causal paths from a detector to an element, an additional flow detector Df : y_3 is attached to junction 1_2. This enforces integral causality at the capacitor if all detectors are

Table 3.4 Listing of causal paths in the DBG in Fig. 3.44

p_1:	$\mathrm{Df}: y_1$	$\xrightarrow{2} 1_1 \xrightarrow{3} \mathrm{R}: R_1 \xrightarrow{3} 1_1 \xrightarrow{1} \mathrm{MSe}: E$
p_2:	$\mathrm{Df}: y_1$	$\xrightarrow{2} 1_1 \xrightarrow{3} \mathrm{R}: R_1$
p_3:	$\mathrm{Df}: y_1$	$\xrightarrow{2} 1_1 \xrightarrow{3} \mathrm{R}: R_1 \xrightarrow{3} 1_1 \xrightarrow{4} 0_1 \xrightarrow{6} 1_2 \xrightarrow{7} \mathrm{R}: R_2$
p_4:	$\mathrm{Df}: y_1$	$\xrightarrow{2} 1_1 \xrightarrow{3} \mathrm{R}: R_1 \xrightarrow{3} 1_1 \xrightarrow{4} 0_1 \xrightarrow{6} 1_2 \xrightarrow{8} 0_2 \xrightarrow{9} \mathrm{R}: R_3$
p_5:	$\mathrm{Df}: y_1$	$\xrightarrow{2} 1_1 \xrightarrow{3} \mathrm{R}: R_1 \xrightarrow{3} 1_1 \xrightarrow{7} 1_2 \xrightarrow{6} 0_1 \xrightarrow{5} \mathrm{C}: C$
p_6:	$\mathrm{Df}: y_1$	$\xrightarrow{2} 1_1 \xrightarrow{3} \mathrm{R}: R_1 \xrightarrow{3} 1_1 \xrightarrow{4} 0_1 \xrightarrow{6} 1_2 \xrightarrow{8} 0_2 \xrightarrow{9} \mathrm{R}: R_3 \xrightarrow{9} 0_2 \xrightarrow{8} 1_2 \xrightarrow{6} 0_1 \xrightarrow{5} \mathrm{C}: C$
p_7:	$\mathrm{Df}: y_2$	$\xrightarrow{2} 1_1 \xrightarrow{4} 0_1 \xrightarrow{6} 1_2 \xrightarrow{8} 0_2 \xrightarrow{10} 1_3 \xrightarrow{12} \mathrm{I}: L$
p_8:	$\mathrm{Df}: y_2$	$\xrightarrow{11} 1_3 \xrightarrow{10} 0_2 \xrightarrow{9} \mathrm{R}: R_3$
p_9:	$\mathrm{Df}: y_2$	$\xrightarrow{11} 1_3 \xrightarrow{10} 0_2 \xrightarrow{9} \mathrm{R}: R_3 \xrightarrow{9} 0_2 \xrightarrow{8} 1_2 \xrightarrow{6} 0_1 \xrightarrow{5} \mathrm{C}: C$
p_{10}:	$\mathrm{R}: R_2$	$\xrightarrow{7} 1_2 \xrightarrow{6} 0_1 \xrightarrow{5} \mathrm{C}: C$
p_{11}:	$\mathrm{R}: R_3$	$\xrightarrow{9} 0_2 \xrightarrow{8} 1_2 \xrightarrow{6} 0_1 \xrightarrow{5} \mathrm{C}: C$
p_{12}:	$\mathrm{Df}: y_2$	$\xrightarrow{11} 1_3 \xrightarrow{10} 0_2 \xrightarrow{9} \mathrm{R}: R_3 \xrightarrow{9} 0_2 \xrightarrow{8} 1_2 \xrightarrow{6} 0_1 \xrightarrow{5} \mathrm{C}: C \xrightarrow{5} 0_1 \xrightarrow{4} 1_1 \xrightarrow{3} 1_1 \xrightarrow{3} \mathrm{R}: R_1$
p_{13}:	$\mathrm{Df}: y_2$	$\xrightarrow{11} 1_3 \xrightarrow{10} 0_2 \xrightarrow{9} \mathrm{R}: R_3 \xrightarrow{9} 0_2 \xrightarrow{8} 1_2 \xrightarrow{6} 0_1 \xrightarrow{5} \mathrm{C}: C \xrightarrow{5} 0_1 \xrightarrow{4} 1_1 \xrightarrow{3} 1_1 \xrightarrow{1} \mathrm{MSe}: E$

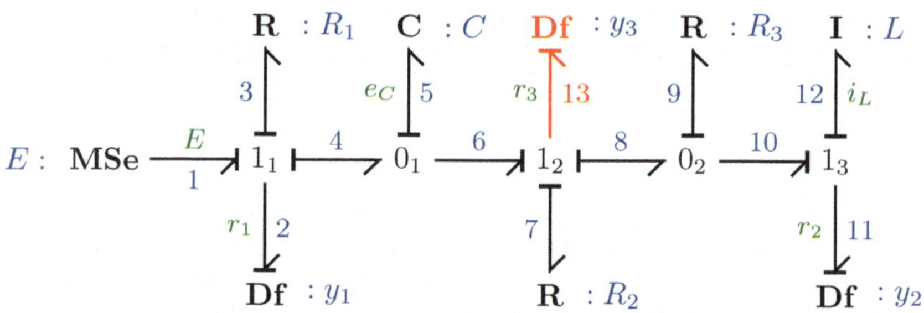

Fig. 3.45 DBG of the RLC circuit in Fig. 3.42 with an additional flow detector $\mathrm{Df}: y_3$

in inverted causality. As a result, the C element has direct causal paths to the two detectors $\mathrm{Df}: y_1$ and $\mathrm{Df}: y_3$ as displayed in the DBG in Fig. 3.45. Accordingly, the two ARRs associated with junctions 1_1 and 1_2 must be differentiated with respect to time to get rid of the initial condition $e_C(0)$ which is unknown in general.

The DBG with three flow detectors in Fig. 3.45 contains the following causal paths from detectors to other elements of the DBG listed in Table 3.5.

Table 3.5 Listing of causal paths in the DBG in Fig. 3.45

p_1':	Df : y_1	$\xrightarrow{2} 1_1 \xrightarrow{1}$ MSe : E
p_2':	Df : y_1	$\xrightarrow{2} 1_1 \xrightarrow{3}$ R : R_1
p_3':	Df : y_1	$\xrightarrow{2} 1_1 \xrightarrow{4} 0_1 \xrightarrow{5}$ C : C
p_4':	Df : y_3	$\xrightarrow{13} 1_2 \xrightarrow{3}$ R : R_2
p_5':	Df : y_3	$\xrightarrow{13} 1_2 \xrightarrow{8} 0_2 \xrightarrow{9}$ R : R_3
p_6':	Df : y_3	$\xrightarrow{13} 1_2 \xrightarrow{6} 0_1 \xrightarrow{5}$ C : C
p_7':	Df : y_2	$\xrightarrow{11} 1_3 \xrightarrow{12}$ I : L
p_8':	Df : y_2	$\xrightarrow{11} 1_3 \xrightarrow{10} 0_2 \xrightarrow{9}$ R : R_3

Table 3.6 Structural fault signature matrix of the DBG in Fig. 3.45 with detectors Df : y_1, Df : y_2 and Df : y_3

Component	ARR_1	ARR_2	ARR_3	D_b	I_b
Se : E	1	0	0	1	0
R : R_1	1	0	0	1	0
C : C	1	0	1	1	1
R : R_2	0	0	1	1	1
R : R_3	0	1	1	1	1
I : L	0	1	0	1	1

Paths p_1' to p_3' pass through junction 1_1 and contribute element parameters E, R_1, C to ARR_1' associated with junction 1_1. Furthermore, paths p_4' to p_6' pass through junction 1_2 and contribute element parameters C, R_2, R_3 to its ARR_3'. Finally, the paths starting from Df : y_2 contribute element parameters R_3 and L to ARR_2'. Causal paths p_3' and p_6' start from different detectors and join only the target bond attached to C : C. Paths p_4' and p_5' have only the bond starting from the same detector in common. The same holds for the two causal paths starting from detector Df : y_2. As a result, all elements except MSe : E and R : R_1 attached to one and the same sensor junction 1_1 can be isolated by means of the three flow sensors. The inspection of causal paths in the DBG with three detectors yields the FSM Table 3.6 without the need to derive equations from the DBG.

All elements have a unique fault signature except MSe : E and R : R_1. The latter two elements are connected in series. The detector Df : y_1 sensing the joint flow through both elements alone is not sufficient to isolate both elements and there are no paths from the other flow detectors to these elements. Non-isolatable parameter faults will be addressed in a subsequent paragraph.

For verification, the following equations are derived from the DBG in Fig. 3.45.

$$r_1 = E - R_1 y_1 - e_c \tag{3.151}$$

$$r_1' := C\dot{r}_1 = C\dot{E} - R_1 C\dot{y}_1 - (y_1 - y_3) \tag{3.152}$$

$$r_3 = e_C - R_2 y_3 - R_3(y_3 - y_2) \tag{3.153}$$

$$r_3' := C\dot{r}_3 = (y_1 - y_3) - CR_2\dot{y}_3 - CR_3(\dot{y}_3 - \dot{y}_2) \tag{3.154}$$

$$r_2 = R_3(y_3 - y_2) - L\dot{y}_2 \tag{3.155}$$

The ARRs derived from the DBG in Fig. 3.45 confirm the structural FSM in Table 3.6 obtained directly from the DBG by following causal paths from a detector to a source or a passive element. In conclusion, for the considered simple RLC circuit, the three flow sensors are a minimal set of sensors that can isolate a maximum of possible parametric faults.

Magnitude of Structurally Isolated Parameter Faults

Once a parametric fault has been isolated, its magnitude can be determined from an ARR. Consider the BG of the RLC circuit (Fig. 3.43). Suppose that the effort source MSe : E and all passive elements may be faulty. If the parameters in the DBG match the faulty parameters in the physical system, then residuals become close to zero. For instance, let residuals r_2 and r_3' deviate significantly from zero, then this means that one of the two parameters R_3 and L is faulty, or both can be simultaneously faulty, i.e. $R_3 := R_3^n + \Delta R_2(t)$, and $L := L^n + \Delta L(t)$, where R_3^n and L^n denote nominal values. Then ARR_2 and ARR_3 are equations that determine the magnitude of these unknown faults.

$$ARR_2 : 0 = (R_3^n + \Delta R_3)(y_3 - y_2) - (L^n + \Delta L)\dot{y}_2 \tag{3.156}$$

$$ARR_3 : 0 = (y_1 - y_3) - R_2^n C^n \dot{y}_3 - (R_3^n + \Delta R_3)C^n(\dot{y}_3 - \dot{y}_2) \tag{3.157}$$

Reformulation yields

$$E_3 := (\Delta R_3)(\dot{y}_3 - \dot{y}_2) = \frac{1}{C^n}\int_0^t (y_1 - y_3)d\tau - R_3^n(y_3 - y_2) - R_2^n y_3 \tag{3.158}$$

$$E_2 := (\Delta L)\dot{y}_2 = R_3^n(y_3 - y_2) + E_3 - L^n\dot{y}_2 \tag{3.159}$$

Equations 3.158, 3.159 may be considered the equations of modulated efforts sinks added to junctions 1_3 and 1_2.

Isolability Matrix

The inspection of causal paths also enables to set up an isolability matrix with a column and a row for each element. A causal path between two passive elements means that their parameters are dependent, i.e. they cannot be isolated from each other. This is indicated by an entry '1' or 'x' in an off-diagonal cell (i, j), $i \neq j$. In the DBG in Fig. 3.44, there are causal paths p_{10}, p_{11} between a resistor and the capacitor. Moreover, there are causal paths from the flow detectors to the source MSe : E and to the inductor I : L, respectively. Accordingly, the causal paths listed in Table 3.4 can be represented by the isolability matrix in Table 3.7.

The additional flow sensor Df : y_3 attached to junction 1_2 reduces the number of causal paths (Table 3.8).

All elements do have a direct causal path to a detector. There are two direct causal paths from detectors Df : y_1 and Df : y_3 to the capacitor. Resistor R : R_3 has two causal paths from detectors Df : y_2 and Df : y_3. All parameters except E and R_1 can be isolated. The latter two elements only have a causal path to detector Df : y_1.

DC Motor Drive

The second illustrative example is a DC motor drive often considered in the literature [5]. Figure 3.46 shows a DBG with two flow detectors which indicate the measurement of the current i_a through the voltage source on the side of the power supply and the angular velocity, ω_l of the load.

Table 3.7 Isolability matrix of the DBG in Fig. 3.44 with two flow sensors Df : y_1 and Df : y_2

Parameter	E	C	R$_1$	R$_2$	R$_3$	L	Df : y_1	Df : y_2
E							1	1
C				1	1		1	1
R_1							1	1
R_2		1					1	
R_3		1					1	1
L								1
Df : y_1	1	1	1	1	1			
Df : y_2	1	1	1		1	1		

Table 3.8 Isolability matrix of the DBG in Fig. 3.45 with three flow sensors Df : y_1, Df : y_2, Df : y_3

Parameter	E	C	R$_1$	R$_2$	R$_3$	L	Df : y_1	Df : y_2	Df : y_3
E							1		
C							1		1
R_1							1		
R_2									1
R_3								1	1
L								1	
Df : y_1	1	1	1						
Df : y_2					1	1			
Df : y_3		1		1	1				

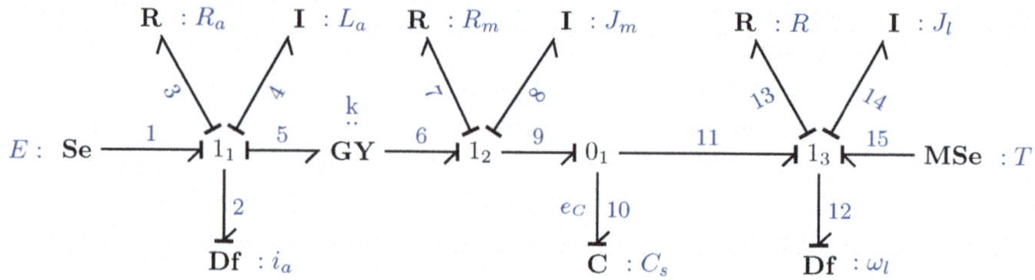

Fig. 3.46 DBG of a DC motor drive

Table 3.9 Listing of causal paths in the DBG in Fig. 3.46

p_1:	$\mathrm{Df}: i_a \overset{2}{\rightarrow} 1_1 \overset{1}{\rightarrow} \mathrm{Se}: E$
p_2:	$\mathrm{Df}: i_a \overset{2}{\rightarrow} 1_1 \overset{3}{\rightarrow} \mathrm{R}: R_a$
p_3:	$\mathrm{Df}: i_a \overset{2}{\rightarrow} 1_1 \overset{4}{\rightarrow} \mathrm{I}: L_a$
p_4:	$\mathrm{Df}: i_a \overset{2}{\rightarrow} 1_1 \overset{5}{\rightarrow} \mathrm{GY} \overset{6}{\rightarrow} 1_2 \overset{9}{\rightarrow} 0_1 \overset{10}{\rightarrow} \mathrm{C}: C_s$
p_5:	$\mathrm{C}: C_s \overset{10}{\rightarrow} 0_1 \overset{9}{\rightarrow} 1_2 \overset{8}{\rightarrow} I: J_m$
p_6:	$\mathrm{Df}: \omega_l \overset{12}{\rightarrow} 1_3 \overset{11}{\rightarrow} 0_1 \overset{9}{\rightarrow} 1_2 \overset{8}{\rightarrow} \mathrm{I}: J_m$
p_7:	$\mathrm{Df}: \omega_l \overset{12}{\rightarrow} 1_3 \overset{11}{\rightarrow} 0_1 \overset{9}{\rightarrow} 1_2 \overset{7}{\rightarrow} \mathrm{R}: R_m$
p_8:	$\mathrm{Df}: \omega_l \overset{12}{\rightarrow} 1_3 \overset{13}{\rightarrow} \mathrm{R}: R$
p_9:	$\mathrm{Df}: \omega_l \overset{12}{\rightarrow} 1_3 \overset{14}{\rightarrow} \mathrm{I}: J_l$
p_{10}:	$\mathrm{Df}: \omega_l \overset{12}{\rightarrow} 1_3 \overset{15}{\rightarrow} \mathrm{MSe}: T$
p_{11}:	$\mathrm{Df}: i_a \overset{2}{\rightarrow} 1_1 \overset{5}{\rightarrow} \mathrm{GY} \overset{6}{\rightarrow} 1_2 \overset{9}{\rightarrow} 0_1 \overset{10}{\rightarrow} \mathrm{C}: C_s \overset{9}{\rightarrow} 1_2 \overset{7}{\rightarrow} \mathrm{R}: R_m$

The shaft between the DC motor and its load is roughly modelled by the capacitor $\mathrm{C} : C_s$ accounting for its mechanical compliance $C_s = 1/k_s$, where k_s denotes the stiffness of the shaft. The modulated source $\mathrm{MSe} : T$ takes into account an external torque $T(t)$ acting on the load.

An inspection of causal paths in the DBG of the motor drive reveals that with the given two flow sensors all possible parameter faults can be detected but none of them can be isolated. Table 3.9 shows a listing of causal paths from detectors to the sources and the passive elements and causals between the latter.

There are four causal paths p_4, p_6, p_7, and p_{11} from a detector to a passive element that include the bond between junctions 1_2 and 0_1. Causal path p_5 between storage elements $\mathrm{C} : C_s$ and $\mathrm{I} : J_m$ also joins that bond. Attaching a third inverted flow detector $\mathrm{Df} : \omega_m$ at junction 1_2 means that the angular velocity, ω_m, of the motor is measured. The additional flow sensor forces the capacitor $\mathrm{C} : C_s$ into integral causality and leads to the causal paths indicated in the DBG in Fig. 3.47 and listed in Table 3.10.

Table 3.11 displays a corresponding FSM.

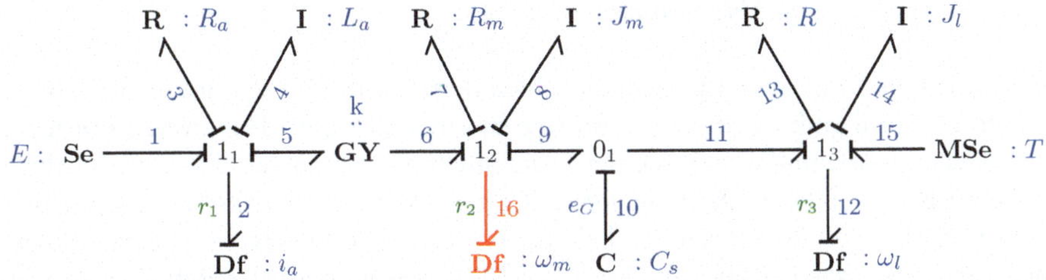

Fig. 3.47 DBG of the DC motor drive with an additional flow detector Df : ω_m

Table 3.10 Listing of causal paths in the DBG in Fig. 3.47

p'_1:	Df : $i_a \xrightarrow{2} 1_1 \xrightarrow{1}$ Se : E
p'_2:	Df : $i_a \xrightarrow{2} 1_1 \xrightarrow{3}$ R : R_a
p'_3:	Df : $i_a \xrightarrow{2} 1_1 \xrightarrow{4}$ I : L_a
p'_4:	Df : $i_a \xrightarrow{2} 1_1 \xrightarrow{5}$ GY : k
p'_5:	Df : $\omega_m \xrightarrow{15} 1_2 \xrightarrow{7}$ R : R_a
p'_6:	Df : $\omega_m \xrightarrow{15} 1_2 \xrightarrow{8}$ I : L_a
p'_7:	Df : $\omega_m \xrightarrow{15} 1_2 \xrightarrow{9} 0_1 \xrightarrow{10}$ C : C_s
p'_8:	Df : $\omega_m \xrightarrow{15} 1_2 \xrightarrow{6}$ GY : k
p'_9:	Df : $\omega_l \xrightarrow{12} 1_3 \xrightarrow{13}$ R : R
p'_{10}:	Df : $\omega_l \xrightarrow{12} 1_3 \xrightarrow{14}$ I : J_l
p'_{11}:	Df : $\omega_l \xrightarrow{12} 1_3 \xrightarrow{15}$ MSe : T

Table 3.11 Structural fault signature matrix of the DBG in Figs. 3.46, 3.47 with detectors Df : i_a, Df : ω_l and Df : ω_m

Component	ARR_1	ARR_3	ARR_2	D_b	I_b
Se : E	1	0	0	1	0
R : R_a	1	0	0	1	0
I : L_a	1	0	0	1	0
GY : k	1	0	1	1	1
I : J_m	0	0	1	1	0
C : C_s	0	1	1	1	1
R : R_m	0	0	1	1	0
R : R	0	1	0	1	0
I : J_l	0	1	0	1	0
MSe : T	0	1	0	1	0

In contrast to the RLC circuit, in this example, the additional sensor for measuring the angular velocity ω_m of the motor still does not enable to isolate element parameters except k and C_s.

Structurally Non-isolatable Parameter Faults

As can be seen from the DBG with the added flow sensor Df : ω_m in Fig. 3.47, there are direct causal paths from a detector to four other elements. However, all of them are attached to the junction of the detector, which means that all of them contribute to the ARR associated with the junction. That is, the single detector at that junction measuring the joint flow through all elements attached to the junction cannot isolate their parameters and causal paths from other detectors to these elements do not exist. In the special case that all element parameters but R_a and R_m can be assumed to be non-faulty, then these two parameters can be isolated with the two flow detectors Df : i_a and Df : ω_m.

As a result, depending on the structure of a DBG, attaching additional detectors to its junctions in certain places cannot always increase the number of disjoint causal paths from detectors to potentially faulty elements and thus cannot increase the number of parametric faults that can be structurally isolated. In order to structurally isolate element parameters in cases such as the DC motor drive it becomes necessary to insert additional junctions and to attach a detector to them. This is illustrated by the DBG fragment displayed in Fig. 3.48.

From the DBG fragment in Fig. 3.48, the following ARRs are obtained.

$$r_1(t) = E - k\omega_m - e_{R_a} - e_{L_a} \tag{3.160}$$

$$r_4(t) = i_a - \frac{1}{R_a} e_{R_a} \tag{3.161}$$

$$r_5(t) = i_a - \frac{1}{L_a}\int_0^t e_{L_a}(\tau)d\tau - i_{L_a}(0) \tag{3.162}$$

If the two additional effort sensors are assumed to be faultless, then a value $r_1(t)$ outside some tolerance bounds as of a time instant t indicates that the voltage supply E has become faulty. Likewise, $r_4 \neq 0$ isolates a faulty parameter R_a.

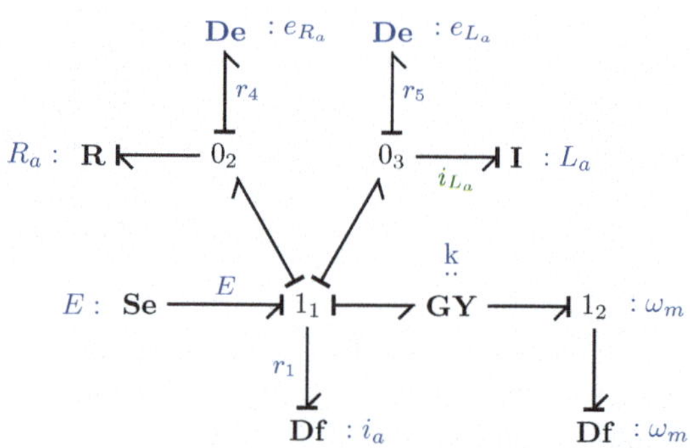

Fig. 3.48 Further parameter isolation by means of additional detectors effort attached to inserted junctions 0_2, 0_3

3.8.2 Faulty Sensors

So far, sensors have been assumed to be faultless. If this is not the case, a faulty sensor can be modelled by a detector of the faultless signal and a modulated sink that provides the faulty signal component. Evidently, a faulty sensor cannot be used for detection and isolation of parametric faults of system components. Therefore, another non-faulty sensor is needed that enables to isolate the sensor fault. For illustration, consider the small passive network in Fig. 3.49 with a faulty sensor for the inductor current and its DBG in Fig. 3.50.

The DBG in Fig. 3.50 reflects that the sensor measures a faulty inductor current $\tilde{i}_L = i_L + \Delta i_L$. As can be seen, there is a direct causal path p_1 from the additional detector De : u_s to the modulated sink MSf : Δi_L and another path p_2 to the inductor I : L which means that the sensor fault Δi_L affects ARR residuals r_2 and r_3. Moreover, there is an indirect causal path p_3 from detector Df : y_1 via R : R_2 to the sink MSf : Δi_L. That is, Δi_L also contributes to ARR$_1$, hence, to all three ARRs.

Moreover, there are the following direct causal paths (Table 3.12).

The last two paths indicate that R_2 contributes to residuals r_1 and r_2. As a result, the sensor fault Δi_L can be structurally isolated. This can be verified by reading

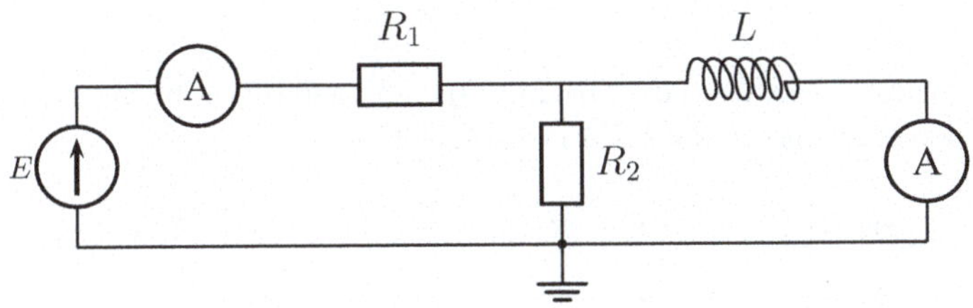

Fig. 3.49 Circuit with a faulty sensor for the inductor current

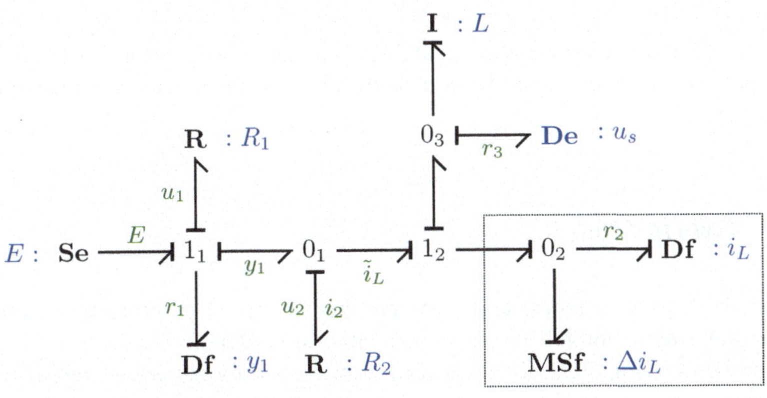

Fig. 3.50 DBG of the circuit in Fig. 3.49

Table 3.12 Listing of causal paths in the DBG in Fig. 3.50

p_4:	Df : $y_1 \rightarrow 1_1 \rightarrow$ Se : E
p_5:	Df : $y_1 \rightarrow 1_1 \rightarrow 0_1 \rightarrow$ R : R_2
p_6:	Df : $i_L \rightarrow 0_2 \rightarrow 1_2 \rightarrow 0_1 \rightarrow$ R : R_2

Table 3.13 Structural fault signature matrix of the DBG in Fig. 3.50

Element	ARR_1	ARR_2	ARR_3	D_b	I_b
Se : E	1	0	0	1	0
R : R_1	1	0	0	1	0
R : R_2	1	1	0	1	1
I : L	0	0	1	1	1
MSf : Δi_L	1	1	1	1	1

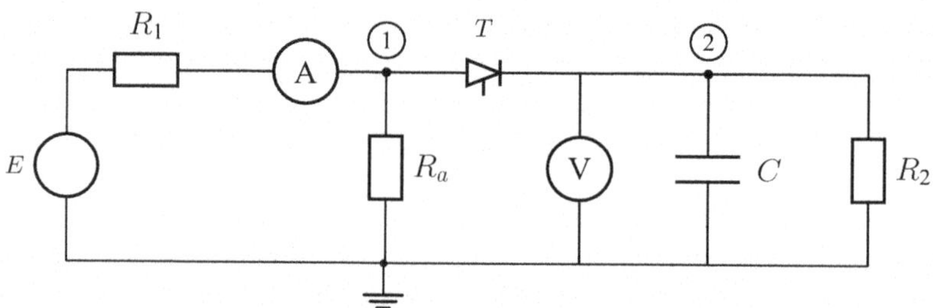

Fig. 3.51 Half-wave rectifier circuit

out the following ARRs from the DBG in Fig. 3.50 and by capturing their structural parameter dependencies in the FSM in Table 3.13.

$$\text{ARR}_1: \quad r_1 = E - R_2(y_1 - (i_L + \Delta i_L)) - R_1 y_1 \tag{3.163}$$

$$\text{ARR}_2: \quad r_2 = R_2(y_1 - (i_L + \Delta i_L)) - u_s \tag{3.164}$$

$$\text{ARR}_3: \quad \dot{r}_3 = \frac{d}{dt} i_L + \frac{d}{dt} \Delta i_L - \frac{1}{L} u_s \tag{3.165}$$

Potentially faulty elements Se : E and R : R_1 cannot be isolated. There are direct causal paths to these elements from detector Df : y_1 but no causal paths from the other detectors.

3.8.3 Hybrid Models

Figure 3.51 displays a simple half-wave rectifier circuit. Depending on the context, the thyristor may be modelled as a switch, either non-ideal or ideal.

In the DBG in Fig. 3.52, the thyristor is represented by the non-standard symbol Sw : m, where $m(t) \in \{0, 1\}$ denotes the switching signal. In a BG with variable causality, the output of a switch is either a zero-value flow, or a zero-value effort

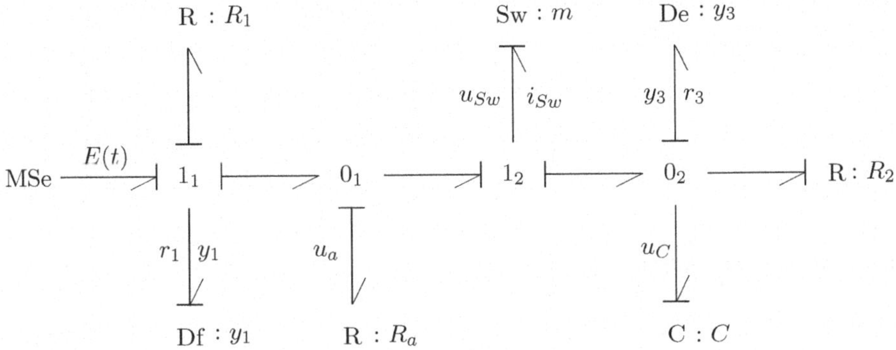

Fig. 3.52 DBG of the half-wave rectifier circuit in Fig. 3.51

depending on its discrete state. As the presented graphical approach to sensor placement and isolation of parametric faults is based on the check of causal paths, a DBG with invariable causalities is more appropriate. According to the current versus voltage characteristic of the thyristor the switch has got fixed conductance causality.

Given the two sensors, it turns out that none of the element parameters can be isolated independently whether the switch is assumed to be ideal or non-ideal.

From the DBG in Fig. 3.52, one obtains

$$r_1 = E - R_1 y_1 - u_a \tag{3.166}$$

$$u_a = R_a(y_1 - i_{sw}) \tag{3.167}$$

$$u_{sw} = u_a - y_3 \tag{3.168}$$

If the switch is considered ideal, it may be described by the implicit equation

$$0 = m\, i_{sw} + (1 - m)\, u_{sw} \tag{3.169}$$

which holds for both switch modes, while its causality in the DBG is kept fixed. Equations 3.167, 3.168, and 3.169 can be solved for the unknown voltage u_a in (3.166)

$$\begin{aligned} r_1 &= E - R_1 y_1 - R_a(y_1 - i_{sw}) \\ &= E - R_1 y_1 - R_a y_1 + R_a \frac{m-1}{m-(1-m)R_a}(R_a y_1 - y_3) \\ &= E - R_1 y_1 - \frac{m}{m-(1-m)R_a} R_a y_1 - R_a \frac{m-1}{m-(1-m)R_a} y_3 \end{aligned} \tag{3.170}$$

Table 3.14 Structural fault signature matrix of the DBG in Fig. 3.52

Element	ARR_1	ARR_3	D_b	I_b
$MSe : E$	1	0	1	0
$R : R_1$	1	0	1	0
$R : R_a$	m	1 − m	m	0
$C : C$	0	1	1	0
$R : R_2$	0	1	1	0

Residual r_3 then reads

$$r_3 = i_{sw} - C\dot{y}_3 - \frac{1}{R_2} y_3 \tag{3.171}$$

$$= \frac{m-1}{m-(1-m)R_a}(R_a y_1 - y_3) - C\dot{y}_3 - \frac{1}{R_2} y_3 \tag{3.172}$$

The structural dependencies of ARR_1 and ARR_3 from the element parameters are captured in the generalised all-mode FSM in Table 3.14.

The result depicted by the all-mode FSM in Table 3.14 is immediately evident by inspection of the circuit in Fig. 3.51. The voltage source $MSe : E$ and resistor $R : R_1$ are in series and cannot be isolated by just the flow detector $Df : y_1$. The capacitor and resistor $R : R_2$ are in parallel. Their parameters cannot be isolated by effort detector $De : y_3$. If the switch is open, resistor $R : R_a$ is in series with $R : R_1$ and belongs to ARR_1, which is the sum of voltage drops in the left mesh. If the switch is closed, $R : R_a$ is in parallel to the capacitor and belongs to ARR_3, which is the sum of currents into node ②.

The same result can be obtained if the switch is considered non-ideal. Its current versus voltage drop characteristic may be replaced by a piecewise linear characteristic and the OFF-conductance may be neglected.

$$i_{sw} = \frac{m}{R_{\text{on}}} u_{sw} \tag{3.173}$$

where R_{on} denotes the ON resistance of the switch.

To separate causal paths, a first step is to attach an additional effort detector $De : y_2$ to junction 0_1. It is interesting to note that an explicit expression for residual r_1 cannot be obtained in the case the switch is assumed ideal because its implicit equation cannot be solved for the current i_{sw}.

$$0 = m\, i_{sw} + (1-m)\, u_{sw} \tag{3.174}$$

$$= m\, i_{sw} + (1-m)\,(y_2 - y_3) \tag{3.175}$$

If a nonlinear switch with a piecewise linear characteristic is used, ARR_1 can be obtained in explicit form.

Finally, in order to isolate parameters R_a, C, and R_2 another sensor for measuring the current through $R : R_2$ is necessary. Figure 3.53 shows the augmented DBG.

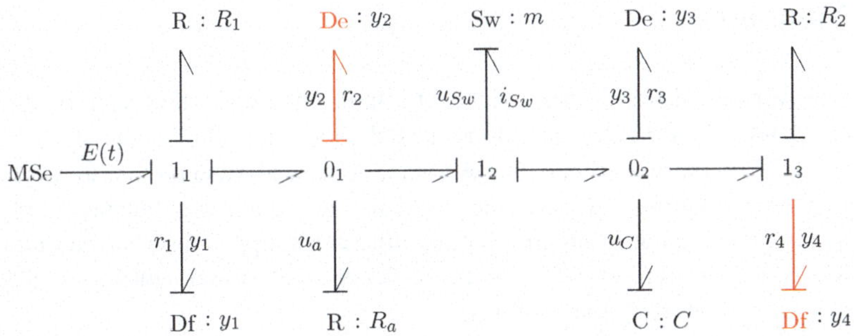

Fig. 3.53 DBG of the half-wave rectifier circuit in Fig. 3.51 augmented by additional detectors De : y_2 and Df : y_4

Table 3.15 Structural fault signature matrix of the augmented DBG in Fig. 3.53

Element	ARR_1	ARR_2	ARR_3	ARR_4	D_b	I_b
MSe : E	1	0	0	0	1	0
R : R_1	1	0	0	0	1	0
R : R_a	0	1	0	0	1	1
R : R_{on}	0	m	m	0	m	m
C : C	0	0	1	0	1	1
R : R_2	0	0	0	1	1	1

As can be seen from the augmented DBG, there are disjoint causal paths from the detectors to the elements. There are only two direct causal paths from detector Df : y_1 to elements MSe : E and R : R_1 which means that the two cannot be isolated. The result can be verified by deriving ARRs from the augmented DBG and by representing the structural relations between ARRs and element parameters in a FSM (Table 3.15).

$$r_1 = E - R_1 y_1 - y_2 \tag{3.176}$$

$$r_2 = y_1 - \frac{1}{R_a} y_2 - i_{sw} \tag{3.177}$$

$$i_{sw} = \frac{m}{R_{on}}(y_2 - y_3) \tag{3.178}$$

$$r_3 = i_{sw} - C\dot{y}_3 - y_4 \tag{3.179}$$

$$r_4 = y_3 - R_2 y_4 \tag{3.180}$$

Evidently, the ON resistance R_{on} of the switch can only be isolated when the switch is closed ($m = 1$).

3.9 Summary

There are various different approaches to fault diagnosis that can be broadly categorised into data-driven and model-based methods, which may be based on Kalman filters, on particle filters, on observers, or on analytical redundancy relations (ARRs). Parity relations [26] are also used in fault diagnosis but have not been addressed in this chapter. Bond Graph modelling can support observer-based as well as ARR-based FDI methods. It is the powerful computational causality concept of Bond Graph methodology that enables

- to set up structural fault signature matrices from causal diagnostic bond graphs for fault detection and isolation without formulating and manipulating equations,
- to derive analytical redundancy relations directly from a diagnostic BG,
- to estimate the magnitude of an isolated fault by following bicausal paths between a faulty element and a detector in a bicausal BG.

A BG model of a system including actuators and sensors can be easily turned in a DBG with storage elements in derivative causality to get rid of initial conditions and detectors in inverted causality as they deliver known measurements into the DBG model. From a DBG, ARRs according to the number of non-redundant sensors can be derived offline. Their online numerical evaluation provides fault indicators. That is, a DBG is a residual generator. If for some fault indicators an ARR in closed symbolic form is not possible, some implicit equations will have to be solved numerically by iteration, or even the entire DBG model will have to be computed.

Entries into ARRs are known control inputs, filtered noisy measurements from the real system, and also time derivatives of measurements. Therefore, filters are of interest that smooth noisy signals and enable their differentiation in discrete time with acceptable accuracy. As an example, Sect. 3.2.1 consideres the Savitzky–Golay (SG) filter in some detail. Section 3.2.2 briefly refers to state variable filters. As to the SG filter, it turns out that the accuracy of the results depends on the type of signal and on the two tuning parameters of the algorithm, i.e. the width of the sliding window and the order of the data fitting polynomial. In general, acceptable results with regard to smoothing and discrete-time differentiation may be obtained for time series that can be represented by continuous time functions with continuous time derivatives. Functions with a discontinuous time derivative, or discontinuous functions such as a clock signal, may pose problems. Differentiation of measurements may be avoided by following a procedure proposed by Benmoussa that leads to time integrals of ARR residuals and has been explained in Sect. 3.5.3.

In order to avoid false alarms as well as misdetections of faults, it is important that ARR residuals are sensitive to faults but robust with regard to parameter uncertainties. In fault detection, two forms of state space models are known, the standard interconnection form and the linear fractional transformation (LFT) form. Both enable to establish adaptive fault thresholds. Bond Graph methodology has developed uncertain BGs, LFT-BGs, and incremental BGs from which both forms of state space models and adaptive fault thresholds can be derived.

Beyond parametric uncertainties, measurement uncertainties, sensor faults, and actuator faults can be easily accounted for in BGs by additional modulated sources as they are additive faults. Their size can be estimated by following bicausal paths from a faulty element to a detector.

As the number of potentially faulty element parameters is higher in general than the number of available sensors, a FSM often indicates that many potential parametric faults can be detected, while only a small number can also be isolated with the given sensors. Section 3.8 therefore addresses the question where to place sensors if technically feasible and how many of them to achieve a maximum of number of isolated potentially faulty element parameters and presents a graphical method that aims at obtaining as much disjoint causal paths from sources and detectors to passive elements. As causal paths in a DBG are analysed, the computational effort is much smaller than the determination of residual parameter sensitivities, which can also be supported by BGs such as sensitivity pseudo BGs (SPBGs), or diagnostic sensitivity pseudo BGs, or by numerical estimation of those parameters which do not have a unique fault signature.

As to fault detection based on a Luenberger observer, Sect. 3.5.1 shows by means of an example that the entries of the observer gain matrix can be directly determined from a BG of the observer. The presentation of observer-based fault detection is preceded by Sect. 3.4 on Kalman filters and particle filters as they are used in FDI as well as in failure prognosis addressed in the following chapter.

References

1. Alem, S., & Benazzouz, D. (2013). Optimal sensor placement for detection and isolation by the structural adjacency matrix. *International Journal of Physical Sciences, 8*(6), 225–230. https://doi.org/10.5897/IJPS12.547.
2. Ali, N., Alam, W., Ur Rehman, A., & Pervaiz, M. (2017). State and disturbance observer based control for a class of linear uncertain systems. In *2017 International Conference on Frontiers of Information Technology (FIT)*, Islamabad (pp. 139–143). https://doi.org/10.1109/FIT.2017.00032.
3. Arulampalam, M. S., Maskell, S., Gordon, N., & Clapp, T. (2002). A tutorial on particle filters for online nonlinear/non-Gaussian Bayesian tracking. *IEEE Transactions on Signal Processing, 50*(2), 174–188.
4. Benmoussa, S. (2013). Approche Bond Graph pour la détectabilité et l'isolabilié algébriques de défaults composants. PhD thesis, L'université Lille 1.
5. Benmoussa, S., Ould Bouamama, B., & Merzouki, R. (2011). Component fault detection and isolation comparison between bond graph and algebraic approach. In *Proceedings of the 5th International Conference on Integrated Modelling and Analysis in Applied Control and Automation* (pp. 214–220).
6. Benmoussa, S., Ould Bouamama, B., & Merzouki, R. (2014). Bond graph approach for plant fault detection and isolation: Application to intelligent autonomous vehicle. *IEEE Transactions on Automation Science and Engineering, 11*(2), 585–593.
7. Björck, A. (1996). *Numerical Methods for Least Squares Problems*. Philadelphia: SIAM.
8. Borutzky, W. (2009). Bond graph model-based fault detection using residual sinks. *Proceedings of the Institution of Mechanical Engineers Part I Journal of Systems and Control Engineering, 223*(3), 337–352.

9. Borutzky, W. (2010). *Bond Graph Methodology – Development and Analysis of Multidisciplinary Dynamic System Models*. London: Springer. ISBN: 978-1-84882-881-0.
10. Borutzky, W. (2015). *Bond Graph Model-Based Fault Diagnosis of Hybrid Systems*. Cham: Springer International Publishing, Switzerland.
11. Borutzky, W. (Ed.) (2016). *Bond Graphs for Modelling, Control and Fault Diagnosis of Engineering Systems*. Springer International Publishing, Switzerland.
12. Borutzky, W. (2018). Sensor placement on diagnostic bond graphs for maximum structural isolation of parametric faults. In J. Granda & D. Karnopp, (Eds.), *Proceedings of 13th International Conference on Bond Graph Modeling and Simulation (ICBGM'2018)* (Simulation Series, Vol. 50(12), pp. 41–49). Bordeaux: SCS.
13. Borutzky, W., Barnard, B., & Thoma, J. (2000) Describing bond graph models of hydraulic components in Modelica. *Mathematics and Computers in Simulation, 53*(4–6), 381–387.
14. Cabanellas, J., Félez, J., & Vera, C. (1995). A formulation of the sensitivity analysis for dynamic systems optimization based on pseudo bond graphs. In F. Cellier & J. Granda (Eds.), *ICBGM'95, International Conference on Bond Graph Modeling and Simulation* (Simulation Series, Vol. 27(1), pp. 135–144). Las Vegas: SCS Publishing.
15. Chen, C. (1999). *Linear System Theory and Design* (3rd ed.). New York: Oxford University Press.
16. Chen, J., Patton, R., & Zhang, H. (1996). Design of unknown input observers and robust fault detection filters. *Journal of Control, 63*(1), 85–105. https://doi.org/10.1080/00207179608921833.
17. Chi, G., & Wang, D. (2015). Sensor placement for fault isolability based on bond graphs. *IEEE Transactions on Automatic Control, 60*(11), 3041–3046.
18. Cox, H. (1965). On the estimation of state variables and parameters for noisy dynamic systems. *IEEE Transactions on Automatic Control, 9*, 5–12.
19. Dai, W., Selesnick, I., Rizzo, J. R., Rucker, J., & Hudson, T. (2017). A nonlinear generalization of the Savitzky-Golay filter and the quantitative analysis of saccades. *Journal of Vision, 17*(9), 10:1–10:15. https://doi.org/10.1167/17.9.10.
20. Daroogheh, N. (2016). Health Monitoring of Nonlinear Systems with Application to Gas Turbine Engines. PhD thesis, Department of Electrical and Computer Engineering, Concordia University, Montréal.
21. Djeziri, M., Merzouki, R., Ould Bouamama, B., & Dauphin-Tanguy, G. (2007). Robust fault diagnosis by using bond graph approach. *IEEE/ASME Transactions on Mechatronics, 12*(6), 599–611.
22. Djeziri, M., Ould Bouamama, B., Merzouki, R., & Dauphin-Tanguy, G. (2009). Optimal sensor placement for fault diagnosis. In *2009 IEEE International Conference on Mechatronics*, Malaga.
23. Djeziri, M. A. (2007). Diagnostic des Systèmes Incrementains par l'Approche Bond Graph. PhD thesis, Ecole Centrale de Lille. http://hal.archives-ouvertes.fr/docs/00/20/00/30/PDF/These-Djeziri-07-12-2007.pdf.
24. Frisk, E., Krysander, M., & Åslund, J. (2009). Sensor placement for fault isolation in linear differential-algebraic systems. *Automatica, 45*, 364–371.
25. Gawthrop, P. (2000). Sensitivity bond graphs. *Journal of the Franklin Institute, 337*, 907–922.
26. Gertler, J. (1997). Fault detection and isolation using parity relations. *Control Engineering Practice, 5*, 653–661.
27. Ghoshal, S., Samantaray, A., & Mukherjee, A. (2005). Improvements to single fault isolation using estimated parameters. In J. Granda & F. Cellier (Eds.), *Proceedings of of the International Conference on Bond Graph Modeling, ICBGM'05* (Simulation Series, Vol. 37(1), pp. 301–306). New Orleans: SCS Publishing. ISBN: 1-56555-287-3.
28. Gordon, N., Salmond, D., & Smith, A. (1993). Novel approach to nonlinear/non-Gaussian Bayesian state estimation. *Radar and Signal Processing, IEE Proceedings F, 140*(2), 107–113.
29. Gorry, A. P. (1990). General least squares smoothing and differentiation by the convolution (Savitzky-Golay) method. *Analytical Chemistry, 62*(6), 570–573.

30. Granda, J. (2007). CAMP-G – User's Manual. Cadsim Engineering, P.O. Box 4083, Davis, Ca 95617. https://www.bondgraph.com.
31. Julier, S. J., & Uhlmann, J. K. (2004). Unscented filtering and nonlinear estimation. *Proceedings of the IEEE, 92*(3), 401–422.
32. Kalman, R. E. (1960). A new approach to linear filtering and prediction problems. *Journal of Basic Engineering, 82*(1), 35–41.
33. Kam, C. (2001). Les bond graphs pour la modélisation des systèmes linéaires incertains. PhD thesis, L'Ecole Centrale de Lille et L'Université des Sciences et Technologies de Lille, Lille.
34. Kam, C., & Dauphin-Tanguy, G. (2001). Bond graph tools for standard interconnection structure determination. In J. J. Granda & G. Dauphin-Tanguy (Eds.), *2001 International Conference on Bond Graph Modeling, and Simulation (ICBGM 2001)* (Simulation Series, Vol. 33(1)). SCS Publishing. ISBN: 1-56555-221-0.
35. Khemliche, M., Ould Bouamama, B., & Haffaf, H. (2004) Optimal sensor placement using bond-graph for FDI design. In *IFAC Proceedings* (Vol. 37(5), pp. 79–84).
36. Kim, D., & Lee, D. (2019). Fault parameter estimation using adaptive fuzzy fading Kalman filter. *Applied Sciences, 9*. https://doi.org/10.3390/app9163329.
37. Kitagawa, G. (1996). Monte Carlo Filter and smoother for non-Gaussian and non-linear state space Models. *Journal of Computational and Graphical Statistics, 5*(1), 1–25.
38. Ljung, L. (1979). Asymptotic behavior of the extended Kalman filter as a parameter estimator for linear systems. *IEEE Transactions on Automatic Control, 34*, 36–50.
39. Merzouki, R., Samantaray, A., Pathak, P., & Ould Bouamama, B. (2013). *Intelligent Mechatronic Systems*. London: Springer.
40. Mukherjee, A., & Samantaray, A. (2001). System modeling through bond graph objects on SYMBOLS 2000. In J. Granda & G. Dauphin-Tanguy (Eds.), *2001 International Conference on Bond Graph Modeling (ICBGM'01), Proceedings of the 2001 Western Simulation Multiconference* (Simulation Series, Vol. 33(1), pp. 164–170). SCS Publishing. ISBN: 1-56555-221-0.
41. Nelder, J., & Mead, R. (1965). A simplex method for function minimization. *Computer Journal, 7*(4), 308–313.
42. Orchard, M. E. (2007). A Particle Filtering-Based Framework for On-Line Fault Diagnosis and Failure Prognosis. PhD Thesis, Georgia Institute of Technology.
43. Ould Bouamama, B., Samantaray, A., Medjaher, K., Staroswiecki, M., & Dauphin-Tanguy, G. (2005). Model builder using functional and bond graph tools for FDI design. *Control Engineering Practice, 13*(7), 875–891.
44. Persson, P., & Strang, G. (2003). Smoothing by Savitzky-Golay and Legrendre Filters. *Communications, Computation, and Finance, 13*, 301–316.
45. Peter, K., & Isermann, R. (1989). Parameter-adaptive Pid-control based on continuous-time process models. In *Proceedings of the 3rd IFAC Adaptive Systems in Control and Signal Processing* (pp. 241–246).
46. Plett, G. L. (2004) Extended Kalman filtering for battery management systems of LiPB-based HEV battery packs: Part 3. State and parameter estimation. *Journal of Power Sources 134*(2), 277–292.
47. Press, W. H., & Teukolsky, S. A. (1990). Savitzky-Golay smoothing filters. *Computers in Physics, 4*, 669–672.
48. Rahmani, A. (1993). Etude structurelle des systèmes linéaires par l'approche bond graph. PhD thesis, L'Université des Sciences et Technologies de Lille, Lille.
49. Ranganathan, A. (2004). The Levenberg-Marquardt Algorithm. http://ananth.in/Notes_les/lmtut.pdf.
50. Ribeiro, M. I. (2004). Kalman and Extended Kalman Filters: Concept, Derivation and Properties. Technical report, Institute for Systems and Robotics Instituto Superior Técnico.
51. Sadeghi, M., & Behnia, F. (2018). Optimum window length of Savitzky-Golay filters with arbitrary order. https://arxiv.org/pdf/1808.10489.
52. Saha, B., Goebel, K., Poll, S., & Christophersen, J. (2009). Prognostics methods for battery health monitoring using a Bayesian framework. *IEEE Transactions on Instrumentation and Measurement* 291–296. https://doi.org/10.1109/TIM.2008.2005965.

53. Salmond, D., & Gordon, N. (2005). An introduction to particle filters. https://pdfs.semanticscholar.org/b36d/076034512a5fa280de39464d03211584c7bf.pdf. INCOMPLETE DRAFT.
54. Samantaray, A., & Ghoshal, S. (2007). Sensitivity bond graph approach to multiple fault isolation through parameter estimation. *Proceedings of the Institution of Mechanical Engineers Part I: Journal of Systems and Control Engineering, 221*(4), 577–587.
55. Samantaray, A., & Ould Bouamama, B. (2008). *Model-Based Process Supervision – A Bond Graph Approach* (Advances in Industrial Control). London: Springer.
56. Samantaray, A., Medjaher, K., Ould Bouamama, B., Staroswiecki, M., & Dauphin-Tanguy, G. (2004). Component based modelling of thermo-fluid systems for sensor placement and fault detection. *SIMULATION: Transaction of the Society for Modeling and Simulation International, 80*(8), 381–398.
57. Samantaray, A., Medjaher, K., Ould Bouamama, B., Staroswiecki, M., & Dauphin-Tanguy, G. (2006). Diagnostic bond graphs for online fault detection and isolation. *Simulation Modelling Practice and Theory, 14*(3), 237–262.
58. Savitzky, A., & Golay, M. J. E. (1964). Smoothing and differentiation of data by simplified least squares procedures. *Analytical Chemistry, 38*(8), 1627–1639.
59. Spuri, F., & Goes, L. (2017). Modeling and parametric identification of a variable-displacement pressure-compensated pump. In *The 15th Scandinavian International Conference on Fluid Power, SICFP'17*, Linköping (pp. 52–60).
60. Terejanu, G. A. (2013). Discrete Kalman Filter Tutorial. https://www.cse.sc.edu/~terejanu/files/tutorialKF.pdf.
61. Touati, Y. (2012). Diagnostic Robuste et estimation de défault à base de Modèle Bond Graph. PhD thesis, l'université Lille 1.
62. Wan, E. A., & van der Merwe, R. (2000). The unscented Kalman Filter for nonlinear estimation. In *The IEEE Conference: Adaptive Systems for Signal Processing, Communications, and Control Symposium 2000. AS-SPCC* (pp. 153–158).
63. WANG, B.-S., & WIDODO, A. (2008). Support vector machine for machine fault diagnosis and prognosis. *Journal of System Design and Dynamics, 2*(1), 12–23. https://doi.org/10.1299/jsdd.2.12, review Paper.

Chapter 4
Failure Prognostic

It's tough to make predictions, especially about the future

Yogi Berra[1]

4.1 Introduction

Beyond fault diagnosis, addressed in the previous chapter, failure prognostic is a constitutive part of Prognostic and Health Management (PHM) [2, 18] and is of imperative importance for safety critical engineering systems and processes, for the supervision, for automation and condition based maintenance (CBM) of industrial processes, for predictive maintenance, and for all kinds of emerging intelligent autonomously operating mobile systems. Predictive anomaly identification based on real-time data, monitoring degradation, detecting precursors to failure, and predicting the remaining useful life (RUL) of faulty components and subsystems becomes more and more important with an increase of system complexity, e.g. in smart manufacturing with communicating and cooperating machines. As more and more systems are equipped with sensors, in situ self-monitoring and prognostic capabilities, they can issue warnings along with confidence intervals when they are about to operate out of admissible tolerances and can automatically set maintenance alarms. Often, changes in selected prognostic features can be mapped to drifting parameters. Vibration sensors, for instance, can be used to monitor the health of rotary machinery, to detect bearing faults, crack developments in gears or fan blades of turbines. Humidity sensors may help to detect contact degradation in electromechanical relays due to corrosion. Failure prognostic enables predictive maintenance of industrial processes and enables a longer lifetime of process

[1] https://www.goodreads.com/author/quotes/79014.Yogi_Berra

W. Borutzky, *Bond Graph Modelling for Control, Fault Diagnosis and Failure Prognosis*, https://doi.org/10.1007/978-3-030-60967-2_4

components, shorter down-times, increased safety, a more efficient use of resources, and a reduction of costs.

Autonomous intelligent mobile systems equipped with fault diagnosis and failure prognosis capabilities can detect precursors of failures such as a detected and ongoing capacitance degradation of the battery of an unmanned aerial vehicle and can check the health state of onboard electronic systems. Based on this information, an onboard prognosis software module can estimate the remaining time left for a safe operation of the system despite the presence of an incipient fault before the increasing affect of the fault on the dynamic system behaviour may lead to a system failure. As a result, an unmanned aerial vehicle such as a taxi, for instance, may change the control, or the mission in order to avoid all kind of possible damage.

Figure 4.1 qualitatively depicts a typical monotonic decrease of the health state of a system due to an incipient fault.

Consider a system with parameter uncertainties that is continuously monitored. As long as parameter variations are within certain bounds, the system is considered to operate normally, i.e. it is assumed that there is no fault. Values indicating the system health state are randomly varying around an acceptable health level in a small envelope. Once an incipient fault has started at time instant t_{oc}, there is some delay until the fault has crossed a fault detection level and is identified as a fault with a magnitude that is increasing with time. As of time point t_P, it is clear that the system state will deteriorate and the task of prognostic starting at this time point is to estimate how long the system can continue to safely operate before the

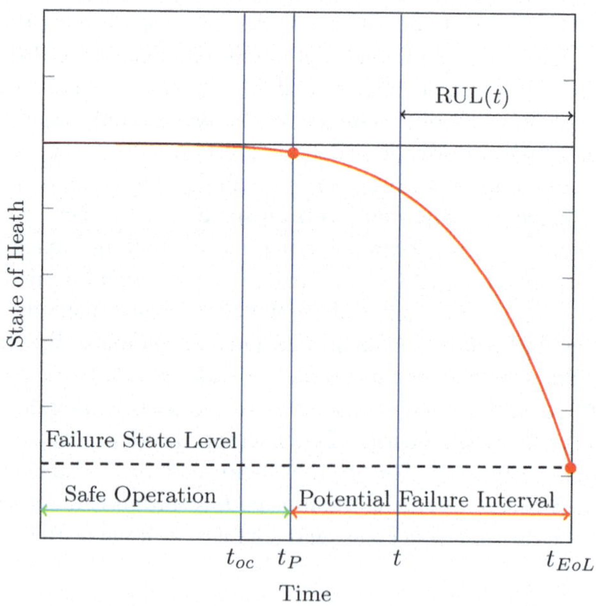

Fig. 4.1 Typical monotonic decrease of the health state of a system due to an incipient fault

progression of the detected fault will cause a functional failure. For illustration, Fig. 4.1 assumes that the system health state qualitatively decreases exponentially. Repeated prediction of the current state into the future, i.e. repeated prediction of the remaining useful life (RUL) at each new time point t, requires that continuous fault diagnosis provides estimates of the increasing fault magnitude.

As the repeated prediction of future states provides random variables with a probability density function (pdf), the time t_{fl} at which the sequence of predicted states reach a failure threshold FT is also uncertain and can only be estimated. Accordingly, the remaining useful life (RUL) as of some current time instant t_c is also a random variable with a pdf and a probability that depend on the preceding condition profile. It is the time left to a system component or to a system that can be used before a failure state level or an intolerable performance degradation level is reached. The failure zone in Fig. 4.1 is the potential failure interval from the time instant t_P to the failure time point t_{EoL}. As the RUL depends on the current time point of observation, its expected value decreases as time progresses.

A failure threshold FT does not necessarily indicate the total failure of a component or the monitored system. It is rather a conservative estimated level beyond which the risk of complete failure would exceed tolerance limits of a hazard zone [37]. In the following, the notion failure alarm threshold will be used. The end of life EoL is a time instant t_{EoL} where the prediction crosses a failure threshold. This time instant is determined through run-to-failure experiments in which a system may be allowed to fail. In general, one does not want to keep a faulty system in operation until a total failure occurs in order to determine the end of life time. That is, the end of life and thus the RUL depend on what is considered a failure and on the predefined failure level.

In mathematical terms, the RUL(t) as of some time instant t can be defined as follows. Assume that a component or the monitored system has not failed before a time instant t' and let $Z(t)$ denote the preceding condition profile, then the conditional lifetime distribution is $F(t \mid t_{\text{fl}} \geq t', Z(t)) = P(t_{\text{fl}} \leq t \mid t_{\text{fl}} \geq t', Z(t))$. The expected residual life as of some time t then is

$$E[RUL(t)] = E[t_{\text{fl}} - t \mid t_{\text{fl}} \geq t', Z(t)] \tag{4.1}$$

where E[RUL(t)] denotes the expectation value of the RUL(t) [21, 44].

The above considerations clearly indicate that constant monitoring of the health of system and a repeated prediction of the time to failure, or the remaining useful life, i.e. failure prognostic are of technical and economical importance. In comparison to fault diagnosis, failure prognosis is still a rather young, still developing research subject.

To estimate the RUL as of a current time instant t_P, it is necessary to know the degradation behaviour of a fault over time. To that end, one may try to develop a model of the process of progressing damage starting from first physical principles. Difficulties, however, may be that the degradation process is not fully understood or that not all needed parameters of a degradation model can be determined.

Other options may be to obtain a degradation model from offline accelerated life tests [15] and to use the results in online health monitoring for the prediction of the RUL [27], or to assume that a potential degradation function candidate is a member of a certain class of functions and to adapt the unknown function parameters by curve fitting. As measured signals carry noise and repeated measurements of one and the same signal may vary, a RUL has to be considered a stochastic quantity. Moreover, as systems may operate in various modes, the degradation behaviour may change from one mode of operation to another making it necessary to change to another class of potential degradation functions [29], [4, Chap. 6].

A review of model-based and data-based prognostic approaches may be found in [1, 13]. In a recent 2016 literature review on prognostic [14], Elattar and his co-authors point out:

> Researchers that have new interest in prognostics need to read hundreds of articles to have a complete picture about prognostics and its relation to other disciplines.
> ...So far, there is no literature that gives a wide and complete vision about prognostics. If one would like to have a clear understanding of all topics related to prognostics, it will be a difficult and time-consuming task. It may take months to just collect articles about prognostics. Reading and understanding all of these articles, especially for new prognostics researchers, is not easy because each article addresses only single or few topics.

4.2 Data-Driven Failure Prognostic

Data-driven failure prognostic exploits measured data provided by the available sensors to learn a behaviour model of the system which includes the degradation behaviour if an incipient fault has happened. To that end, first, some conditioning of raw signals, i.e. amplification, attenuation, filtering must be done. Moreover, features must be selected or extracted from the available data, or the amount of features must be reduced. A model, solely based on the obtained data, is used to estimate the system's current state and then to recursively predict its future health state in order to estimate the remaining useful life (RUL) of a system with an incipient fault until some failure criterion is met, i.e. when the fault magnitude or a health index crosses an alarm threshold. Due to various uncertainties such as manufacturing variability, measurement uncertainties, unforeseen future system loads or environmental changes, uncertainty with regard to an initial state, estimation of the current state, prediction of future states, and consequently estimation of the RUL carries some uncertainty.

Data-driven prognostic methods are not based on a mathematical system model derived from physical principles but instead use dynamic Bayesian networks (DBNs) [3, 26], hidden Markov models (HMMs), neural networks, or statistical methods. Probabilistic prognostic considers system states and consequently RUL estimates as random variables with a probability density distribution (pdf) and a challenge of prognostic is to minimise the uncertainty on the RUL estimation.

Fig. 4.2 A typical flow of data-driven failure prognostic

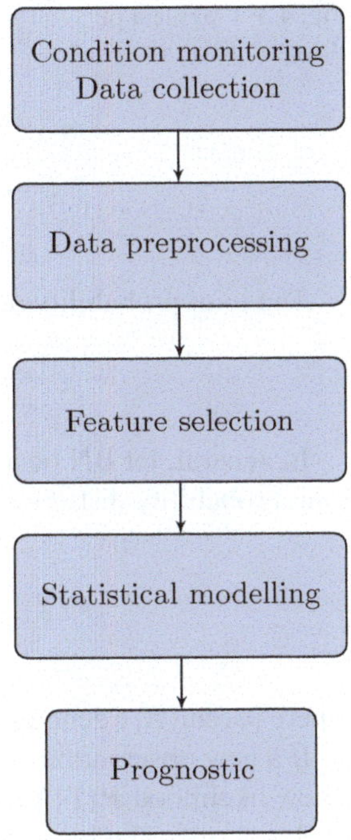

Figure 4.2 depicts the typical flow of data-driven failure prognostic for the case that statistical methods are used.

4.2.1 Stochastic Data-Driven Methods

In the following, some basic ideas of Bayesian-based data-driven approaches are briefly summarised.

Bayesian Networks

A Bayesian network (BN), also known as belief network, is a directed, acyclic graph, in which the vertices represent random variables. Oriented edges lead from a cause variable to an effect variable. In fault diagnosis, they represent fault-symptom relations. Each node is associated with a conditional probability distribution, except for root nodes, for which a probability defined by prior knowledge is specified. The edges in a Bayesian network represent the joint probability distribution of the connected variables. Consider the BN in Fig. 4.3.

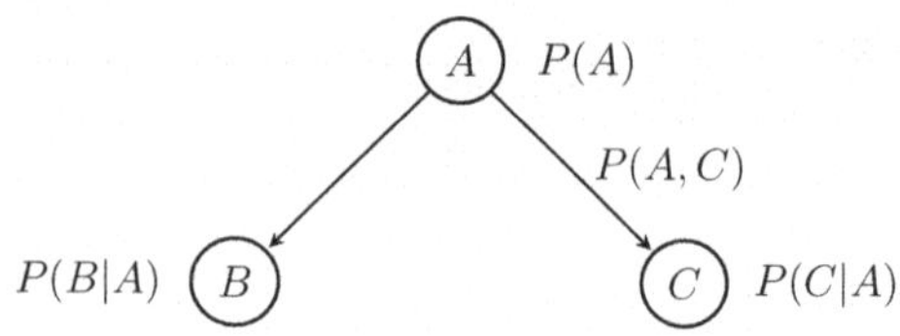

Fig. 4.3 Fragment of Bayesian network

The joint probability distribution P(B,A) of the edge (B,A) is

$$P(B, A) = P(B|A) \cdot P(A) \tag{4.2}$$

In general, let BN be a Bayesian network with nodes $X_1, X_2, \ldots, X_n$. Then the joint probability distribution is

$$P(X_1, X_2, \ldots, X_n) = \prod_{i=1}^{n} P(X_i|\text{parent}(X_i)) \tag{4.3}$$

where parent(X_i) denotes the parent set of node X_i.

If a new observation Y related to the node X has become available giving rise to a new likelihood $P(Y|X)$, then the prior probability $P(X)$ can be updated by means of Bayes rule. The result is the posteriori probability $P(X|Y)$ of node X.

$$P(X|Y) = \frac{P(Y|X) \cdot P(X)}{P(Y)} \tag{4.4}$$

where $P(Y) = P(Y|X) \cdot P(X) + P(Y|\bar{X}) \cdot P(\bar{X})$ is a normalising factor indicating the total probability of the evidence and $\bar{X}$ denotes the complement of X.

The effect of newly available evidence can be propagated through the BN by updating the probabilities of the adjacent nodes of an edge. By repeating this process at each new time instant, the evolution of the health state of a system can be estimated and can be used for prognosis. In a system with an incipient fault, the time evolution of the health state indicates some degradation growth over time.

A BN is a static model that represents a joint probability distribution at a fixed point. A dynamic Bayesian network (DBN) is the temporal extension of a static BN. It allows variables to be represented at multiple discrete time instances within the same network structure. That is, a static BN is repeated at discrete time points and its variables carry a time index. In addition to the conditional dependency in the static BN, there are arcs in a DBG denoting temporal dependencies between static BN instantiations [12].

Other Bayesian-based stochastic approaches to data-driven prognostic estimate the state of a system by means of particle filters, or Kalman filters revisited in Sect. 3.4.

Particle filters use Bayes' rule to update the prior pdf $p(\boldsymbol{x}_k|z_{k-1})$ of the state $\boldsymbol{x}_k$ at time t_k by means of the newly available measurement z_k (p. 69, Fig. 3.10) and approximate the posteriori conditional state probability distribution $p(\boldsymbol{x}_k|z_k)$ by means of drawn samples. In order to estimate the RUL, particle filtering uses the equation of the deteriorating state recursively to propagate the posterior PDF estimate defined by the set of N particles $\boldsymbol{x}_p^i$ with associated weights w_p^i, $\{<\boldsymbol{x}_p^i, w_p^i>: i = 1, \ldots, N\}$, in time until $\boldsymbol{x}_p^i$ no longer meets the failure criteria at some end of life time t_{EoL}^i. The RUL distribution is then the probability $p(t_{\text{EoL}}^i - t_P)$ given by the distribution of w_p^i, where t_p denotes the start of the prediction. Figure 4.4 shows a flowchart of particle filter based RUL prediction.

In contrast to autoregressive moving average (ARMA) methods considered in the following section, particle filtering enables long term predictions as model parameters can be included as part of the state vector. That is, state estimation is combined with model parameter adaption so that RUL prediction can start with a tuned model. In [33], it is demonstrated that given some restrictions, the problem of an exponential growth of computational complexity with state dimension can be avoided in particle filter based state estimation combined with parameter adaption.

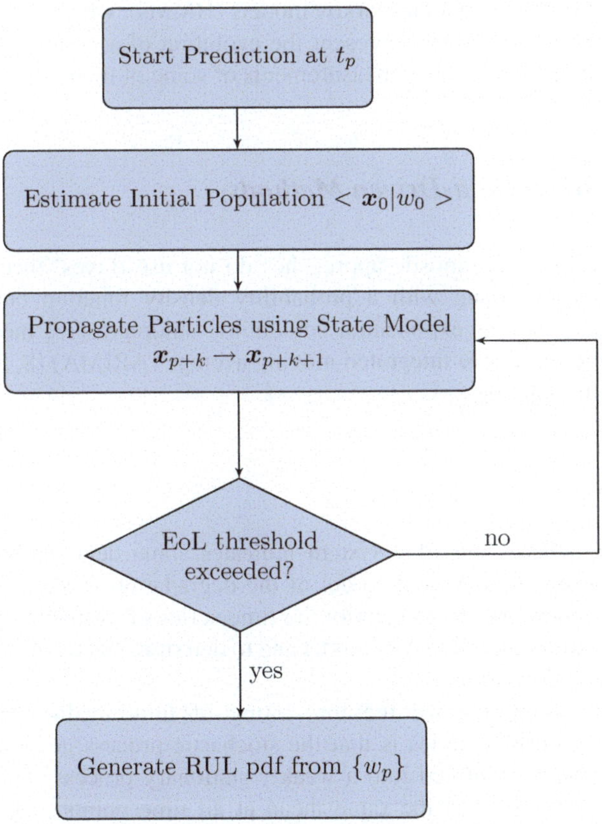

Fig. 4.4 Flowchart for RUL prediction by means of a particle filter [33]

In [22] and [4, Chap. 7], Jha proposes an approach to system health monitoring that is based on bond graph modelling and particle filtering. Bond graphs in LFT form are used for robust detection of the onset of an incipient fault. The failure prognostic part then uses particle filtering and assumes that the incipient fault has been isolated and that a degradation model of the prognostic candidate is known a priori.

Danes et al. report an approach that uses ARRs for fault detection in systems represented by a hybrid model [47] and particle filtering for failure prognosis [11]. Unknown degradation trends are assumed to be linear or exponential with respect to time. The state vector used in particle filtering is augmented by the coefficients of a set of potentially faulty elements and degradation rates. Let F^1, F^2 be coefficients that describe faulty parameter dynamics, and let r^1, r^2 denote degradation rates, k_F the fault occurrence time, k_D the time instant of fault detection, and T the sampling period, then the exponential degradation model used in [11] reads

$$F_k = F^1 \exp(-r^1(k - k_F)T) + F^2(1 - r^2(k - k_F)T) \qquad k \geq k_F \quad (4.5)$$

Still another option is hidden Markov models (HMMs), which are a special case of Bayesian networks. HMMs represent the evolution of a system. The true state variables are hidden. Only noisy measurements of some of them are observable.

4.2.2 *Statistical Data-Driven Methods*

Statistical data-driven prognostic approaches do not use Bayes' theorem for estimating future states along with a probability density function but apply time series techniques such as exponential smoothing, autoregressive moving average (ARMA), or autoregressive integrated moving average (ARIMA) [8, 50] for degradation trend extrapolation.

Linear Regression

If the monitored behaviour of a system indicates some degradation and if it is difficult to develop an analytical model of the degradation process from physical principles, an option may be to capture the time series of available measured data by means of an autoregressive (AR) model and to determine its unknown parameters by least square optimisation.

A common assumption for the forecasting of future values and for trend estimation based on time series is that the stochastic process generating the time series data can be transformed into a weakly stationary process. This means that the transformed process has the same mean at all time points, that the variance is finite for all time instances, and that the covariance between the values at any two time points only depends on the difference between the two times, and not on their location on the time axis. The most common transformation is differencing, i.e. subtracting the previous value from the current value and considering the time series of the differences instead of the original one. Sometimes, differencing must be applied a number of times.

Given a time series of measured data, linear regression assumes that future values are linearly dependent on past observations and random errors.

$$y(t) = \sum_{j=1}^{p} \theta_j \phi_j(t) + \epsilon(t) \tag{4.6}$$

where $y(t)$ is the data of a time series at time instant t, θ_j are the regression weights to be determined accounting for the influence of preceding values on $y(t)$, $\phi_j(t) = y(t-j)$ the regressors, and $\epsilon(t)$ a random error. The latter is commonly assumed a Gaussian white noise process. Equation 4.6 is an autoregressive model with lag p denoted in the literature as AR(p).

An ARMA(p,q) model is obtained by adding a moving average model (MA) to an autoregressive model.

$$y(t) = \underbrace{\sum_{i=1}^{p} \theta_i y(t-i)}_{\text{autoregressive part}} + \underbrace{\sum_{i=1}^{q} \theta_i \epsilon(t-i)}_{\text{moving average part}} \tag{4.7}$$

If the process generating the time series data is not stationary, differencing is applied repeatedly until near-stationary is obtained. The resulting time series is then used for an ARMA model. If differencing is needed d times, the notion ARIMA(p,d,q) model is used in the literature.

The number p of lags in the autoregressive part of the model may be determined by inspection of the partial autocorrelation plot, the number q of terms in the moving average part required to remove any autocorrelation in the series made stationary may be found by inspection of the autocorrelation plot. For the determination of the parameters (p,d,q) of an ARIMA model, the open source Python statsmodels package may be used [39].

For n data values at time points $t_i = t - i, i = 1, \ldots, n$, one obtains from (4.6)

$$\underbrace{\begin{bmatrix} y(t_1) \\ \vdots \\ y(t_n) \end{bmatrix}}_{\boldsymbol{y}} = \underbrace{\begin{bmatrix} \phi_1(t_1) & \ldots & \phi_p(t_1) \\ \vdots & \ddots & \vdots \\ \phi_1(t_n) & \ldots & \phi_p(t_n) \end{bmatrix}}_{\boldsymbol{\Phi}} \underbrace{\begin{bmatrix} \theta_1 \\ \vdots \\ \theta_p \end{bmatrix}}_{\boldsymbol{\theta}} + \underbrace{\begin{bmatrix} \epsilon(t_1) \\ \vdots \\ \epsilon(t_n) \end{bmatrix}}_{\boldsymbol{\epsilon}} \tag{4.8}$$

or

$$\boldsymbol{y} = \boldsymbol{\Phi} \boldsymbol{\theta} + \boldsymbol{\epsilon} \tag{4.9}$$

where $\boldsymbol{\Phi}$ denotes the known observation data matrix and $\boldsymbol{\theta}$ the parameter vector to be estimated.

In the case $n > p$, i.e. more data are available than unknown parameters, the least square solution for the vector of unknown regression weights $\boldsymbol{\theta}$ is the estimate

$$\hat{\boldsymbol{\theta}} = \underbrace{\left(\Phi^T \Phi\right)^{-1} \boldsymbol{\Phi}^T}_{\boldsymbol{\Phi}^+} \boldsymbol{y} \tag{4.10}$$

where $\boldsymbol{\Phi}^+$ is the Moore–Penrose pseudo-inverse of $\boldsymbol{\Phi}$. The weighted sum

$$\hat{y}(t) = \sum_{j=1}^{p} \hat{\theta}_j y(t-j) \tag{4.11}$$

of p past measurement values $y(t-j)$ is then an approximation of $y(t)$ with least square error $||\boldsymbol{\epsilon}||^2 = \boldsymbol{\epsilon}^T \boldsymbol{\epsilon}$.

Recursive Least Square Method

There is an extension of the above linear regression called *recursive* least square (RLS) estimation that enables to predict future approximations $\hat{y}$ by proceeding in time and by using weights computed in a prior step in the update [19].

Equation 4.6 may be written in the form

$$y(t) = \underbrace{[y(t-1)\, y(t-2) \ldots y(t-k)]}_{\boldsymbol{\Psi}^T(t-1)} \boldsymbol{\theta}(t-1) + \epsilon(t) \tag{4.12}$$

The determination of a parameter vector $\boldsymbol{\theta}(t)$ that minimises the square error at the current time instant t then yields:

$$\hat{\boldsymbol{\theta}}(t) = \hat{\boldsymbol{\theta}}(t-1) + \boldsymbol{K}(t)[y(t) - \hat{y}(t)] \tag{4.13}$$

where

$$\mathbf{P}(t) = \mathbf{P}(t-1) - \frac{\mathbf{P}(t-1)\boldsymbol{\Psi}(t-1)\Psi^T(t-1)\mathbf{P}(t-1)}{1 + \boldsymbol{\Psi}^T(t-1)\mathbf{P}(t-1)\boldsymbol{\Psi}(t-1)} \tag{4.14}$$

$$\boldsymbol{K}(t) = \mathbf{P}(t)\boldsymbol{\Psi}(t-1) \tag{4.15}$$

Equations 4.13 and 4.14 are recursive and need initial values. If no prior information about the parameters is available, their initial values are commonly set to zero. The initial value for the recursive computation of matrix $\mathbf{P}$ is commonly $\mathbf{P}(0) = \delta\mathbf{I}$ where $\delta \gg 1$ a positive constant and $\mathbf{I}$ the unity matrix. It can be shown that $\mathbf{P}(t)$ is proportional to the covariance matrix of the parameter vector $\boldsymbol{\theta}(t)$ [19]. The factor $\boldsymbol{K}(t)$ in (4.13) is known as the Kalman gain vector.

Suppose that the values of a time series till the time instant t have been fitted by linear regression, then time series forecasting into the future until a predefined failure alarm threshold is reached may be performed by recursive least square estimation. However, the application of an ARMA model assumes that the stochastic process generating the time series data is stationary. Therefore, and due to the uncertainty with regard to the initial conditions of the recursion, in general, ARMA models are more suited for short-term predictions.

Forgetting Factor Recursive Least Square Method

The so-called exponentially weighted least squares method, or *forgetting factor recursive least square* (FFRLS) method uses a forgetting factor $0 < \lambda < 1$ that gradually reduces the affect of old observations on the error. New data are more weighted than old ones so that the method can respond to changes in the observation data and may be used for non-stationary processes. The factor increases the sensitivity of the algorithm and prevents that the entries in the matrix $\mathbf{P}$ become too small.

$$||\epsilon||^2 = J(\boldsymbol{\theta}(t)) = \sum_{i=1}^{t} \lambda^{t-i}[\boldsymbol{\Psi}^T(i-1)\boldsymbol{\theta}(i) - y(i)]^2 \tag{4.16}$$

The forgetting factor in the squared error results in a new matrix $\bar{\mathbf{P}}$.

$$\bar{\mathbf{P}}(t) = \lambda^{-1}\mathbf{P}(t) \tag{4.17}$$

The choice of a constant value for the forgetting factor is difficult. For $\lambda = 1$, the matrix $\mathbf{P}(t)$ and the Kalman gain vector $\boldsymbol{K}(t)$ become progressively insensitive to parameter changes. Conversely, for $\lambda < 1$, $\mathbf{P}(t)$ may be more sensitive to parameter changes but also more sensitive to noise. In practice, a typical value is $\lambda = .98$. In view of this difficulty, RLS algorithms with a variable adaptive forgetting factor have been proposed in the literature, for instance in [40, 51].

Sliding Window Recursive Least Squares ARMA Parameter Estimation

Instead of using a weighting factor that gradually reduces the influence of past data, the sum of squared errors may be minimised over a set of L recent observation samples in a sliding window so that the estimated parameter vector $\hat{\boldsymbol{\theta}}(t)$ at time instant t and the estimate $\hat{y}(t)$ are sufficiently accurate [52, 53]. A lengthy mathematical development yields modifications of the expressions for the matrix $\mathbf{P}(t)$ (4.14) and the estimate $\hat{\boldsymbol{\theta}}(t)$ of the parameter vector (4.13) that account for the L lasted observation data only [49]. The computational effort is twice the one required by the standard RLS algorithm.

In [49], the parameter estimates in the current sliding window are used to set up a state space model in phase-variable form.

$$\hat{\boldsymbol{x}}_{k+1} = \begin{bmatrix} \hat{x}_{1(k+1)} \\ \hat{x}_{2(k+1)} \\ \vdots \\ \hat{x}_{p(k+1)} \end{bmatrix} = \underbrace{\begin{bmatrix} 0 & 1 & 0 & 0 & \dots & 0 \\ 0 & 0 & 1 & 0 & \dots & 0 \\ \vdots & \vdots & \vdots & \vdots & \vdots & \vdots \\ 0 & 0 & 0 & 0 & \dots & 1 \\ \hat{\alpha}_0 & \hat{\alpha}_1 & \hat{\alpha}_2 & \hat{\alpha}_3 & \dots & \hat{\alpha}_{p-1} \end{bmatrix}}_{\mathbf{A}} \underbrace{\begin{bmatrix} \hat{x}_{1(k)} \\ \hat{x}_{2(k)} \\ \vdots \\ \hat{x}_{p(k)} \end{bmatrix}}_{\hat{\boldsymbol{x}}_k} + \underbrace{\begin{bmatrix} 0 \\ 0 \\ \vdots \\ 0 \\ \hat{\beta}_0 \end{bmatrix}}_{\mathbf{B}} u_k \quad (4.18)$$

As the entries in the matrices are obtained only from data in the latest sliding window by least square estimation, the state space model can adapt to changes of the statistical features of the AR model.

Let

$$\hat{x}(k+1) = \sum_{j=0}^{p-1} \hat{\alpha}_j \underbrace{x(k+1-p+j)}_{=:\, \hat{x}_{j+1}(k)} + \hat{\beta}_0 u(k) \quad (4.19)$$

be an estimation of $x(k+1)$ and $\hat{\alpha}_j$, $\hat{\beta}_0$ the least square estimated parameters. As

$$\hat{x}(k+1-p+j) = \hat{x}_j(k+1) = \hat{x}_{j+1}(k), \quad j = 0, 1, \dots, p-1 \quad (4.20)$$

$$\hat{x}(k+1) = \hat{x}_p(k+1) \quad (4.21)$$

the above state space model (4.18) is obtained for the vector

$$\hat{\boldsymbol{x}}(k+1) = [\hat{x}_1(k+1)\ \hat{x}_2(k+1)\ \dots\ \hat{x}_p(k+1)]^T \quad (4.22)$$

Combining Identification of a System with Deteriorating Behaviour and Failure Prognostic Based on Kalman Filtering

In [49], the state space model (4.18) is then given to a Kalman filter for a N-step ahead prediction of the current state into the future with $N = 1, 2, 3 \dots$ till the time instant t_{fl} at which a damage relevant feature reaches a failure alarm threshold. Subtracting the current time from t_{fl} gives a predicted RUL. The time is then advanced by one step and the procedure is repeated.

The equations of the discrete Kalman filter (3.20), (3.21) read for the case of a N-step ahead prediction:

$$\hat{\boldsymbol{x}}_{k+N|k} = \mathbf{A}_{k+N|k}\,\hat{\boldsymbol{x}}_{k|k} + \mathbf{B}_k \boldsymbol{u}_k \tag{4.23}$$

$$\mathbf{P}_{k+N|k} = \mathbf{A}_{k+N|k}\mathbf{P}_{k|k}\mathbf{A}^T_{k+N|k} + \mathbf{Q}_{k+N|k} \tag{4.24}$$

with

$\hat{\boldsymbol{x}}_{k\|k}$	:	updated filter estimate at t_k
$\hat{\boldsymbol{x}}_{k+N\|k}$	:	predicted estimate of $\boldsymbol{x}$ at time instant t_{k+N} given all measurements through the current time instant t_k
$\mathbf{P}_{k\|k}$	:	error covariance matrix associated with the prediction $\hat{\boldsymbol{x}}_{k\|k-1}$
$\mathbf{P}_{k+N\|k}$	:	error covariance associated with the predicted estimate $\hat{\boldsymbol{x}}_{k+N\|k}$
$\mathbf{A}_{k+N\|k}$	:	transition matrix from step k to $k+N$
$\mathbf{Q}_{k+N\|k}$	:	covariance of the cumulative effect of Gaussian white noise from step k to $k+N$

This prognostic approach based on a combination of system identification by means of sliding RLS estimation and Kalman filtering can adapt to changes, e.g. of ambient conditions, can adapt to future load uncertainties, to input data uncertainties, such as the initial state of the system, variability in material properties, and manufacturing variability.

However, it is to be noted that the covariance matrices $\mathbf{Q}$ and $\mathbf{R}$ accounting for process noise and measurement noise respectively must be specified *in advance* before the Kalman filter is used. Thus, changes of the noise characteristics are not taken into account by the filter unless the matrices are adapted, but matrix $\mathbf{Q}$ affects the estimation of the state error covariance matrix $\mathbf{P}$ (cf. (3.21), p. 64), and matrix $\mathbf{R}$ the measurement prediction covariance matrix $\mathbf{S}$, and by consequence the Kalman gain matrix $\mathbf{K}$ (cf. (3.22), p. 64). Moreover, as process noise may include disturbances and modelling errors, the entries of matrix $\mathbf{Q}$ are difficult to determine. These uncertainties will result in uncertainties of the RUL estimates. What may be measured is the accuracy of the ARMA model, which affects the accuracy of the state space matrices built from prior measurements as the system behaviour deteriorates and used by the Kalman filter in the prediction of future states.

4.2.3 Neural Networks

Other approaches to data-driven failure prognostic use artificial neural networks (ANNs) which have become quite popular because they may outperform traditional data-driven approaches, e.g. those based on regression. ANNs require few model assumptions, are self-adaptive, and are well suited for practical problems when data but little knowledge about the system under study is available. Their functionality is inspired by the biological mechanisms of the human brain. The nodes of a neural network (NN), usually called neurons, are the basic processing units. They build groups on an input layer, on one or more hidden layers, and on an output layer and interact via connections to which an *adjustable* weight is assigned. Given some inputs, the purpose of a NN is to learn to produce desired outputs which requires quite a number of training data sets to learn about the relationships between inputs and outputs. Learning means to adapt the weights so that the network sufficiently accurately captures the relationships between inputs and outputs. The neurons of the input layer multiply the inputs by weights in parallel. The sum of weighted inputs along with an added bias is then fed forward to neurons of the first hidden layer. Each neuron on the hidden layers and the output layer is equipped with a nonlinear activation function which converts the weighted inputs into its output. Common activation functions are symmetric hard limit, linear threshold, symmetric sigmoid, or hyperbolic tangent [45]. The same happens between the first hidden layer and a possibly subsequent hidden layer as well as between the final hidden layer and the output layer.

A NN architecture in which the inputs to the neurons of a particular layer only depend on the outputs of the neurons on the preceding layer is called a feedforward network (FFNN). In static networks, the response of the network to a set of inputs is independent of its last outputs. Besides this forward propagation of signals from input to output nodes, the errors between the training data and the outputs of the network are also propagated *backward* for updating the weights in order to minimise the errors. Such NN architectures are often called backward-propagation networks (BPNNs) in the literature. Dynamic or recurrent neural networks (RNNs) with feedback connections between output and input nodes enable a backward flow of signals and a modelling of the time evolution of a dynamic system.

The training of NNs with back propagation, i.e. the adaption of the weights, is often based on a minimisation of the least square error by means of the Levenberg–Marquardt (LM) algorithm.

Let $\boldsymbol{w}$ be the vector of the weights of a neural network, d_{kp} the desired value of the kth output and the pth pattern, o_{kp} the actual value of the kth output and the pth pattern, P the number of patterns, and K the number of the network outputs. The so-called performance index $F(\boldsymbol{w})$ to be optimised is then defined as

$$F(\boldsymbol{w}) = \boldsymbol{E}^T \boldsymbol{E} \tag{4.25}$$

where

$$\boldsymbol{E} = [e_{11} \ldots e_{K1}\ e_{12} \ldots e_{K2}\ \ldots\ e_{1p} \ldots e_{KP}]^T \tag{4.26}$$

and

$$e_{kp} = d_{kp} - o_{kp}, \qquad k = 1, \ldots, K, \qquad p = 1, \ldots, P \tag{4.27}$$

The NN weights are then updated according to the equation

$$\boldsymbol{w}_{t+1} = \boldsymbol{w}_t - \left(\mathbf{J}_t^T \mathbf{J}_t + \mu_t \mathbf{I}\right)^{-1} \mathbf{J}_t^T \boldsymbol{E} \tag{4.28}$$

where $\mathbf{I}$ denotes the identity matrix, μ the learning factor, and $\mathbf{J} = \partial \boldsymbol{e}/\partial \boldsymbol{w}$ the Jacobian of the output errors with respect the weights of the NN. As can be seen, the LM algorithm requires the inversion of a $N \times N$ matrix at each time step. Therefore, for NNs of large size, the LM algorithm is not practical. For the training of NNs with one hidden layer, Wilamowski et al. proposed a modification of the LM algorithm with an improved convergence rate [48].

Figure 4.5 [45] shows a static feedforward neural network with one hidden layer. The annotations used have the following meaning.

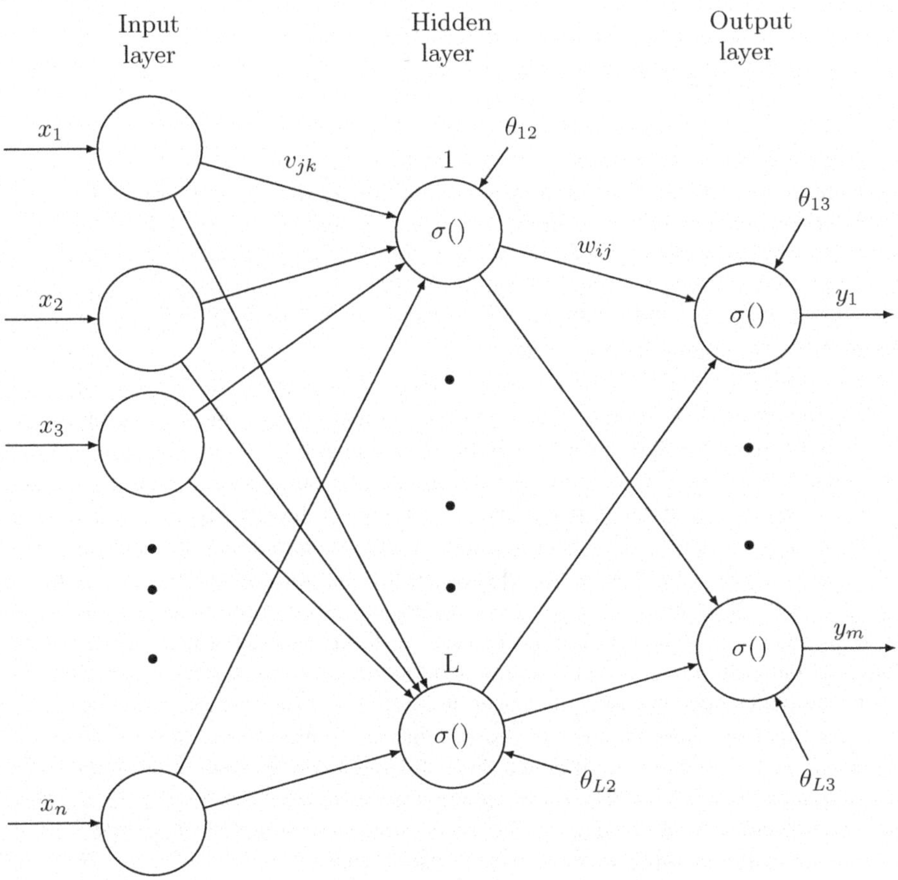

Fig. 4.5 A static feedforward neural network with one hidden layer

x_k	:	inputs into the NN
y_i	:	outputs of the NN
$\sigma()$	:	nonlinear activation function of the neurons on the hidden layer and the output layer
v_{jk}	:	hidden layer weights
w_{ij}	:	output layer weights
θ_{vj}	:	hidden layer thresholds
θ_{wi}	:	output layer thresholds
L	:	number of neurons on the hidden layer

With these notations, the function of the static feedforward NN can be expressed as follows [45]:

$$y_i = \sum_{j=1}^{L} w_{ij}\sigma\left(\sum_{k=1}^{n} v_{jk}x_k + \theta_{v_j}\right) + \theta_{wi} \qquad i = 1, \ldots, m \tag{4.29}$$

After a NN has been trained, it can be used for fault diagnosis, forecasting, and failure prognostic. The nonlinear action functions and the fast and simple calculations enable a NN to compute a complex system behaviour without having a model for the relations between inputs and outputs. In prognostic, NNs can detect abnormalities between healthy and deteriorated measured signals by assuming early samples in each time series being healthy and latest samples as deteriorated. On this basis, NNs can predict the RUL [20].

The design of a NN, however, is not trivial. For a start, the number of input nodes, the number of hidden layers, and the number of neurons on each layer has to be defined. If the training of an initial NN architecture does not approximate sufficiently accurate the relation between inputs and outputs of a complex system with a deteriorating dynamic behaviour, the number of hidden layers and of hidden nodes may be increased. The determination of an optimal number of nodes has been subject of various research works reported in the literature [24]. However, a more complex BPNN architecture with more nodes and more connections means more weights to be adjusted, which requires more training data and entails the problem that a global optimum is difficult to find and that the convergence rate depending on initial estimates becomes low. Moreover, noise and bias on the input data caused by sensors affect the determination of the weights. An option to account for noise on the input data is to define confidence limits for the error between training data and the output of the NN [28]. The problem with uncertain initial values for the weights may be overcome by bootstrapping, i.e. by re-running the training of the NN several times with different subsets of the training data [16].

4.3 Model-Based Failure Prognostic

Physics-Based Failure Prognostic

A dynamic model of a real world system developed from first principles of physics can not only be used for the simulation of the dynamic behaviour but also for fault diagnosis and even for failure prognostic if the fundamental physics of a degradation process taking place in the system is well enough understood and if a model of the degradation mechanism is incorporated into the dynamic system model. Such a model accounting for an inherent degradation process enables a prediction of long term deterioration of the system behaviour. As the model development is based on assumptions and makes use of approximations, the result needs to be validated before used for failure prognostic. However, the development of a degradation model (DM) can be difficult and costly, if even not impossible as the number of model parameters increases with the model complexity and not all parameters may be available. Therefore, often, a model of some type is assumed according to the degradation process and its parameters are estimated.

A well known example of a degradation model is the law of Paris–Erdogan [25] for crack growth in mechanical structures due to fatigue.

$$\frac{d\theta}{dn} = C(\Delta K)^{\gamma} \tag{4.30}$$

where θ is the crack length, n the number of cycles, $d\theta/dn$ the crack growth rate, ΔK the stress intensity factor, $0 < C \ll 1$ and $2 < \gamma < 4$ are material constants.

In [23], Kulkarni develops physics-based degradation models for electrolytic capacitors under conditions such as electrical overstress and thermal overstress to track their degradation. RUL prediction is performed by Bayesian inference using a Kalman filter approach.

Hybrid Failure Prognostic

As physics-based modelling of a non-deterministic degradation process in a complex system subject to various operational conditions such as environmental impacts and changing loading conditions is difficult, or impractical, model-based and data-driven techniques are combined. In addition to a system model developed from first principles of physics such as a BG model, a model of the damage progression can be obtained by fitting the parameters of a suitable mathematical model to a set of available degradation data. The degradation function with parameters to be determined may be linear, logarithmic, exponential, or a polynomial. Once a degradation model of sufficient accuracy has been learned this way, it can be used in conjunction with the system model for recursive estimation of the current state of health and for predicting the RUL [10, 34].

Jha [22], [4, Chap. 7] combines bond graph-based fault diagnosis for detecting degradation initiation and particle filtering for online estimation of the current state of a system with a deteriorating dynamic behaviour, i.e. for estimating the damage and for subsequently RUL prediction. Under some assumptions, the actual value of a single faulty parameter $\theta^d(t)$ is implicitly given by an ARR derived from a BG in LFT form and included in the state vector of a faulty model in state space form that is used in the prognostic module.

4.4 Determination of a Degradation Model from ARRs

Section 3.5.2 shows that constraints between known inputs and measured signals (ARRs) can be derived from a diagnostic BG and that their evaluation can be used for FDI. If unknowns can be eliminated, ARRs are of the explicit form (3.87)

$$r_k(t) = g_k(\boldsymbol{u}_m(t), \boldsymbol{y}_n(t), \boldsymbol{\theta}_p, \boldsymbol{b}(t)) \tag{4.31}$$

where $\boldsymbol{u}_m(t) = [\boldsymbol{u}(t)\,\dot{\boldsymbol{u}}(t)\,\ldots, \boldsymbol{u}^{(m)}(t)]^T$, $\boldsymbol{y}_n(t) = [\boldsymbol{y}(t)\,\dot{\boldsymbol{y}}(t)\,\ldots\,\boldsymbol{y}^{(n)}(t)]^T$, and $\boldsymbol{\theta}_p = [\theta_1 \ldots \theta_p]^T$.

Assume that at time instant t_0 parameter θ_1 has been detected and isolated as a single incipient fault. If the value of the unknown varying faulty parameter $\tilde{\theta}_1(t)$ at time instances $t > t_0$ was known and if the nominal parameter value $\theta_1^n(t)$ is replaced by the faulty one, then the residual of the ARR containing $\tilde{\theta}(t)$ vanishes.

$$0 = r_1(t) = g_1(\boldsymbol{u}_m(t), \boldsymbol{y}_n(t), \tilde{\theta}_1(t), \theta_2, \ldots, \theta_p, \boldsymbol{b}(t)) \tag{4.32}$$

Equation 4.32 may be considered an implicit relation for the degradation profile versus time that can be solved for $\tilde{\theta}_1(t)$ if the conditions of the implicit function theorem hold (cf. Appendix C). That is, there exists a real valued function $\psi()$ such that

$$\tilde{\theta}_1(t_k) = \psi_1(\theta_2, \ldots, \theta_p, \boldsymbol{u}_m(t_k), \boldsymbol{y}_n(t_k), \boldsymbol{b}(t_k)) \tag{4.33}$$

The values for $\tilde{\theta}_1(t_k)$ are obtained sequentially from sampled inputs and measurements and time derivatives of both signals. Even if the raw signals corrupted with noise are filtered before their values are used as arguments of function $\psi()$ with some time delay, the value of $\tilde{\theta}_1(t_k)$ used in the estimation of the next system state carries some uncertainty. Hence, all state of health estimations and thus the RUL prediction result in uncertain values. Figure 4.6 displays a flowchart of the ARR-based procedure.

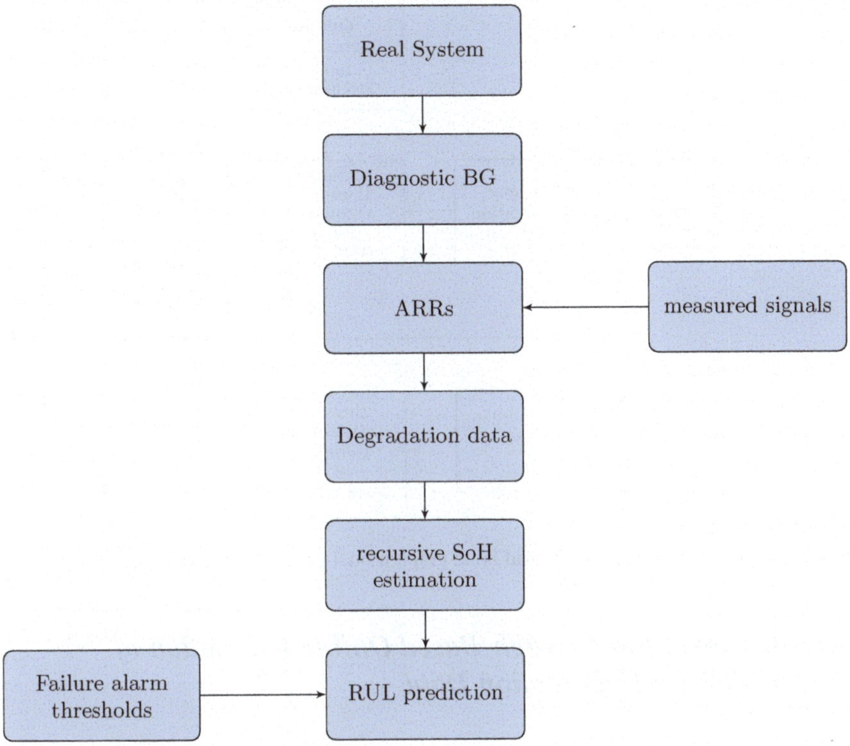

Fig. 4.6 Flowchart of an ARR-based prognostic

4.5 A Hybrid Bond Graph Model-Based Data-Driven Approach

This section presents a novel hybrid bond graph model-based data-driven approach to failure prognostic. The physics model-based part of the hybrid approach uses a diagnostic BG for the acquisition of degradation data. To that end, known system input signals and measured signals are inserted into equations, which have been derived offline from a DBG, to estimate the numerical values of a faulty parameter at sampling time points in a sliding time window of fixed size and to use them in the data-driven part of the approach to failure prognosis.

For each of the overlapping instances of the sliding time window, the data-driven part of the approach identifies a mathematical function over the time window and projects it into the future to obtain an estimate of the current RUL. As a result, a sequence of RUL estimates is obtained with values that tend to zero as the considered faulty component reaches its end of life [6]. Figure 4.7 displays the scheme of the hybrid bond graph model-based data-driven method.

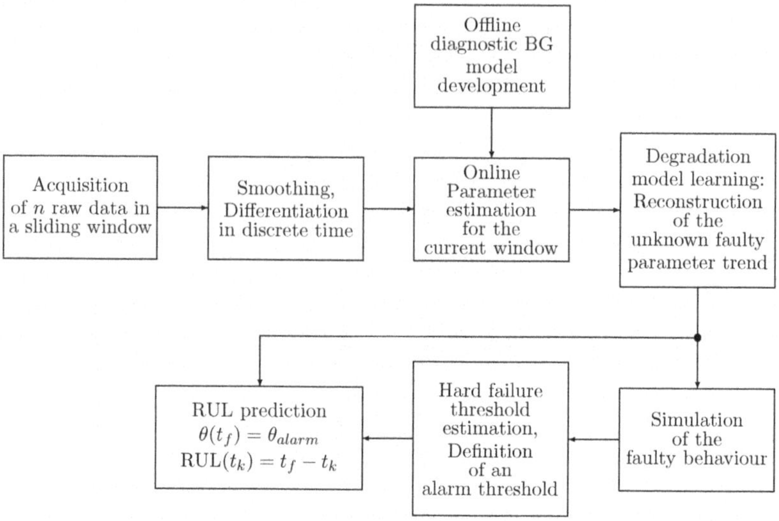

Fig. 4.7 Scheme of the proposed hybrid bond graph model-based data-driven approach

4.5.1 Bicausal Bond Graph-Based Online Estimation of Unknown Degradation Data

Parametric degradation means that the deterioration of the dynamic behaviour of an engineering system can be attributed to the fact that some of its parameters increasingly deviate from their nominal values with time following a function of which an exact analytical expression is mostly unknown. The approach in this section determines a time series for an *unknown* degradation function by parameter estimation.

As shown in Sect. 2.5, bicausal BGs allow that both power co-variables, effort and flow, of a bond attached to a power port of an element, may be inputs into the element. Accordingly, they may be used for parameter estimation and thus can be used for setting up an equation that determines the degradation function $\Phi_\theta(t)$ of a faulty element parameter $\Theta(t_i) = \Theta_n + \Phi_\theta(t_i)$ at sampling time instances t_i, where Θ_n denotes the nominal parameter value.

In a bicausal diagnostic BG, detectors representing sensors are in inverted causality because they provide known measured signals into the diagnostic model. The two conjugate power signals into the port of a faulty element can be determined by following causal paths from sources providing known input signals and from detectors in inverted causality to the port of an identified faulty element. That is, both power variables in the element's constitutive equation are known, while its faulty trending parameter at the current sampling time point t_i is unknown. The constitutive element law results in a possibly implicit equation that determines the unknown parameter value $\Theta(t_i)$.

Example: Boost Converter

Consider the circuit schematic of the open loop boost converter in Fig. 4.8. It is assumed that the converter used, e.g. in power generation plants, operates in continuous conduction mode (CCM) with a sensor for the inductor current $\tilde{i}_L$ and a sensor for the output voltage $\tilde{V}$. A fault in this electronic component may lead to a failure in a power distribution system and to a degradation of its performance.

If the MOSFET transistor and the diode, D, are modelled as two conversely commutating ideal switches $\mathrm{Sw} : s_i,\ i = 1, 2$, then the circuit immediately transforms into the DBG in Fig. 4.9 with two detectors in inverted causality that are assumed to be non-faulty.

It can be shown that the connection of the two ideal oppositely commutating switches in the dashed box can be replaced by a modulated transformer. Note that the causality at the ports of the submodel in the dashed box does not change when the causalities at the switch ports are inverted. Disregarding the switch port causalities indicated in Fig. 4.9, the behaviour of the switches shall be described by an *implicit* equation that accounts for their two discrete states.

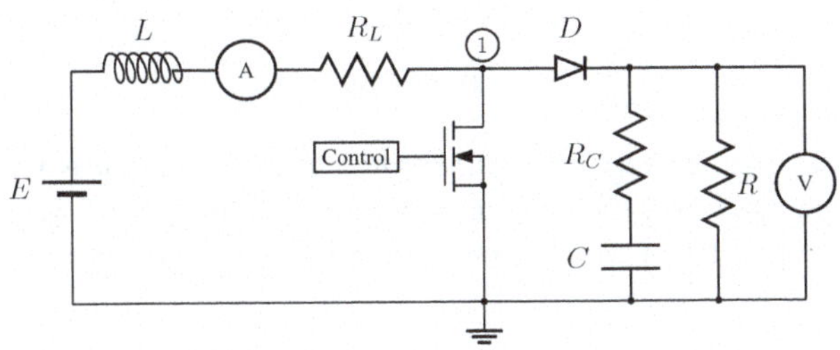

Fig. 4.8 Circuit schematic of an open loop boost converter with a load resistor *R*

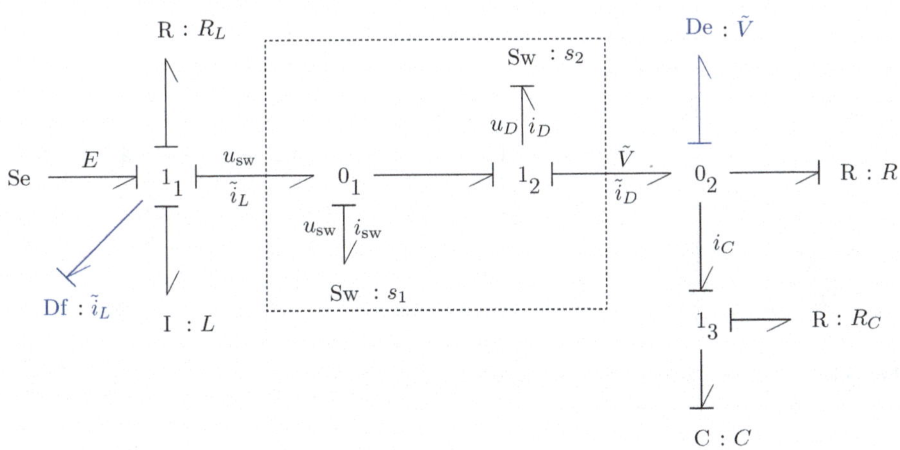

Fig. 4.9 DBG of the boost converter in Fig. 4.8

$$0 = s_1 u_{\mathrm{sw}} + \bar{s}_1 i_{\mathrm{sw}} \tag{4.34}$$

$$0 = s_2 u_{\mathrm{D}} + \bar{s}_2 i_{\mathrm{D}} \tag{4.35}$$

From the BG, one gets

$$\tilde{i}_L = i_{\mathrm{sw}} + i_{\mathrm{D}} \tag{4.36}$$

$$\tilde{V} = u_{\mathrm{sw}} - u_{\mathrm{D}} \tag{4.37}$$

Substitution into the implicit switch equations yields two linear equations for two unknowns u_{sw}, u_{D}.

$$s_1 u_{\mathrm{sw}} - \bar{s}_1 i_{\mathrm{D}} = -\bar{s}_1 \tilde{i}_L \quad | \cdot \bar{s}_2 \mid \cdot s_2 \tag{4.38}$$

$$s_2 u_{\mathrm{sw}} - \bar{s}_2 i_{\mathrm{D}} = s_2 \tilde{V} \quad | \cdot \bar{s}_1 \mid \cdot s_1 \tag{4.39}$$

Observing $s_2 = \bar{s}_1$ one obtains for the unknowns after some steps the equations of a modulated transformer with the modulus $s_2(t) \in \{0, 1\}$.

$$u_{\mathrm{sw}} = s_2 \tilde{V} \tag{4.40}$$

$$i_{\mathrm{D}} = s_2 \tilde{i}_L \tag{4.41}$$

If the small equivalent series resistance, R_C, of the capacitor is neglected and if variables are averaged over the switching period, then the circuit may be presented by the DBG in Fig. 4.10, in which d denotes the duty ratio of the signal controlling the transistor.

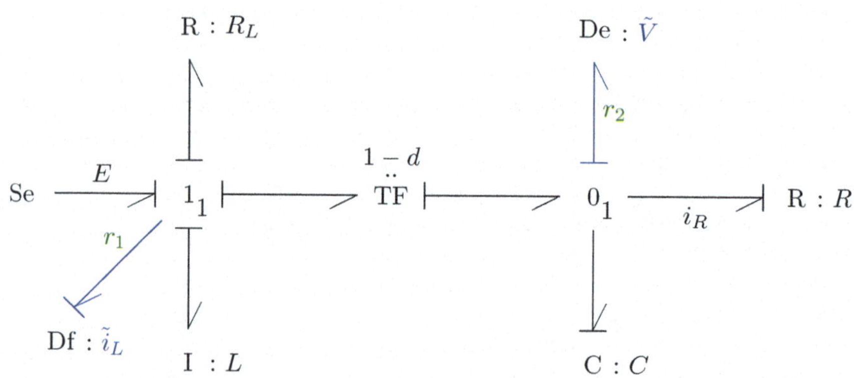

Fig. 4.10 Averaged DBG of the boost converter in Fig. 4.8

Fault Detection and Isolation

First, ARRs are derived from the DBG in Fig. 4.10 in order to check which potential parameter faults can be detected and isolated by means of the two sensors. From the DBG in Fig. 4.10, the following two ARRs are easily derived:

$$\text{ARR}_1 : r_1 = E - R_L - L\frac{d}{dt}\tilde{i}_L - (1-d)\tilde{V} \tag{4.42}$$

$$\text{ARR}_2 : r_2 = (1-d)\tilde{i}_L - C\frac{d}{dt}\tilde{V} - \frac{1}{R}\tilde{V} \tag{4.43}$$

Their structural dependencies from element parameters are represented by the FSM in Table 4.1. As can be seen, all parametric faults can be detected by means of the two sensors but none can be isolated except from a faulty duty ratio d.

As can be seen from the FSM in Table 4.1, the two storage elements do have a different fault signature. That is, they could be isolated if all other elements could be assumed to be non-faulty. In the following, a single fault hypothesis is adopted and two cases are considered. At first, it is assumed that the parameter of the resistive load element is deteriorating with time. The second scenario assumes that the parameter of one of the two storage elements progressively deviates from its nominal value. For both cases it is shown how the numerical values of the respective unknown degradation function can be estimated by means of known system inputs and measured values or simulated measurements.

Estimating the Unknown Degradation of the Load Resistance

Assume that the cause for a monitored abnormal dynamic behaviour of the boost converter circuit has been isolated and can be attributed to the time-variant load resistance $R(t)$ that is increasingly deviating from its nominal value R_n with time, i.e. $R(t) = R_n + \Phi_R(t)$, where $\Phi_R(t)$ denotes the unknown degradation function. Given monitored measurements, the task is to estimate the time-varying resistance $R(t)$. To that end, the bond attached to the port of the R-element is replaced by a bicausal bond as depicted in Fig. 4.11.

Table 4.1 Fault signature matrix of the DBG in Fig. 4.10

Element	r_1	r_2	D_b	I_b
Se : E	1	0	1	0
R : R_L	1	0	1	0
I : L	1	0	1	0
TF : d	1	1	1	①
C : C	0	1	1	0
R : R	0	1	1	0

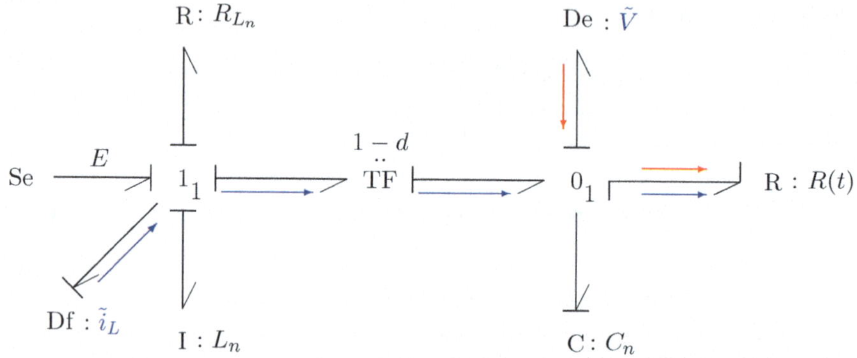

Fig. 4.11 Averaged bicausal BG of the boost converter with a trending resistance $R(t)$

As can be seen from the bicausal DBG in Fig. 4.11, there is a causal path from the flow detector Df : $\tilde{i}_L$ and another one from the effort detector De : $\tilde{V}$ to the power port of the R-element. The tilde denotes (filtered) measurements obtained from the real system or simulated measurements. As both port variables are determined by real measurements or simulated data provided by sensors into the DBG model, the time evolution of the resistance $R(t)$, i.e. numerical values of the degradation function $\Phi_R(t)$ can be computed. From the bicausal BG, one obtains

$$\tilde{i}_R = (1-d)\tilde{i}_L - C_n \frac{d\tilde{V}}{dt} \tag{4.44}$$

$$\tilde{V} = R(t)\tilde{i}_R = (R_n + \Phi_R(t))\tilde{i}_R \tag{4.45}$$

and as a result, an implicit algebraic equation for the unknown degradation function $\Phi_R(t)$.

$$\tilde{i}_R \Phi_R(t) = \tilde{V} - R_n \tilde{i}_R \tag{4.46}$$

Note that the computation of the degradation function values requires the differentiation in discrete time of the measured voltage $\tilde{V}$ (Eq. 4.44).

The decomposition of the load resistance into a nominal part R_n and a degradation part $\Phi_R(t)$ can be directly expressed in the bicausal BG as shown in Fig. 4.12.

By following the causal paths from the detectors to the trending part of the load resistor one obtains for the degradation function $\Phi_R(t)$

$$[(1-d)\tilde{i}_L - C_n \dot{\tilde{V}}]\, \Phi_R(t) = \tilde{V} - [R_n((1-d)\tilde{i}_L - C_n \dot{\tilde{V}})] \tag{4.47}$$

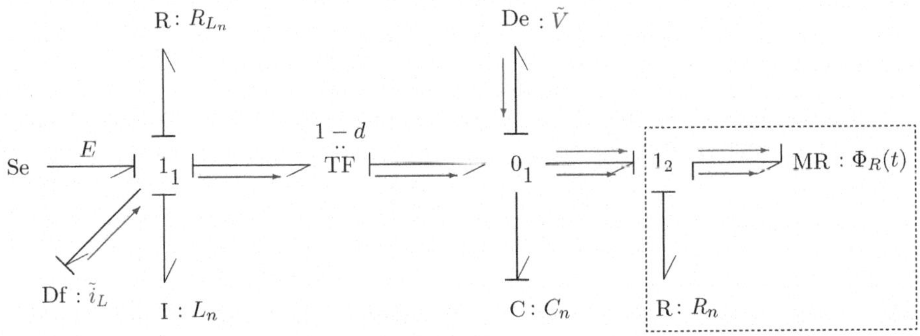

Fig. 4.12 Averaged bicausal BG of the boost converter with a decomposed trending resistance $R(t)$

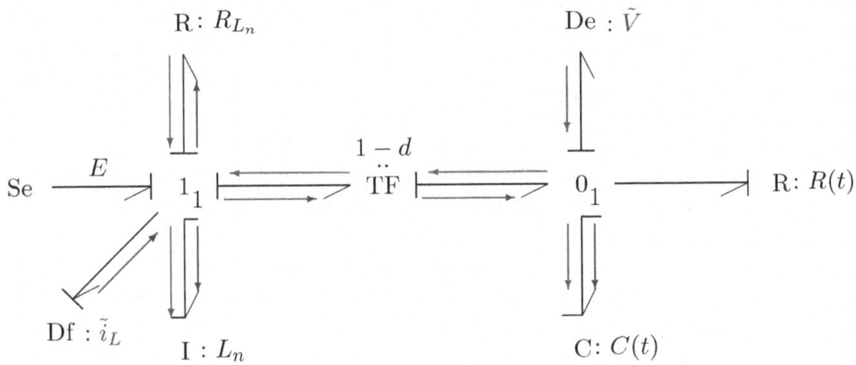

Fig. 4.13 Averaged bicausal BG of the boost converter with trending parameters $C(t)$, $L(t)$

Estimating the Unknown Parameter Degradation of a Storage Element

In the bicausal BG of Fig. 4.13, the bond attached to the power port of the C-element and the I element, respectively, has also been turned into a bicausal bond.

As a result, there are causal paths from the two detectors to the C-port so that the numerical values of a decaying capacitance $C(t)$ could be determined. First, it is assumed that only the capacitor as one of the two storage elements is faulty and that its capacitance trend follows an unknown degradation function, i.e. $C(t) = C_n + \Phi_C(t)$.

$$q_C = C(t)\tilde{V} = (C_n + \Phi_C(t))\tilde{V} \tag{4.48}$$

The causal paths from the flow detector Df : $\tilde{i}_L$ and from the effort detector De : $\tilde{V}$ to the C-element in the DBG of Fig. 4.13 yield

$$\dot{q}_C = (1 - d)\tilde{i}_L - \frac{\tilde{V}}{R_n} \tag{4.49}$$

Substitution of (4.49) into (4.48) gives the following implicit equation for the degradation function $\Phi_C(t)$:

$$\tilde{V}(t)\Phi_C(t) = \int_{t_{f_1}}^{t} \dot{q}_C(\tau)d\tau + \Phi_C(t_{f_1})\tilde{V}(t_{f_1}) - C_n\tilde{V} \tag{4.50}$$

where t_{f_1} denotes the time instant when the incipient fault exceeds an (adaptive) fault threshold and thus is detected. That is, $\Phi_C(t) \neq 0$ for $t > t_{f_1}$. Below that threshold the value of the capacitance may vary. A robust fault detection is insensitive to small parameter deviations from their nominal values.

As can be seen from the bicausal BG in Fig. 4.13, there are another two causal paths from the two detectors in inverted causality to the inductor with a faulty inductance $L(t) = L_n + \Phi_L(t)$. If just the inductor becomes faulty as of a time instant t_{f_2}, then similar to the computation of $\Phi_C(t)$ above, one obtains for the unknown degradation function $\Phi_L(t)$ from the bicausal BG

$$\begin{aligned}\tilde{u}_L = \frac{d}{dt}(Lf_L(\tilde{i}_L)) &= \frac{d}{dt}[(L_n + \Phi_L(t))f_L(\tilde{i}_L)] \\ &= E - R_{L_n}\tilde{i}_L - (1-d)\tilde{V}\end{aligned} \tag{4.51}$$

or

$$\Phi_L(t)\tilde{i}_L(t) = \int_{t_{f_2}}^{t} \tilde{u}_L(\tau)d\tau + \Phi_L(t_{f_1})\tilde{i}_L(t_{f_2}) - L_n f_L(\tilde{i}_L) \tag{4.52}$$

The integration in (4.50) and (4.52) may be performed numerically e.g. by means of the trapezoidal rule. Note that in the case of a storage element no measured signals need to be differentiated.

If the constitutive equation of an element is not linear with regard to its faulty time-varying parameter $\Theta(t)$, a nonlinear equation is to be solved by iteration at each sampled time point t_i. The solution is then stored in the buffer for the sliding window.

$$0 = \Phi(\Theta(t_i), \tilde{f}(t_i), \tilde{e}(t_i)) \tag{4.53}$$

where the two conjugate power port variables $\tilde{e}(t)$ and $\tilde{f}(t)$ are obtained by following causal paths from sources and detectors in inverted causality.

4.5.2 *ARR-Based Estimation of Degradation Data on Two DBGs*

In this section, a second, alternative BG-based approach to the development of equations for the estimation of trending parameters is presented. The numerical values of an unknown degradation function are estimated by using a first stage and a second stage diagnostic bond graph model [5]. Inputs into the first stage DBG with *nominal* parameters are measurements from a real system or simulated measurements. Both DBGs are used to generate mode-dependent ARRs that may include discrete switch states. Evaluation of the ARRs from the first stage DBG provides residuals that enable to detect the onset of incipient faults. In the following, residuals of ARRs derived from a first stage DBG are indicated by an upper index. In addition, it is assumed that faults can be isolated. The second stage DBG model accounts for parametric degradation by means of an *unknown* function. ARRs derived from the second stage DBG make use of the residuals of the first stage ARRs and constitute an implicit relation for the unknown degradation function. The computation can take place online concurrently to the monitoring of the real system and the measurement of its output signals.

Let ARRs be derived from a first stage DBG.

$$\boldsymbol{r}^{1}(t) = \mathbf{f}_1(\boldsymbol{u}_n(t)\,,\ \tilde{\boldsymbol{y}}_m(t)\,,\boldsymbol{\Theta}_n, b(t)) \tag{4.54}$$

where $\boldsymbol{u}_n(t)$ denotes the vector of input signals and of all its derivatives up to the order of n, $\tilde{\boldsymbol{y}}_m(t)$ the vector of measured signals and of all its derivatives up to the order of m, $\boldsymbol{\Theta}_n$ the vector of nominal physical system parameters, and $b(t) \in \mathbb{N}$ the system mode. The vector of all discrete switch states $\boldsymbol{\sigma}(t)$ is mapped onto $b(t)$ which is constant between two consecutive discrete events.

As can be seen from (4.54), a deviation of ARR residuals from values within defined bounds may be due to a faulty switch state change leading to an invalid pattern of switch states, i.e. to a physically inadmissible system mode of operation. In some works reported in the literature, such abrupt changes in the system structure are categorised as *hard* faults [41]. If the mode of operation of a monitored system has been identified and is admissible, then the cause for ARR residuals outside given thresholds are parameters that have deviated from their nominal values beyond parameter uncertainties. Faults due to continuous parameter variations are sometimes called *soft* faults.

Now, a second DBG is used that accounts for unknown parameter degradations $\boldsymbol{\Phi}(t)$, i.e. $\boldsymbol{\Theta}(t) = \boldsymbol{\Theta}_n + \boldsymbol{\Phi}(t)$. If the latter were known and if measured outputs $\tilde{\boldsymbol{y}}(t)$ of the faulty system are inputs into the DBG of the second stage, then the evaluation of the ARRs would give residuals close to zero. (The right parameter values causing the faulty system outputs would be used in the ARRs.)

$$\mathbf{0} \approx \mathbf{f}_2(\boldsymbol{u}_n(t)\,, \tilde{\boldsymbol{y}}_m(t)\,,\ \boldsymbol{\Theta}(t), b(t)) \tag{4.55}$$

Observing (4.54) and (4.55) yields an implicit relation for the unknown degradation functions and can be used to determine time series of the latter ones.

$$\mathbf{0} = \mathbf{f}_3(\boldsymbol{u}_n(t)\,,\ \tilde{\boldsymbol{y}}_m(t)\,,\ \boldsymbol{\Theta}_n\,,\ \boldsymbol{\Phi}(t)\,,\ b(t)\,,\ \boldsymbol{r}^1(t)) \tag{4.56}$$

Since the DBG of the second stage represents a linear time-variant model and storage elements are in derivative causality, ARRs are linear or linear after reformulation with respect to the unknown degradation functions. In case parameter fault signatures are unique, then the degradation function $\Phi_i(t)$ of time-varying parameter $\Theta_i(t)$ is given by a unique implicit relation. The system of ARRs can be solved so that the degradation $\Phi_i(t)$ of a time-varying parameter $\Theta_i(t)$ can be expressed as a mode-dependent function $\mathbf{g}_i()$ of the first stage ARR residuals, the inputs and outputs of the non-faulty system model, and the nominal parameters.

$$\Phi_i(t) = \boldsymbol{g}_i(\boldsymbol{r}^1(t), \boldsymbol{u}_n(t), \boldsymbol{y}_m(t), \boldsymbol{\Theta}_n, b(t)) \tag{4.57}$$

Figure 4.14 displays the scheme of the two stage DBG model-based online determination of unknown parameter degradation functions.

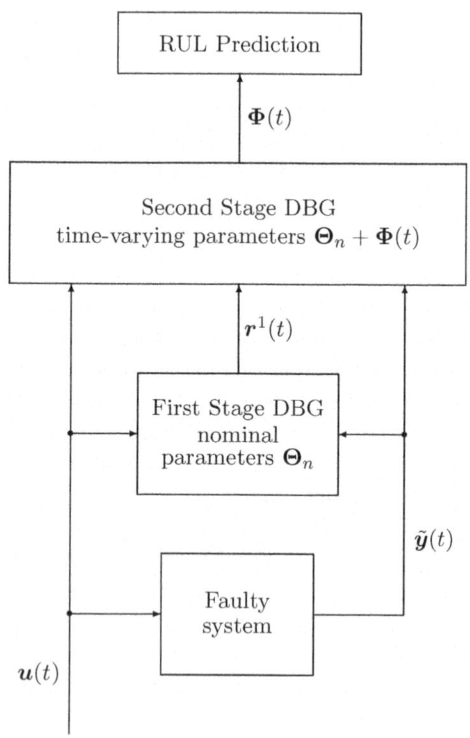

Fig. 4.14 Scheme of the ARR-based online determination of parameter degradation functions

Example: Boost Converter

The example of the boost converter depicted in Fig. 4.8, p. 151 is used again for illustration of this ARR-based approach to a determination of the values of an unknown degradation function.

Derivation of ARRs from the First Stage DBG

Two sensors are available. Accordingly, the following two ARRs can be directly obtained from the DBG of the first stage:

$$r_1 = E - L\frac{d\tilde{i}_L}{dt} - R_L\tilde{i}_L - u_{sw} \tag{4.58}$$

$$r_2 = i_D - i_C - \frac{\tilde{V}}{R} \tag{4.59}$$

Furthermore, as to the capacitor current, i_C, the following two equations are derived from the DBG:

$$i_C = C\dot{u}_C \tag{4.60}$$

$$\tilde{V} = R_C i_C + u_C \tag{4.61}$$

Combining (4.59), (4.60), and (4.61) yields another residual r_2'.

$$\begin{aligned} r_2' &:= r_2 + C R_C\, \dot{r}_2 \\ &= i_D - \frac{\tilde{V}}{R} - C(1 + \frac{R_C}{R})\dot{\tilde{V}} + R_C\, C\frac{di_D}{dt} \end{aligned} \tag{4.62}$$

In the following, the small equivalent series resistance, R_C, of the capacitor is neglected.

Disregarding the causality at the ports of the two switches, an *implicit* equation is set up for each of them.

$$\text{Sw}: \quad 0 = \bar{s}_1 u_{sw} + s_1 i_{sw} \tag{4.63}$$

$$\text{D}: \quad 0 = \bar{s}_2 u_D + s_2 i_D \tag{4.64}$$

where s_1, s_2 denote the discrete switch states and $\bar{s}_i = 1 - s_i$ $(i = 1, 2)$. Taking into account the equations

$$i_{sw} = \tilde{i}_L - i_D \tag{4.65}$$

$$u_D = u_{sw} - \tilde{V} \tag{4.66}$$

obtained from the DBG of the first stage and observing $\bar{s}_s = s_1$, the above implicit switch equations can be solved so that the switch variables can be expressed by measured variables. As a result, the two mode-dependent ARRs obtained from the DBG of the first stage read

$$r_1^1(t) = E - L_n \frac{d\tilde{i}_L}{dt} - R_L^n \tilde{i}_L(t) - s_1(t)\tilde{V}(t) \tag{4.67}$$

$$r_2^1(t) = s_1(t)\tilde{i}_L(t) - C_n \dot{\tilde{V}} - \frac{\tilde{V}(t)}{R_n} \tag{4.68}$$

where the index n indicates nominal parameters.

If the inductor current, $\tilde{i}_L(t)$, and the voltage across the capacitor, $\tilde{V}(t)$, are obtained by measurements of the faulty system, then at least one of the above ARR residuals differs significantly from zero as of some time instant t_1.

Determination of Degradation Functions from ARRs of the Second Stage DBG

If the inductance is deteriorating as of time instant t_1, then only residual $r_1^1(t)$ deviates from zero. The second stage ARR containing the time-varying inductance $L(t) = L_n + \Phi_L(t)$ then reads

$$\begin{aligned} 0 = r_1^2(t) &= E - \frac{d}{dt}(L(t)\,\tilde{i}_L) - R_L^n \tilde{i}_L - s_1(t)\tilde{V} \\ &= \underbrace{E - L_n \frac{d\tilde{i}_L}{dt} - R_L^n \tilde{i}_L(t) - s_1(t)\tilde{V}(t)}_{r_1^1(t)} - \tilde{i}_L(t)\frac{d\Phi_L}{dt} - \left(\frac{d\tilde{i}_L}{dt}\right)\Phi_L \end{aligned} \tag{4.69}$$

Since $\tilde{i}_L(t)$ and $\tilde{V}(t)$ are known as measured outputs of the faulty system model, (4.69) is a linear time-variant ODE for the unknown degradation function $\Phi_L(t)$ that accounts for the system mode.

$$\tilde{i}_L(t)\frac{d\Phi_L}{dt} + \left(\frac{d\tilde{i}_L}{dt}\right)\Phi_L = r_1^1(t) \tag{4.70}$$

If the cause for an abnormal system behaviour is a change of the load resistance, then only residual $r_2^1(t)$ deviates from values close to zero and becomes time-varying as of some time t_2, i.e. $R(t) = R_n + \Phi_R(t) \quad \forall t > t_2$,

From the DBG of the second stage, the following mode-dependent constraint can be derived:

$$0 = r_2^2(t) = \underbrace{s_1(t)\tilde{i}_L(t) - C_n\dot{\tilde{V}} - \frac{1}{R_n}\tilde{V}}_{r_2^1(t)} + \frac{\Phi_R(t)}{R_n(R_n + \Phi_R(t))}\tilde{V} \tag{4.71}$$

If the degradation function $\Phi_R(t)$ were known, the second stage ARR residual r_2^2 would vanish. If $r_2^1 \neq 0$, then (4.71) determines the unknown degradation function $\Phi_R(t)$. Reformulation of (4.71) gives an implicit linear time-varying algebraic equation for $\Phi_R(t)$ that depends on the first stage residual $r_2^1(t)$.

$$\left(\frac{\tilde{V}(t)}{R_n} + r_2^1(t)\right)\phi_R(t) = -r_2^1(t)R_n \tag{4.72}$$

4.5.3 Learning a Mathematical Degradation Model

As illustrated in previous Sect. 4.5.1, equations for estimating the discrete values of a time series of an unknown degradation function can be directly derived from a bicausal BG by following causal paths from sources and detectors in inverted causality to the port of a faulty element. That is, inputs to these equations are only known control inputs and measurements. Numerical values of an unknown parameter degradation function can be computed at sample time points online concurrently to the health monitoring of a real system and the continuous measurement of signals. As soon as n measured values of each needed signal are available and are stored in a buffer holding values in a sliding window, the (filtered) trend of a faulty parameter $\Theta(t)$ can be approximated up to a time instant t_c^k at the end of the current window w_k and can be projected into the future to see when it would intersect with a failure or safety critical threshold. This means that online failure prognostic starts with some delay when some earlier data has been collected and stored in the buffer so that regression of the data available in the current window w_k can take place.

Once computed numerical values of an unknown function of the degradation of parameter Θ_i at sample time points are available, they may be treated like degradation data of a feature extracted from measurement data. Direct measurement of degradation is often not possible without being invasive or destructive. The concurrent computation of a time series of degradation data by an evaluation of equations derived offline from a bicausal diagnostic BG can be considered as the data acquisition phase.

Given n available degradation data $\Theta(t_i^k) = \Theta_n + \Phi_\Theta(t_i^k)$, $(i = 1, \ldots, n)$ obtained from filtered real measurements or simulated measurements in the current k^{th} sliding time window w^k that are stored in a buffer of fixed size, a number of basic mathematical models, i.e. linear, exponential, or power models with parameters to be determined may be tested to see which one fits best the data in the current window w_k. This task of learning a mathematical model can be carried out e.g.

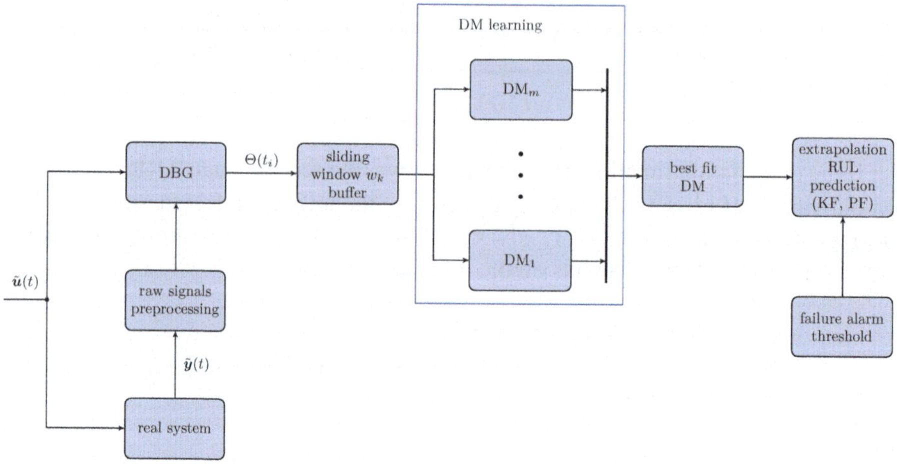

Fig. 4.15 Simultaneous training of various potential degradation models (DMs)

by open source mathematical software such as GNU Octave or by commercial software such as Weibull++ [31] or by the Matlab Predictive Toolbox™ [42] and can be performed in parallel on a multiprocessor, multicore computer. The data set for all tests is the same. For this simultaneous training of various degradation models, a commercial product such as the Matlab Parallel Toolbox™ [7] may be used. As an evaluation criterion for the best fit, the root mean square error (RSME) may be used. Figure 4.15 illustrates the simultaneous training of various potential degradation models. The best fitting model is given to the projection and RUL estimation module.

4.5.4 *Projection and RUL Estimation*

The mathematical function $\Theta(t_i^k) = \Theta_n + \Phi_\Theta(t_i^k)$ approximating the degradation data obtained by parameter estimation can then be projected into the future by means of a Kalman filter to determine a time point t_f^k at which the time evolution of the faulty trending parameter $\Theta(t)$ intersects with a given failure level threshold. The time span from the current time t_c^k (current age of the system) to the time instant t_f^k where the time evolution of the parameter $\Theta(t)$ obtained from degradation data in the kth window w^k intersects with a failure level threshold or a safety critical threshold, gives an estimate of the remaining useful life RUL^k.

$$\text{RUL}^k := t_f^k - t_c^k \tag{4.73}$$

With progressing time new filtered deteriorated values of parameter Θ_i become available while some older values drop out at the tail of the fixed length buffer. The latter holds values of a sliding time window. That is, time windows overlap. For a new time window w^k, the two steps, i.e. the determination of the best fitting degradation model and its extrapolation are repeated. That is, RUL prediction is updated online as new data arrive. Determination of the best fit degradation model for the current time window w^k means that the parameters of the latest degradation model are just updated, or that the degradation function used in the previous window is dismissed and replaced by a member of another class of functions.

As time advances, i.e. the system becomes older, t_c^k takes a new value and the intersection with the failure level threshold gives a new time to failure value. As a result, one obtains a new value for the RUL. Repeating these steps while time is progressing results in a sequence of RUL estimates $\mathrm{RUL}^k(\Theta_i)$, which ultimately converge to zero as a faulty component reaches its EoL. This two step prognosis procedure consisting of a regression analysis of the degradation data in a sliding window w^k and an extrapolation that provides an estimate of the time to failure can be carried out concurrently to the monitoring of a real process, simultaneously for multiple degradation mechanisms that do not compete, and in parallel on a multicore, multiprocessor computer. The global system-level RUL is then the infimum of all component RULs.

Figure 4.16 illustrates the regression analysis of the degradation data in the current sliding window w^k and the extrapolation as of current time instant t_c^k into the future until the parameter degradation trend $\Theta(t)$ intersects with an alarm threshold Θ_{alarm}. The error bars in the current window w_k indicate that the degradation data are obtained by parameter estimation from uncertain measurements. The parameters of the best fitting analytical function are therefore random with a mean and a

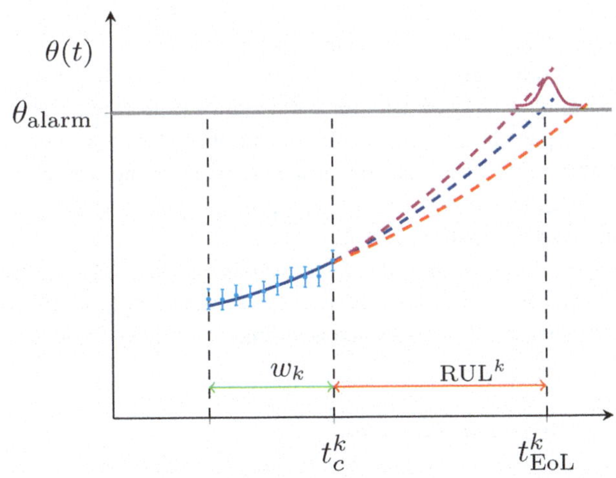

Fig. 4.16 Regression and extrapolation of estimated values of a trending parameter $\theta(t)$

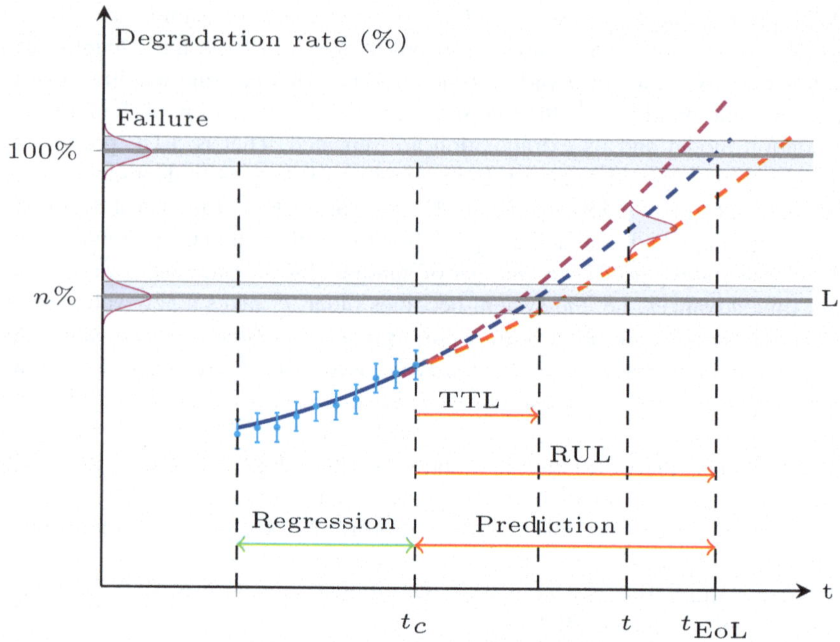

Fig. 4.17 Uncertainties in projection and RUL estimation

variance. This affects the extrapolation such that the RUL is also a random variable. The dashed continuation of the graph indicates that the results of a long-term extrapolation become increasingly uncertain. However, with a moving window, the recursive regression and extrapolation gradually turns into a short-term prediction with more accurate values of the RUL. The bell-shaped pdf above the alarm threshold indicates that the EoL times are random variables.

The bell-shaped pdf at time instant t in Fig. 4.17 emphasises that the prediction of a trending parameter $\theta(t)$ or a feature at a future time point t is uncertain. The result is a random parameter with a pdf. Furthermore, a degradation rate level L below the failure threshold is also uncertain as well as the associated time span, $\text{TTL} \leq \text{TTF} = \text{RUL}$, from the current age, t_c, of the system to the time point where the degradation rate reaches the level L.

Figure 4.18 [13] displays the pdf of the remaining time TTL to the degradation rate level L. The shaded area below the curve represents the probability that TTL is still ahead of a given time T with some confidence level, say of 90%.

Accuracy of Regression and Prediction

The determination of numerical values of an unknown degradation function $\Phi_{\Theta}(t)$ at sample time points and the projection of the degradation behaviour of a time-

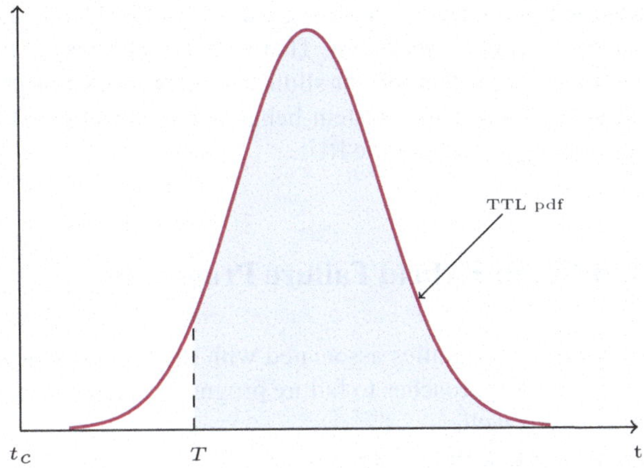

Fig. 4.18 Uncertainty of the remaining time to the degradation rate level L

varying parameter $\Theta(t) = \Theta_n + \Phi_\Theta(t)$ into the future require a sufficient number of degradation data in the current window w^k in order to accurately identify the parameters of a potential degradation model. The amount of available degradation data, i.e. the size of the sliding window, affects the uncertainty in the values of the degradation model parameters and has an effect on the estimation of the time to failure. Commercial software such as Weilbull++ [31] can compute upper and lower bounds for the RUL with a certain confidence level. In order to meet given accuracy requirements for the parameters of the degradation model to be fitted, the size of the sliding window may vary. The boundaries for the RUL become more narrow with progression of time, they build a converging cone as the sliding time window moves on, i.e. the identification of a best fitting degradation model, its extrapolation, and the prediction of the time to failure become more accurate as a faulty component approaches its EoL.

Failure Prognostic for Hybrid Systems

An advantage of a *repeated* identification of a mathematical model for the degradation process is that in the case of a hybrid model for each system mode of operation a possibly different degradation behaviour can be taken into account. In systems represented by a hybrid model, degradation of a component in its ON mode may stop when the component switches into OFF mode. An example may be the mass flow through an increasingly clogged valve. As long as the valve is open, its discharge coefficient, c_d, decreases with time and this increase of the valve's hydraulic resistance can be revealed from continuously sampled measurement data. Evidently, when the valve is switched off, i.e. when this faulty system component becomes inactive, then the last value of the discharge coefficient before closure is retained, degradation is stopped as long as the valve is in OFF mode, i.e. the decline of the time evolution $c_d(t)$ becomes zero. That is, extrapolating the time evolution of the faulty parameter from the current sliding window does not provide a RUL estimate. In that system mode, the system behaviour is not affected by the faulty valve and nothing can be said about the RUL.

4.6 Uncertainties in Hybrid Failure Prognostic

There are a number of uncertainties associated with the two presented bond graph model-based, data-driven approaches to failure prognostic that are common to other prognostic approaches as well.

As Provan points out in [30] back in 2003:

> ... uncertainty is central to any definition of prognosis. This is because a prognosis involves a projection into the future, and we argue that all such future projections must contain some uncertainty, since the future cannot be predicted with certainty.

Modelling Uncertainties

At first, a bicausal BG model from which equations for the determination of numerical values of unknown degradation functions are derived relies on modelling assumptions. Some features of a system or a process may not be fully understood, approximated, neglected, or modelled in a simplified manner. Moreover, model parameters may be uncertain, and numerical computation may start from an uncertain initial system state.

Measurement Uncertainties

Inputs into the constitutive element equations derived from a DBG are noisy, random measurements and known control inputs. Sensors delivering the measurements may be biased, may show drift, or may be incorrectly calibrated. That is, signal preprocessing as discussed in Sect. 3.2, i.e. appropriate filtering of acquired raw measured data is needed that retains essential information and enables feature extraction.

Statistical and Environment Uncertainties

Repeated measurements provide sets of values of a sensor signal at sampling points t_i with a certain distribution that is often not symmetric. Unforeseen environmental changes, future load conditions may entail variations of operating conditions which in turn may result in future faults. Also, a degradation process identified from measurements may change stochastically. Therefore, confidence limits of a predicted RUL may be of more importance than a specific RUL value [37].

Degradation Model Uncertainties

As the fault estimation of a trending parameter $\Theta(t)$ at sample time points t_i in the kth sliding time window w^k uses measurements, the values of an incipient fault, i.e. the degradation values $\Phi_\theta(t)$ at each sample time instant are random with a probability density function (pdf). Accordingly, the parameters of a best fitting degradation function are also to be considered random parameters with a mean value and a pdf, which affects the prediction of the kth failure time point t_f^k and the estimation of RUL^k. The question then is which probability density functions are suited for approximation and how to choose their parameters.

For instance, in the commercial Matlab Predictive Maintenance Toolbox™, the exponential degradation model fitting data of a health indicator $h(t)$ extracted from measured data is defined as

$$h(t) = \phi + \alpha \ \exp\left(\beta t + \epsilon - \frac{\sigma^2}{2}\right) \tag{4.74}$$

where

$h(t)$	:	health indicator as a function of time t
Φ	:	intercept term considered constant
α	:	lognormal-distributed parameter
β	:	Gaussian-distributed parameter
ϵ	:	Gaussian white noise : $\mathcal{N}(0, \sigma^2)$

so that the expectation value $E[h(t)]$ is

$$E[h(t)|\alpha, \beta] = \Phi + \alpha \ \exp(\beta t) \tag{4.75}$$

The distribution of the random parameters α, β of this exponential degradation model is updated at each time step t to the posteriors based on the latest observation of the health indicator. If no historical data is available values for the expectation value and the variance of their priors have to be chosen [43].

Prediction Uncertainties

The result of a long term RUL prediction is not a single value but a random variable with an expectation value and a pdf. A pdf of initial conditions is propagated forward. Let $\varphi_\theta(x \mid t)$ be the pdf of the trending parameter $\theta(t)$ at time instant t, then the probability $\pi[\theta'] := P(\theta(t) \leq \theta')$, i.e. the probability that the magnitude of $\theta(t)$ at time instant t is less than a value θ', is

$$\pi[\theta'] = \int_0^{\theta'} \varphi_\theta(x \mid t)dx \; < \; \int_0^{\theta_{FT}} \varphi_\theta(x \mid t)dx \tag{4.76}$$

where θ_{FT} is the failure threshold.

As has been shown in [35], the RUL pdf may not be a normal distribution even if the trending parameter at sample time points t_i is normally distributed.

Prognostic Metrics

In prognostic, various performance metrics are in use. They enable to quantify the prediction quality, can be used for comparison of the performance of different RUL prediction algorithms, incorporate probability distribution information, and allow for a convenient and intuitive visualisation of algorithm performance. In the following, only the so-called $\alpha - \lambda$ accuracy metric is considered. A more comprehensive presentation addressing also other performance metrics may be found, for instance, in [36, 37], [17, Chap. 5], [45, Chap. 7].

The binary prognostic metric called $\alpha - \lambda$ accuracy determines whether the accuracy of the RUL prediction is within α % of the actual RUL at a specific time instant t_λ expressed as a fraction of the time interval between the prediction time point t_p and the actual failure time point t_{EoL} [17, Chap. 5], [36].

$$\alpha - \lambda \text{ Accuracy} := \begin{cases} 1 & (1-\alpha)\cdot r_*(t_\lambda) \leq r(t_\lambda) \leq (1-\alpha)r_*(t_\lambda) \\ & \text{if} \\ 0 & \text{otherwise} \end{cases} \tag{4.77}$$

where

$\alpha \in [0, 1]$	:	accuracy modifier
$\lambda \in [0, 1]$	:	time window modifier: $t_\lambda = t_p + \lambda(t_{EoL} - t_p)$
t_p	:	time point of the first prediction
$r(t_\lambda)$	:	predicted RUL at t_λ
$r_*(t_\lambda)$	:	ground truth or true RUL

Let $\phi_r(x|t_\lambda)$ denote the pdf of the predicted RUL r at t_λ and

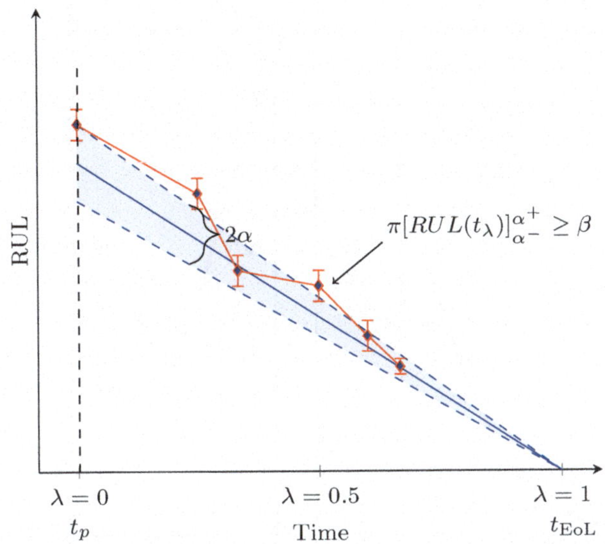

Fig. 4.19 $\alpha - \lambda$ Accuracy

$$\pi[r(t_\lambda)]_{\alpha^-}^{\alpha^+} = \int_{\alpha^-}^{\alpha^+} \phi_r(x \mid t = t_\lambda)dx \tag{4.78}$$

the total probability within the bounds α^- and α^+.

The $\alpha - \beta$ accuracy indicates whether the prediction accuracy at t_λ falls within desired α bounds. Figure 4.19 [36] visualises the $\alpha - \beta$ accuracy where

$\pi[r(t_\lambda)]_{\alpha^-}^{\alpha^+}$: total probability of the predicted $r(t_\lambda)$ between $\alpha^- := r_*(t_\lambda) - \alpha \cdot r(t_\lambda)$ and $\alpha^+ := r_*(t_\lambda) + \alpha \cdot r(t_\lambda)$

β: minimum acceptable probability mass

Figure 4.20 [38] shows the α bounds and the β parameter on a RUL pdf. The normal Gaussian distribution has only been used for illustration.

The tolerated error gets smaller with more narrow α bounds. The higher the lower limit β for the total probability of the predicted RUL between the bounds, the more confident is the prediction.

Risk Assessment

As a RUL is not a single value but is random, risk management introduces a so-called *maximum allowable probability of failure* (PoF), i.e. a maximum acceptable level of risk of equipment failure in order to support maintenance decisions and corrective actions (Fig. 4.21, [9]). The choice of a PoF value depends on the plant to

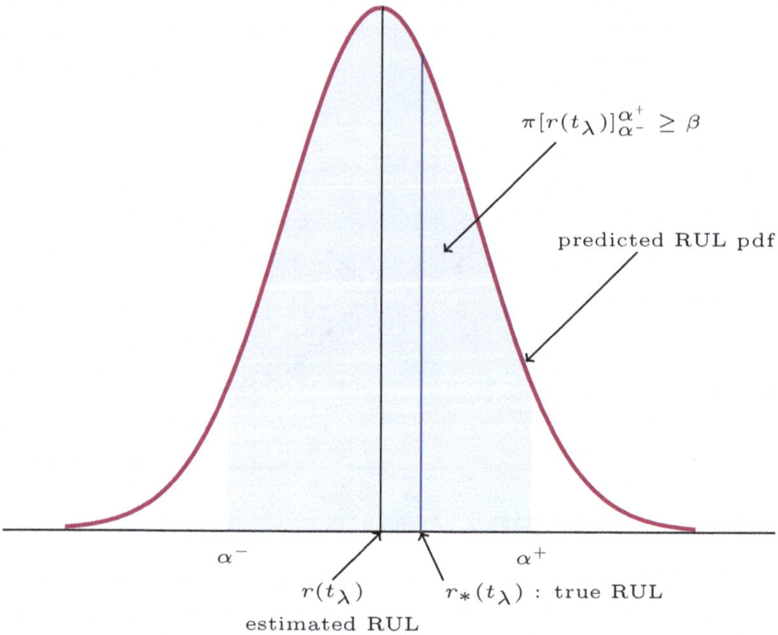

Fig. 4.20 Illustration of the α bounds and the β parameter on a RUL pdf

be operated and on various aspects. The time instant at which this level is reached is called *Just-in-Time-Point* t_{JITP} [9] and gives rise to the introduction of the so-called lead-time interval

$$t_{LTI} := t_{JITP} - t_P \tag{4.79}$$

where t_P denotes the time at which prediction starts. Corrective maintenance actions must be taken before $t_{LTI} < E[RUL]$ elapsed. Such maintenance actions make sure that a plant does not operate beyond a maximum allowable PoF.

Failure Threshold

Prediction of the failure time depends on the failure threshold that has been set. With insufficient a priori knowledge, the choice of a failure alarm threshold below the EoL failure threshold ensuring a safety margin is uncertain so that for a failure threshold a pdf has to be assumed. This is indicated in Fig. 4.17 by a grey envelop around the failure threshold which may be called hazard zone [46]. A proper choice of a failure alarm threshold is crucial as the intersection of an extrapolated

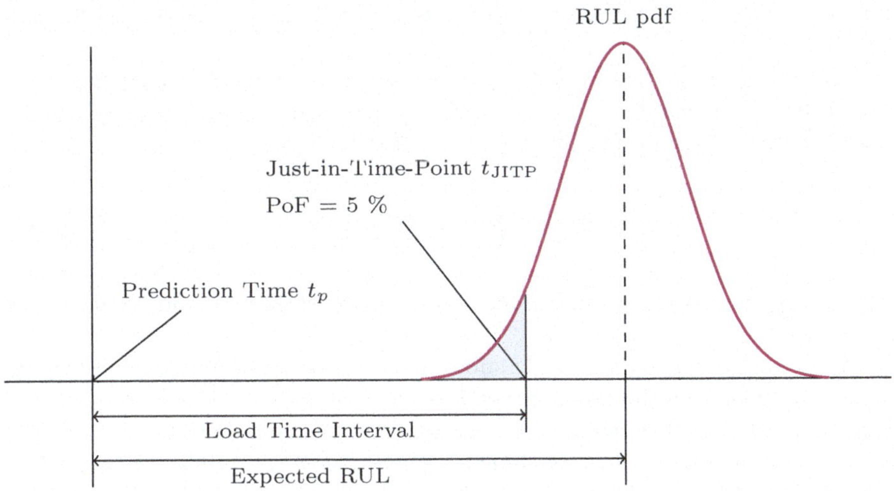

Fig. 4.21 Maximum allowable probability of failure (PoF) and Just-in-Time-Point t_{JITP}

degradation trend provides a time instant t_{alarm} at which a decision on the action to be taken must be made.

Once a best fitting mathematical model $r\theta^1(t) = \theta_n^1 + \Phi_\theta^1(t)$ for the trend of an incipient fault $\theta^1(t)$ has been identified, the faulty behaviour of the system can be simulated. To that end, θ_n^1 is replaced by $r\theta^1(t)$ in the computation of the nominal state space model. Let $y(t)$ be an output signal that indicates the failure of a component or of the system. For instance, if in a hydraulic system a valve is completely clogged, then there is no outflow which may entail that the system does no longer perform its intended function. In electronic systems, a current through a component persistently equal to zero as of some time instant t_{EoL}, or a battery voltage that has reached a critical minimal value indicates its failure. As a result, the system functionality may be reduced or may even cease. Let

$$\dot{\boldsymbol{x}}(t) = \mathbf{f}(\boldsymbol{x}(t), \boldsymbol{u}(t), \boldsymbol{\theta}) \tag{4.80}$$

$$y(t) = g(\boldsymbol{x}(t), \boldsymbol{u}(t), \boldsymbol{\theta}) \tag{4.81}$$

be a state space model derived from a bond graph in integral causality, where $\boldsymbol{x}(t)$ denotes the state, $\boldsymbol{u}(t)$, the vector of known control input signals, and $\boldsymbol{\theta} = (r\theta^1(t), \theta_n^2, \ldots, \theta_n^p)$ the parameter vector. Then a condition

$$y(t_{\mathrm{EoL}}) = y_0 \,, \tag{4.82}$$

where y_0 denotes a value characterising a failure, determines the time instant of failure, t_{EoL}, and $r\theta^1(t_{\mathrm{EoL}})$ is an estimate of the hard failure threshold θ_{FT}^1. The time point t_{EoL} implicitly given by condition (4.82) can be determined through simulation.

Accordingly, an alarm threshold θ_{alarm} that is reached at the decision point $t_{\mathrm{DP}} < t_{\mathrm{EoL}}$ with a safety margin to θ^1_{FT} can be chosen and used in online prognostic of the real engineering system. The probability $\pi_{t_{\mathrm{DP}}}$ for $\theta < \theta_{\mathrm{FT}}$ at t_{DP} is

$$\pi_{t_{\mathrm{DP}}} = \int_0^{\theta_{\mathrm{FT}}} p_\theta(a \mid t_{\mathrm{DP}}) da \tag{4.83}$$

where $p_\theta(a \mid t_{\mathrm{DP}})$ is the pdf of θ at the future time point $t_{\mathrm{DP}} > t_c$ and t_c denotes the current age.

In addition to the above considerations, it should be taken into account that the value of a failure threshold may depend on the system mode of operation and the impact of possibly changing environmental conditions on a system. As a result, a failure threshold may not be just a time independent constant with some uncertainty but may also change with time.

Onset of the Degradation and Start of the Prediction

Furthermore, there is a time delay between the occurrence at time instant t_{oc} and the detection of an incipient parametric fault at t_D, and the start of the prediction at t_P. In [32], a corrected time t_C is obtained by subtracting a margin from t_D that guarantees that the system is in a fault mode at time t_C.

A parameter value $\theta(t)$ deviating from its nominal value θ_n must not only touch constant fault thresholds, $a = \theta_n \pm 3\sigma_0$, with σ_0 denoting the standard deviation of θ, before the onset of degradation, or touch time dependent adaptive fault thresholds, but must increasingly deviate from these boundaries with time in order to be identified as a progressive fault. Collecting sampled data of a trending parameter $\theta(t)$ in a buffer for a sliding time window can start when a fault threshold is touched. Values are checked whether they are outside of fixed or adaptive fault thresholds and whether they show a steady incline or decline. If this is not the case, no incipient fault has happened. Otherwise, regression and extrapolation of an identified degradation trend can take place when the buffer is filled, which means some delay for failure prognosis.

Some Advantages of the Presented Hybrid Method

The estimation of numerical values of an unknown degradation function in the data acquisition phase by evaluating an equation derived from a bicausal BG at sample time points or obtained by using ARRs derived from two DBGs can be performed in parallel for multiple simultaneous parametric faults concurrent to the continuous monitoring of a real process as long as a current fault does not undo a fault that started earlier, in other words, as long as incipient faults can still be detected and isolated.

At the stage of an online learning of a degradation model, the coefficients of a mathematical function that best fit measured data in the current sliding time window are considered random with a mean value and a probability density function. This affects the projection of a learned parameter degradation trend into the future towards the intersection with a failure threshold.

For the fitting of measured degradation data in each consecutive window, w^k, pertaining to a faulty component, the mean value of random parameters of various potentially appropriate basic mathematical degradation functions can be computed in parallel.A criterion such as the root mean square error (RSME) can single out the best fitting function among a set of potential candidates.

The repeated identification of a best fit degradation model for consecutive time windows w^k enables to account for possible changes of the degradation behaviour from window to window that may be due to changes of the system mode of operation or may be caused by system load changes, or changes in the system's environment.

Extrapolating a faulty parameter trend $\theta(t)$ from each time window w^k results in a sequence of RUL values $\mathrm{RUL}(\theta, t_k)$ that tend to zero for $t_k \rightarrow t_{\mathrm{EoL}(\theta)}$.

4.7 Summary

Failure prognostic builds on fault diagnosis and needs some knowledge about the degradation process or at least degradation data extracted from measurements at sampling points that can be projected into the future until a predefined failure alarm threshold is reached.

Data-driven as well as model-based methods for failure prognostic do have their pros and cons. This suggests to combine the benefits of both approaches.

This chapter presents two new hybrid BG-based, data-driven approaches to failure prognosis that do not need to know a mathematical model of the damage process nor assume a priori that the degradation behaviour follows a certain mathematical function for which the coefficients are to be fitted. Instead, discrete numerical values of the unknown degradation model at sampling points are generated by parameter estimation on a bicausal BG, or are obtained by evaluating ARRs derived from two stage one and stage two DBGs. The needed equations are set up offline. Parameter estimation based on a bicausal BG as well as the determination of numerical values of an unknown degradation function based on ARRs is illustrated by means of a boost converter.

The online evaluation of constitutive element equations, or ARRs provides discrete values of a faulty parameter that increasingly deviates from its nominal value with time. The computed values are stored in a buffer for a sliding time window and can be used for *simultaneous training of various degradation models*. A criterion such as the RSME decides on the best fitting degradation model (DM).

Extrapolation of the latter one can be used to compute the intersection with a predefined failure alarm threshold. The learning of a best fitting DM and its extrapolation is repeated while the window is moving forward in time so that a sequence of RUL estimates is obtained.

Section 4.6 addresses some uncertainties such as random parameters in degradation models and probability density functions in RUL prediction. As to the various performance metrics used in prognostic, figures are reproduced that illustrate the $\alpha - \lambda$ accuracy metric.

The chapter concludes with a brief summary of the advantages of the presented hybrid BG model-based, data-driven method for failure prognostic.

References

1. An, D., Kim, N., & Choi, J. (2014). Practical options for selecting data-driven or physics-based prognostics algorithms with reviews. *Reliability Engineering and System Safety, 133*, 223–236. https://doi.org/10.1016/j.ress.2014.09.014.
2. Atamuradov, V., Medjaher, K., Dersin, P., Lamoureux, B., & Zerhouni, N. (2017). Prognostics and health management for maintenance practitioners – review, implementation and tools evaluation. *International Journal of Prognostics and Health Management, 2017*(060). https://www.semanticscholar.org/paper/Prognostics-and-health-management-for-maintenance-Atamuradov-Medjaher/e4c6211ab54ae83c7f1763f37451bf8e1b130dcb.
3. Bartram, G.W. (2013). System Health Diagnosis And Prognosis Using Dynamic Bayesian Networks. Ph.D. Thesis, Vanderbilt University, Nashville.
4. Borutzky, W. (Ed.). (2016). *Bond Graphs for Modelling, Control and Fault Diagnosis of Engineering Systems* (2nd ed.). Switzerland: Springer International Publishing. https://doi.org/10.1007/978-3-319-47434-2.
5. Borutzky, W. (2018). Determination of a function for a degradation process by means of two diagnostic bond graphs. In *Proceedings of the 10th IFAC Symposium on Fault Detection, Supervision and Safety for Technical Processes*, IFAC, Warsaw.
6. Borutzky, W. (2020). A hybrid bond graph model-based – data driven method for failure prognostic. *Procedia Manufacturing, 42*, 188–192. Open-access article.
7. Boston University. (n.d.) Matlab parallel computing toolbox tutorial. https://www.bu.edu/tech/support/research/training-consulting/online-tutorials/matlab-pct/, Online-tutorial.
8. Box, G., Jenkins, G., Reinsel, G., & Ljung, G. (2015). *Time Series Analysis: Forecasting and Control* (5th ed.). Hoboken: Wiley.
9. Butler, S. (2012). Prognostic Algorithms for Condition Monitoring and Remaining Useful Life Estimation. Ph.D. thesis, National University Of Ireland, Maynooth.
10. Celaya, J., Kulkarni, C., Biswas, G., Saha, S., & Goebel, K. (2011). A model-based prognostics methodology for electrolytic capacitors based on electrical overstress accelerated ageing. In *Proceedings of Annual Conference of the Prognostics and Health Management Society*, Open-access article.
11. Danes, M., Ramezani, A., & Moghaddam, J.Z. (2015). Enhanced prognosis of hybrid systems with unknown mode changes. *Modares Journal of Electrical Engineering, 15*(2), 21–26.
12. Dean, T., & Kanazawa, K. (1989). A model for reasoning about persistence and causation. *Computational Intelligence, 5*, 142–150.
13. Dragomir, O.E., Gouriveau, R., Dragomir, F., Minca, E., & Zerhouni, N. (2009). Review of prognostic problem in condition-based maintenance. In *Proceedings of European Control Conference ECC'09*, Budapest (pp. 1585–1592). http://hal.archives-ouvertes.fr/docs/00/41/87/61/PDF/ECC09_0917_FI-1.pdf.

14. Elattar, H.M., Elminir, H.K., & Riad, A.M. (2016). Prognostics: A literature review. *Complex Intelligent Syststems, 2*, 125–154. https://doi.org/10.1007/s40747-016-0019-3, open access.
15. Escobar, L.A., & Meeker, W.Q. (2006). A review of accelerated test models. *Statistical Science, 21*(4), 552–577. https://doi.org/10.1214/088342306000000321.
16. Franke, J., & Neumann, M. (2000). Bootstrapping neural networks. *Neural Computation, 12*, 463–484. https://doi.org/10.1162/089976600300015204.
17. Goebel, K., Saxena, A., Saha, S., Saha, B., & Celaya, J. (2012). Prognostic performance metrics. In *Machine Learning and Knowledge Discovery for Engineering Systems Health Management* (1st ed., pp. 149–171). New York: Chapman & Hall/CRC. https://doi.org/10.1201/b11580.
18. Gouriveau, R., Medjaher, K., & Zerhouni, N. (2016). *From Prognostics and Health Systems Management to Predictive Maintenance 1: Monitoring and Prognostics* (Vol. 4). Wiley. https://doi.org/10.1002/9781119371052.
19. Haykin, S. (2014). *Adaptive Filter Theory* (5th ed.). Pearson, out of print.
20. Heimes, F. (2008). Recurrent neural networks for remaining useful life estimation. In *IEEE 2008 Prognostics and Health Management conference*. https://doi.org/10.1109/PHM.2008.4711422.
21. Jardine, A.K.S., Lin, D., & Banjevic, D. (2006). A review on machinery diagnostic and prognostics implementing condition-based maintenance. *Mechanical Systems and Signal Processing, 20*(7), 1483–1510. https://doi.org/10.1016/j.ymssp.2005.09.012.
22. Jha, M. (2015). Diagnostics and Prognostics of Uncertain Dynamical Systems in a Bond Graph Framework. Ph.D. Thesis, École Centrale de Lille, Université Lille Nord-de-France.
23. Kulkarni, C. (2013). A Physics-Based Degradation Modeling Framework for Diagnostic and Prognostic Studies in Electrolytic Capacitors. Ph.D. Thesis, Graduate School of Vanderbilt University, Nashville
24. Lawrence, S., Giles, C., & Tsoi, A. (1996) What Size Neural Network Gives Optimal Generalization? Convergence Properties of Back-propagation. Technical Report UMIACS-TR-96-22 and CS-TR-3617, Institute for Advanced Computer Studies, University of Maryland
25. Luo, M., Namburu, M., Pattipati, K., & Qiao, L. (2003). Model-based prognostic techniques. In *Proceedings of IEEE Autotestcon 2003* (pp. 330–340)
26. McNaught, K., & Zagorecki, A. (2010). Using dynamic Bayesian networks for prognostic modelling to inform maintenance decision making. *IEEM 2009 – IEEE International Conference on Industrial Engineering and Engineering Management* (pp. 1155–1159). https://doi.org/10.1109/IEEM.2009.5372973.
27. Medjaher, K., & Zerhouni, N. (2013). Hybrid prognostic method applied to mechatronic systems. *International Journal of Advanced Manufacturing Technology 69*(1–4), 823–834. https://doi.org/10.1007/s00170-013-5064-0.
28. Personnaz, I.R.R. (2000). Construction of confidence intervals for neural networks based on least squares estimation. *Neural Networks, 13*, 463–484.
29. Prakash, O., Samantaray, A., Bhattacharyya, R., & Ghoshal, S. (2018). Adaptive prognosis for a multi-component dynamical system of unknown degradation modes. In *Proceedings of 10th IFAC Symposium on Fault Detection, Supervision and Safety for Technical Processes*, IFAC, Warsaw
30. Provan, G. (2003). Prognosis and condition-based monitoring: An open systems architecture. In *5th IFAC Symposium on Fault Detection, Supervision and Safety of Technical Processes* (pp. 81–86). Washington, D.C.: IFAC.
31. ReliaSoft Corporation. (n.d.) Life data analysis reference. Chapter 21: Degradation data analysis. http://reliawiki.org/index.php/Degradation_Data_Analysis.
32. Rozas, H., Claveria, R.M., Orchard, M.E., & Medjaher, K. (2018). Residual-based scheme for detection and characterization of faults in lithium-ion. In *Proceedings of 10th IFAC Symposium on Fault Detection, Supervision and Safety for Technical Processes*. Warsaw: IFAC.
33. Saha, B., & Goebel, K. (2011). Model adaptation for prognostics in a particle filtering framework. *International Journal of Prognostics and Health Management*, Open-access article.

34. Saha, B., Goebel, K., Poll, S., & Christophersen, J. (2009). Prognostics methods for battery health monitoring using a bayesian framework. *IEEE Transactions on Instrumentation and Measurement*, 291–296. https://doi.org/10.1109/TIM.2008.2005965.
35. Sankararaman, S., & Goebel, K. (2013). Why is the remaining useful life prediction uncertain? In *Annual Conference of the Prognostics and Health Management Society* (pp. 1–13), open-access article.
36. Saxena, A., Celaya, J., Saha, B., Saha, S., & Goebel, K. (2009). On applying the prognostic performance metrics. In *Proceedings of the Annual Conference of the Prognostics and Health Management Society*, Open-access article.
37. Saxena, A., Celaya, J., Saha, B., Saha, S., & Goebel, K. (2010). Metrics for offline evaluation of prognostic performance. *International Journal of Prognostics and Health Management, 1*(4), 2153–2648, Open-access Article.
38. Saxena, A., Roychoudhury, I., Celaya, J., & Goebel, K. (2012). Requirements flowdown for prognostics and health management. In *AIAA Infotech at Aerospace Conference and Exhibit 2012*. https://doi.org/10.2514/6.2012-2554.
39. Seabold, S., & Perktold, J. (2010). Statsmodels: Econometric and statistical modeling with python. In *Proceedings of the 9th Python in Science Conference*.
40. Sun, X., Ji, J., Ren, B., Xie, C., & Yan, D. (2019). Adaptive forgetting factor recursive least square algorithm for online identification of equivalent circuit model parameters of a Lithium-Ion battery. *Energies, 12*, 2242. https://doi.org/10.3390/en12122242, Open-access article.
41. Tanwani, A., Domínguez-García, A., & Liberzon, D. (2011). An inversion-based approach to fault detection and isolation in switching electrical networks. *IEEE Transactions on Control Systems Technology, 19*(5), 1059–1074.
42. The Mathworks Inc. (n.d.) Three ways to estimate remaining useful life. https://www.mathworks.com/products/predictive-maintenance.html, white paper.
43. The Matlab Inc. (n.d.) RUL estimation using RUL estimator models – related topics: Wind turbine high-speed bearing prognosis. https://uk.mathworks.com/help/predmaint/ug/rul-estimation-using-rul-estimator-models.html.
44. Usynin, A.V. (2007). A Generic Prognostic Framework for Remaining Useful Life Prediction of Complex Engineering Systems. Ph.D. Thesis, University of Tennessee. https://trace.tennessee.edu/utk_graddiss/319.
45. Vachtsevanos, G., Lewis, F.L., Roemer, M., Hess, A., & Wu, B. (2006). *Intelligent Fault Diagnosis and Prognosis for Engineering Systems*. Hoboken: Wiley.
46. Vachtsevanos, G., Saxena, A., Orchard, M., Clements, S., Goebel, K., Gorospe, G., Kulkani, C.H., Daigle, M., Poll, S., Roychoudhury, I., Sankararaman, S., & Teubert, C. (n.d.) *Introduction to Prognosis*. https://www.phmsociety.org/sites/phmsociety.org/files/PROGNOSTICS_TUTORIAL.pdf, online available.
47. Wang, D., Yu, M., Low, C., & Arogeti, S. (2013). *Model-Based Health Monitoring of Hybrid Systems*. New York: Springer.
48. Wilamowski, B., Iplikci, S., Kaynak, O., & Efe, M. (2001). An algorithm for fast convergence in training neural networks. In *Proceedings of the International Joint Conference on Neural Networks*, IEEE (Vol. 3, pp. 1778–1782). https://doi.org/10.1109/IJCNN.2001.938431.
49. Wileman, A. (2016). An Investigation into the Prognosis of Electromagnetic Relays. Ph.D. Thesis, Cranfield University.
50. Wu, W., Hu, J., & Zhang, J. (2007). Prognostics of machine health condition using an improved ARIMA-based prediction method. In *2007 2nd IEEE Conference on Industrial Electronics and Applications* (pp. 1062–1067)
51. Yung, W., & Man, K. (1993). Optimal selected forgetting factor for RLS estimation. In *IFAC 12th Trienrual World Congress* (pp. 331–334). Sydney: IFAC.
52. Zhang, J.J.J. (2004). A novel variable-length sliding window blockwise least-squares algorithm for on-line estimation of time-varying parameters. *International Journal of Adaptive Control Signal Processing, 18*(6), 505–521.
53. Zhao, K., Ling, F., Levari, H., & Proakis, J. (1994). Sliding window order-recursive least-squares algorithms. *IEEE Transactions on Signal Processing, 42*, 1961–1972.

Chapter 5
Fault Tolerant Control

5.1 Introduction

Fault tolerant control (FTC) builds up on fault diagnosis and is of major importance for systems of systems such as commercial aircrafts and autonomously operating systems. The task is to react to a fault that has been detected and isolated so that the system can continue its operation in the presence of a component fault and to ensure safety at the same time. Required decisions and steps to change the control in response to the detection of a fault are also closely related to failure prognosis. Imagine that a crack in the fuselage of an aircraft has happened or that the battery of a battery powered unmanned aerial vehicle has suffered some degradation of its state of charge, or one of its actuators has failed. Then the severity of the fault and a RUL estimate decide on how the control is changed and whether even the flight mission must be changed.

Fault tolerant strategies are usually categorised into two classes depending on whether they use a passive or an active approach [23]. Passive FTC relies on a robust controller of fixed structure that enables to cope with a set of faults taken into account at the stage of system design. Active FTC (AFTC) is a challenging task because the parameters of a control algorithm or even the algorithm itself is to be changed online by a supervision system. The integration of FDI and reconfigurable control for active fault tolerant control systems have been addressed in [30]. Figure 5.1 [8] displays the general scheme of active FTC.

The following more formal definitions are recalled by using the notation in [27].

Definition 5.1 (Control Problem) A control problem is defined as a triple $< O, C(\boldsymbol{\theta}), U >$, where O is a set of objectives, $C(\boldsymbol{\theta})$ a set of constraints, and U a set of admissible control laws. The parameter vector $\boldsymbol{\theta}$ accounts for uncertainties on which the constraints may depend. The task is to achieve one or more of the objectives while the system behaviour must satisfy given constraints.

W. Borutzky, *Bond Graph Modelling for Control, Fault Diagnosis and Failure Prognosis*, https://doi.org/10.1007/978-3-030-60967-2_5

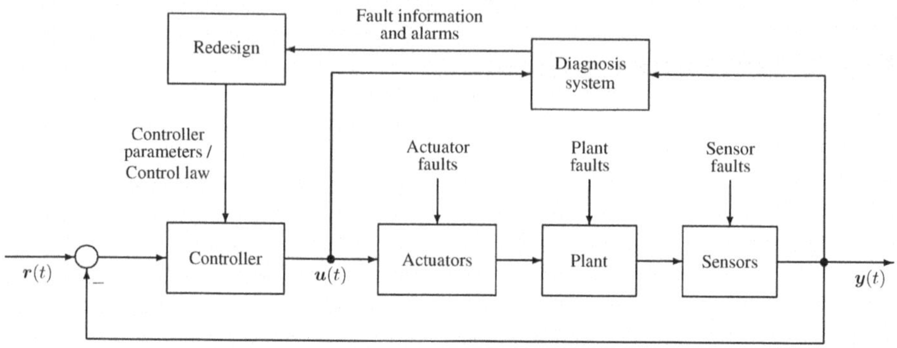

Fig. 5.1 Conceptual scheme of active FTC

Remark 5.1 Faults do change the constraints on a control problem. They transform a control problem $< O, C_n(\boldsymbol{\theta}_n), U >$ with nominal constraints and parameters $C_n(\boldsymbol{\theta}_n)$ into a problem $< O, C_f(\boldsymbol{\theta}_f), U >$, where $C_f(\boldsymbol{\theta}_f)$ denotes a set of constraints and parameters associated with the faulty system. Note that faults do not change the objectives.

Definition 5.2 (Recoverable Fault) A fault is said to be *recoverable* if the objectives that have been set can still be achieved after the faulty control problem $< O, C_f(\boldsymbol{\theta}_f), U >$ has been solved.

Definition 5.3 (Fault Accommodation) Assume that fault detection reveals that the problem to be solved no longer is $< O, C_n(\boldsymbol{\theta}_n), U_n >$. Furthermore, fault isolation has determined the subset of constraints that are unchanged and the subset of control laws $U_f \in U_n$ that can still be used. Let $\hat{C}_f(\boldsymbol{\theta}_f), \hat{U}_f$ be an estimate of the faulty impact. Fault accommodation is then a fault tolerant strategy that solves the problem $< O, \hat{C}_f(\boldsymbol{\theta}_f), \hat{U}_f >$.

Remark 5.2 Fault accommodation means that the faulty system has the same structure as the healthy one. Only the control law is changed. Instead of the initial problem $< O, C_n(\boldsymbol{\theta}_n), U_n >$, the problem $< O, \hat{C}_f(\boldsymbol{\theta}_f), \hat{U}_f >$ is solved. Therefore, fault accommodation is also sometimes called *control reconfiguration*. Fault accommodation is only possible if a fault has been detected, isolated, and estimated. Even if a solution exists, the faulty system described by $C_f(\boldsymbol{\theta}_f), U_f$ may fail to satisfy the objectives O although the available model of the faulty system does.

Definition 5.4 (System Reconfiguration) Suppose that a fault has been detected and isolated but cannot be estimated. System reconfiguration is a fault tolerant control strategy that switches off the detected and isolated faulty system components and tries to achieve the objectives by using only the remaining healthy part of the system. Let $C'_n(\boldsymbol{\theta})$ denote the subset of constraints and U'_n the subset of control laws associated with the healthy part of the system. System reconfiguration then tries to solve the problem $< O, C_n'(\theta), U_n' >$.

Remark 5.3 Unlike fault accommodation, system reconfiguration does not need an estimation of a detected and isolated fault. Once a fault has been detected and isolated, system reconfiguration chooses a new configuration from a set of possible ones and an associated control law. The appropriate control low for a number of possible configurations may have been computed offline and implemented in a bank of controllers so that online only a switching to a control appropriate for the detected and isolated fault may take place. Fault accommodation instead requires an *online* computation of the new control law and a real-time solution of the control problem associated with the faulty system. As long as a solution has not been found, the nominal control is still effective and the fault is not compensated.

Definition 5.5 (Supervision Problem) If neither fault accommodation nor system reconfiguration can solve a control problem, clearly, the system goal must be changed, i.e. the set of possible control objectives O in the triple $< O, C(\boldsymbol{\Theta}), U >$ must be changed, where $\boldsymbol{\Theta}$ denotes a set of parameters that is supposed to be known and comprises nominal and uncertain parameters. This reconfiguration of the objectives is a *supervision problem*. That is, FTC is associated with a decision problem.

Remark 5.4 Often, new objectives are defined by human operators. For instance, depending on the severity of an event, a pilot of an aeroplane may decide to continue the flight on a lower altitude, or to change the mission and to fly to the nearest airport. In an industrial process, the solution of a supervision problem may be the decision that a component needs maintenance.

Bond Graph Modelling and FTC

Bond graph modelling can not only support fault diagnosis and failure prognosis as presented in Chaps. 3 and 4 but can also serve FTC. A bond graph approach to passive fault tolerant control (PFTC) of systems described by continuous time model has been presented in [17]. In active FTC, the change of the controller law after a fault has occurred requires *system inversion*, i.e. to find an input so that the faulty system produces a desired output. One way to decide whether a model is invertible and to determine an input required to produce a desired output is to assign *bicausalities* to a BG (Sect. 2.7) [11, 12, 19, 20].

Samantaray and Bouamama address fault tolerant control of systems represented by a continuous time model and related issues such as system inversion in Chapter 11 of their book on model-based process supervision [24]. Loureiro presents in his PhD thesis [12] a bond graph approach to diagnosis and FTC and applies it to an intelligent autonomous vehicle. A fault tolerant control approach to switched LTI systems has been proposed in [29], which uses bond graphs for mode identification and an observer-based reliable state feedback control focusing on the time delay between FDI and fault accommodation during which the original control law is still applied to the faulty system. The AFTC strategy proposed by Allous and Zanzouri in [1] makes use of an inverse BG model and a Luenberger observer. In Chapter 5 of [9], the author proposes an integration of bond graph-based fault diagnosis and fault accommodation through inverse simulation. A comprehensive presentation of fault tolerant control may be found in [7, 14]. An excellent survey is given in [2, Chap. 2].

5.2 Fault Accommodation Using an Inverse Faulty System Model

Fault accommodation aims at an automatic controller reconfiguration in order to prevent that a severe fault that has been detected by monitoring and fault diagnosis leads to a system or a component failure and to make sure that the dynamic behaviour of the closed loop system remains close to a required reference dynamic behaviour despite a persistent fault. Fault accommodation is an active Fault Tolerant Control (FTC) technique that changes the controller law after a fault has occurred and relies on a faultless operation of the sensors and the actuators [7, Appendix 5], [6, 27]. That is, a system that has become faulty as of some time instant is not repaired immediately and remains in operation due to a modified control signal. The control objective is still achieved, the performance of the faulty system may have, however, deteriorated somewhat.

If some sensors fail, the system becomes partially unobservable and information needed by the controller gets lost. On the other hand, some actuator faults may make the system partially uncontrollable. As a result, a real-time re-adjustment of the controller law linking the same controller inputs and outputs in use before the fault event is not possible any more. Some sensor information fed into the controller as well as some actuators may not be available any more, that is, some control loops are broken so that both a change of the control-loop structure and a new control law is required that accounts for the changed configuration. Reconfigurable FTC requires a fast fault diagnosis and that the system remains observable and controllable with the sensors and actuators still in operation. Nevertheless, it always takes some time to detect and to isolate a fault. Accordingly, there is a time span, the time to reconfiguration, in which the system behaviour is faulty but the system input required to compensate the fault has not been reconstructed yet. That is, the control law for the healthy system is still effective so that a signal is sent to one of the actuators that has become faulty which may affect the stability of the closed loop system. In this chapter, it is assumed that sensors and actuators do not fail. Faults may only occur in the plant.

For closed loop systems with a nominal feedback controller, FDI (robust with regard to uncertain parameters) is not an easy task as its objectives are opposite to those of the controller so that there is a trade-off between good closed loop performance and good fault detection. The problem of FDI in closed loop systems has been addressed, for instance, in [4, 21].

Small multiplicative faults in the plant and additive faults in its actuators as well as small disturbances on the system are compensated by a feedback controller and will not be relevant as long as the control error is forced to values close to zero with time. Slightly faulty monitored values of a controlled variable due to a small sensor offset are also not detected in case there are no redundant sensors so that a voting among multiple sensors of the same signal is not possible. The feedback controller will make a controlled variable provided by a faulty sensor equal to the reference.

Large faults either in the plant or in its sensors and actuators may exceed the robustness of a feedback controller and can lead to a degradation of the normal dynamic closed loop behaviour and can impair safety or reliability so that an active FTC is required for a fast recovery from a severe fault and for maintaining stability and accuracy. Once a large fault has occurred, a nominal closed loop controller aims at compensating the fault which takes some time. During this time, a fault indicator will exceed a fault threshold so that there is time to detect the fault. However, it is important that fault detection, isolation, and input reconstruction are fast.

Now, suppose that a large fault has happened, that its effect on the dynamic behaviour of a closed loop system has been detected and that the fault has been isolated. The input required for its accommodation then can be determined by means of an *inverse model* that uses the identified faulty parameter and the desired system output as input. The block diagram in Fig. 5.2 [9] displays the scheme of fault accommodation based on an inverse model.

As long as there is no fault, switch 2 is open and the output of the real system $\tilde{\boldsymbol{y}}$ equals the desired output $\boldsymbol{y}_{\text{des}}(t)$ of the healthy system model $\tilde{\boldsymbol{y}}(t) = \boldsymbol{y}_{\text{des}}(t)$ which serves as a reference. (The output of the inverse model equals the command variable.) When an abrupt fault, f, has happened in the system at time instant t_1, it takes some time Δt to detect and to isolate it. Meanwhile, the dynamic output behaviour of the faulty system, $\tilde{\boldsymbol{y}}(t)$, deviates increasingly from the desired output behaviour $\boldsymbol{y}_{\text{des}}(t)$ and the nominal feedback controller aims at compensating the error. Once the fault has been detected and isolated, the faulty parameter, $\tilde{p}$, is fed into the inverse system model so that the latter one becomes an inverse faulty system model. With the information about the fault and the output of the healthy system model as input, $\boldsymbol{y}_{\text{des}} = \boldsymbol{u}^*$, the inverse faulty system model provides an output signal $\boldsymbol{y}^* = \boldsymbol{u}_{\text{req}}$. By closing switch 2, this signal is added to the output of the nominal feedback controller at time instant $t_2 > t_1$ (cf. [22]). At this time instant, the error has already been somewhat reduced by the nominal controller. The sum of both inputs into the faulty system drives the error to zero and forces the faulty system to behave as the healthy system despite the fault. The forward model of the healthy system and the inverse faulty system model constitute an additional inversion-based feedforward controller which becomes active in the case of a fault and provides $\boldsymbol{u}_{\text{req}}$ as an input into the faulty system. The fault compensated system then can be treated as a normally operating system. In case there is a small disturbance on the faulty

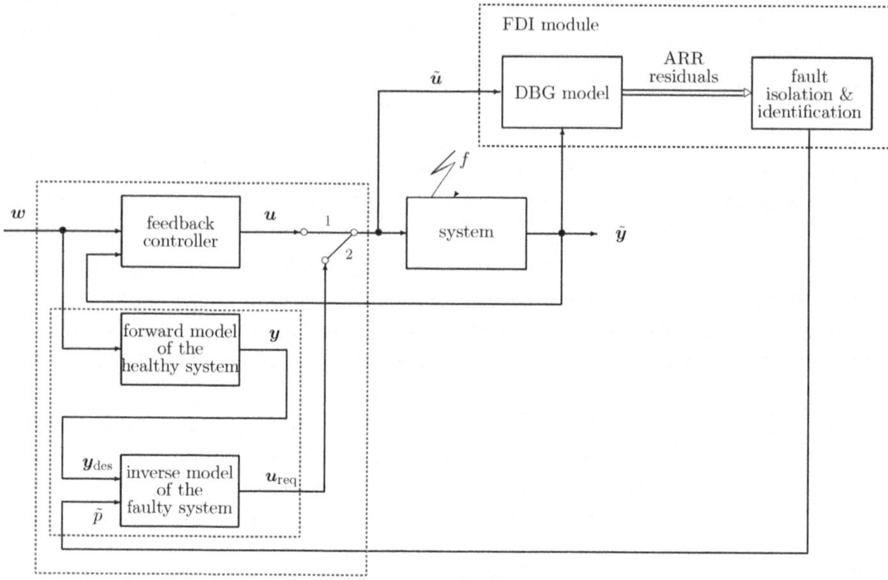

Fig. 5.2 Fault accommodation based on an inverse model

system after the fault has been accommodated, it is compensated by the feedback controller.

There are various approaches to system inversion. The input to be reconstructed can be obtained by

- designing a feedback system (proper inversion) [10, 16],
- numerical solution of a DAE system [28],
- inverse simulation [9, Chap. 5], [13]
- application of bicausality to a bond graph [11, 12, 18]

A review of inverse simulation methods may be found in [15, Chap. 4].

5.3 Implicit System Inversion

Given a LTI SISO forward model of a system, a well known classical approach to determine its inverse model *implicitly* is to place the forward model into a high gain feedback loop as depicted in Fig. 5.3.

In Fig. 5.3, the signal block K denotes the feedback gain and $G_f(s) = 1/D_f(s)$ with a polynomial $D(s)$ in $s \in \mathbb{C}$ the transfer function of a low pass filter which is necessary for a proper inverse model. Let $G_p(s) = N(s)/D(s)$ be the transfer function of the system. Then, its inverse $G_{\text{inv}} = D(s)/N(s)$ cannot usually be implemented as the order of the polynomial $D(s)$ is greater than the one of the

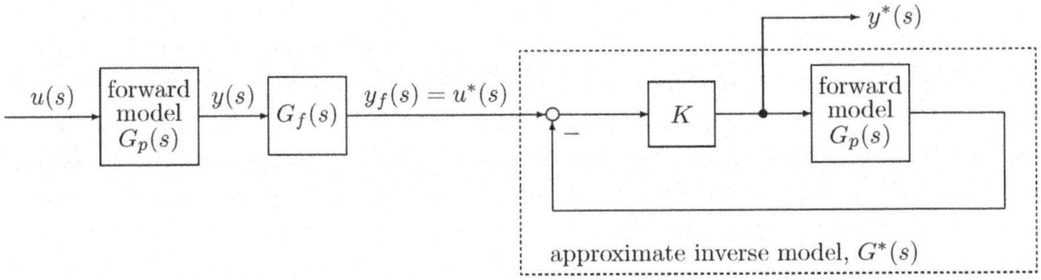

Fig. 5.3 Implicit system inversion scheme

polynomial $N(s)$. That is, the inverse system is not proper. The transfer function of the low pass filter then adds some high frequency poles to the inverse system making it proper.

From the block diagram in Fig. 5.3 one obtains immediately for a finite feedback gain K an approximate inverse $G^*(s) := y^*(s)/u^*(s)$ with

$$y^*(s) = \frac{1}{\frac{1}{K} + G_p(s)} u^*(s) = \frac{G_p(s)\, G_f(s)}{\frac{1}{K} + G_p(s)} u(s) \tag{5.1}$$

For $K \to \infty$, the output, $y^*(s)$, of the approximate inverse system equals the filtered input $u_f(s) := G_f(s)u(s)$ and $G^*(s) \to G_p^{-1}(s) = G_{\text{inv}}(s)$.

For a finite gain K

$$\begin{aligned} y^*(s) &= G^*(s)u^*(s) = G^*(s)G_f(s)y(s) \\ &= \frac{K D(s)}{D(s) + K N(s)} \cdot \frac{1}{D_f(s)} y(s) \end{aligned} \tag{5.2}$$

That is, the transfer function of the prefilter makes the inverse system proper.

To make sure that $y^*(t) = u(t)$ for $t \to \infty$, the prefilter and a finite feedback gain K must fulfil the condition

$$G_p(0)\, G_f(0)\, G^*(0) = G_f(0)\, G_p(0) \frac{K}{1 + K G_p(0)} = 1 \tag{5.3}$$

where $G^*(0)$ is the stationary gain of $G^*(s)$. Thus,

$$G_f(0) = 1 + \frac{1}{K G_p(0)} \tag{5.4}$$

For $K \to \infty$, the gain of the prefilter, $G_f(0)$, approaches 1, which means that for a sufficiently high feedback gain K, the output of the inverse system is sufficiently close to the input signal of the system anyway regardless of the prefilter.

The implicit inversion approach is also applicable to LTI MIMO systems. The scalar feedback gain then becomes a matrix. The forward model used by the implicit inversion scheme may be a BG in preferred integral causality. In the case of an LTI MIMO system $\Sigma(\mathbf{A}, \mathbf{B}, \mathbf{C}, \mathbf{D})$, the matrices of a state space model can be derived from the causal BG. Once the matrices of the forward model are available, the matrices of the implicitly determined inverse model can be easily obtained analytically. For a proper LTI MIMO system ($\mathbf{D} = \mathbf{0}$), the matrices of the inverse model marked by a star read [10]

$$\mathbf{A}^* = \mathbf{A} - \mathbf{B}\,\mathbf{K}\,\mathbf{C} \tag{5.5a}$$

$$\mathbf{B}^* = \mathbf{B}\,\mathbf{K} \tag{5.5b}$$

$$\mathbf{C}^* = -\mathbf{K}\,\mathbf{C} \tag{5.5c}$$

$$\mathbf{D}^* = \mathbf{K} \tag{5.5d}$$

The matrices in (5.5) may be computed by a mathematical software package such as GNU Octave. Alternatively, the implicit inversion scheme with a BG forward model can be directly used in an offline simulation to validate a control law reconstructed for fault accommodation.

It is convenient that the implicit inversion scheme uses the forward model of a system. A disadvantage may be that a high feedback gain introduces high frequencies into the inverse model, i.e. increases its bandwidth. High frequencies effects on the output introduced by the high feedback gain are, however, reduced by the low pass prefilter.

5.4 Input Reconstruction from a Bicausal Bond Graph of the Inverse Faulty System

Fault accommodation as depicted in Fig. 5.2 needs an inverse model in which a faulty parameter isolated and estimated by the FDI module has been inserted. One option to determine the input required for compensating a parametric fault is to derive the equations of a forward model from a behavioural BG and to consider a component $u_{\text{req}}(t)$ of the input vector $\boldsymbol{u}(t)$ as unknown and a component $y_{\text{des}}(t)$ of the output vector $\boldsymbol{y}(t)$ as known. In general, the result is an implicit DAE system. It may happen that some states of the faulty system can be expressed as a function of the remaining states, the input $\boldsymbol{u}_{\text{req}}$ to be determined, the desired output $\boldsymbol{y}_{\text{des}}$, and its time derivatives. Depending on the structure of the DAE system and the nonlinearities involved, all unknown states of the faulty system may even be eliminated. As a result, the required input $\boldsymbol{u}_{\text{req}}$ can be expressed by the desired output $\boldsymbol{y}_{\text{des}}$ and its derivatives [9, Chap. 5].

Another option that does not need to isolate and to estimate a parametric fault is to solve ARRs derived from a DBG of the forward model for the required input

u_{req}. When a fault happens, the residual of one of the ARRs differs significantly from zero. This ARR may depend on more than one component parameter so that the fault is detected but not isolated. However, for input reconstruction it is sufficient just to know the value of the ARR residual. The faulty behaviour may be caused by one of the elements bonded directly or indirectly to the sensor junction for which the ARR has been set up [9, Chap. 5].

A different active FTC approach reported in [1] uses a Luenberger observer and inserts output residuals into the control loop and by that way also does not need fault isolation and estimation for input recovery.

Example: Input Reconstruction on the Bicausal BG of a DC Motor Drive

This section illustrates input reconstruction on a bicausal BG by means of the open loop DC motor drive taken from [9, Chap. 5]. It is assumed that a single parametric fault has been detected, isolated, and estimated. As demonstrated in Sect. 2.7, an inverse model with respect to an input, output pair of signals can be obtained from a bicausal BG.

Figure 5.4 reproduces the schematic and Fig. 5.5 shows an averaged forward BG model, where $R := R_L + R_{on}$ and R_{on} denotes the ON-resistance of the switch modelling the transistor Q. The signal $m(t)$ switches the transistor on and off with a duty cycle d.

As can be seen from the forward BG in Fig. 5.5, there is an I/O power line associated with an I/O causal path from the voltage source Se : E to the flow detector Df : ω and another power line from the modulated source MSe : T to the flow detector Df : ω. These I/O power lines are not disjoint; the two of them pass through junction 1_3. However, if the load torque T is observed and an effort detector is inserted, then the I/O power lines are disjoint. The power lines from E to the detector De : T and from T to ω are not considered as they are not of minimal length. The BG fragment of the DC motor drive in Fig. 5.6 indicates the two disjoint I/O power lines from Se : E to SS : ω and from MSe : T to SS : T highlighted in red and in blue, respectively.

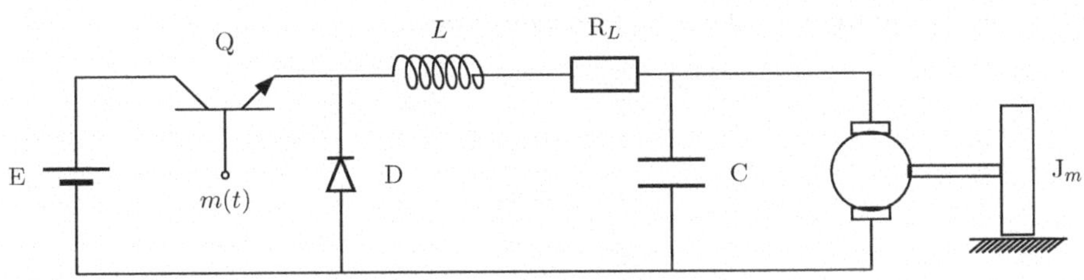

Fig. 5.4 Schematic of an open loop DC motor drive [9]

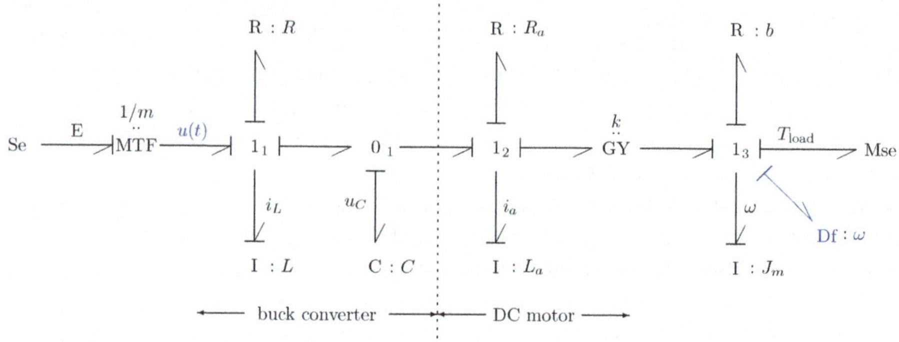

Fig. 5.5 Averaged forward BG model of the open loop DC motor drive in Fig. 5.4

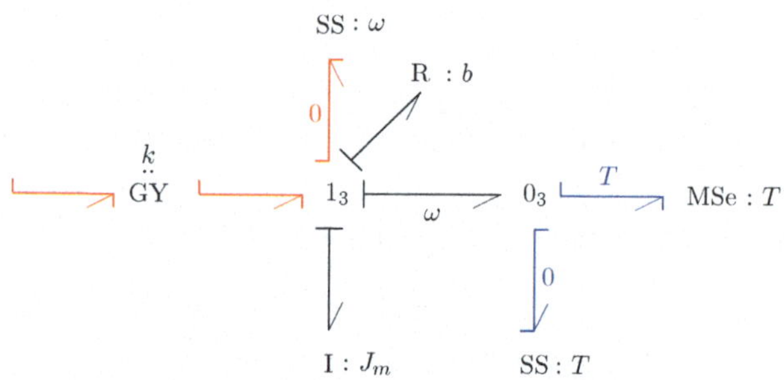

Fig. 5.6 Disjoint I/O power lines from Se : E to SS : ω and from MSe : T to SS : T

That is, the forward model of the DC motor drive is *structurally* invertible with respect to the signal pair E, ω. Algebraic invertibility of a general forward LTI model depends on its parameters so that a theorem such as the one of Moylan (Theorem 2.4), p. 34 or the one of Sain-Massey (Theorem 2.5), p. 35 must be checked.

In the bicausal BG of the faulty system with a faulty armature resistance $\tilde{R}_a = R_a^n + \Delta R_a$ depicted in Fig. 5.7, the effort source Se : E and the flow detector Df : ω are replaced by the non-standard source sensor element SS.Following the bicausal path from the flow source sensor SS : ω to the effort source sensor SS : E, the subsequent equations can be derived, in which state variables carry a tilde to distinguish them from the states of the healthy system.

$$1_3 : \qquad e_2 = b\,\omega_{\text{des}} + J_m \dot{\omega}_{\text{des}} + T_{\text{load}} \tag{5.6}$$

$$GY : \qquad e_3 = k\,\omega_{\text{des}} \tag{5.7}$$

$$\tilde{i}_a = f_3 = \frac{1}{k} e_2 \tag{5.8}$$

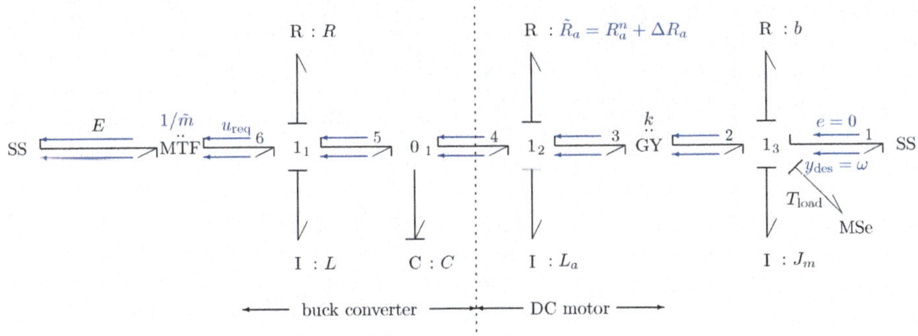

Fig. 5.7 Bicausal DBG of the buck-converter-DC motor

$$1_2 : \tilde{u}_C \;=\; e_4 = \tilde{R}_a\, i_a + L_a \frac{di_a}{dt} + e_3 \tag{5.9}$$

$$0_1 : \;\tilde{i}_L \;=\; f_5 = f_3 + C\dot{e}_4 \tag{5.10}$$

$$1_1 : \qquad e_6 = R\, f_5 + L\frac{df_5}{dt} + e_5 \tag{5.11}$$

$$MTF : \qquad u_{\text{req}} = \tilde{m}\, E \tag{5.12}$$

Let $y_{\text{des}} = \omega_{\text{des}} = \omega = y$, $u := mE = (1-d)E$ in case $\Delta R_a = 0$, $u_{\text{req}} = \tilde{m}E = (1-\tilde{d}(t))E$ in case $\Delta R_a \neq 0$, $\tilde{x}_1 := \tilde{i}_L$, $\tilde{x}_2 := \tilde{u}_C$, $\tilde{x}_3 := \tilde{i}_a$, $\tilde{x}_4 := \tilde{\omega} = \omega$. Then the equations derived from the bicausal BG may be formulated as an implicit DAE for the inverse faulty model.

$$\begin{bmatrix} L & 0 & 0 & 0 & 0 \\ 0 & C & 0 & 0 & 0 \\ 0 & 0 & L_a & 0 & 0 \\ 0 & 0 & 0 & J_m & 0 \\ 0 & 0 & 0 & 0 & 0 \end{bmatrix} \begin{bmatrix} \dot{\tilde{x}}_1 \\ \dot{\tilde{x}}_2 \\ \dot{\tilde{x}}_3 \\ \dot{\tilde{x}}_4 \\ \dot{u}_{\text{req}} \end{bmatrix} = \begin{bmatrix} -R & -1 & 0 & 0 & 1 \\ 1 & 0 & -1 & 0 & 0 \\ 0 & 1 & -\tilde{R}_a & -k & 0 \\ 0 & 0 & k & -b & 0 \\ 0 & 0 & 0 & 1 & 0 \end{bmatrix} \begin{bmatrix} \tilde{x}_1 \\ \tilde{x}_2 \\ \tilde{x}_3 \\ \tilde{x}_4 \\ u_{\text{req}} \end{bmatrix} + \begin{bmatrix} 0 & 0 \\ 0 & 0 \\ -1 & 0 \\ 0 & 1 \end{bmatrix} \begin{bmatrix} T_{\text{load}} \\ -y_{\text{des}} \end{bmatrix} \tag{5.13}$$

As the assignment of bicausality along the power line from ω to E forces all storage elements into derivative causality, the inverse model is stateless so that the required input u_{req} only depends on the inputs and their time derivatives. In fact, in this example, all unknown states of the faulty system can be eliminated and the input u_{req} required for compensation of the fault can be expressed by the desired output y_{des} and its time derivatives, where y_{des} equals the angular velocity of the healthy system, and by the load torque T_{load} and its time derivatives.

Assume that it is the armature resistance that has abruptly increased as of some time instance, i.e. $\tilde{R}_a = R_a + \Delta R_a$ so that a change of the duty cycle d of the signal controlling the transistor Q is necessary to change the output voltage of the buck converter and to accommodate the fault. This fault cannot be isolated just by structural inspection of a FSM even if a detector in inverted causality is connected to every junction of the DBG in Fig. 5.7.

From (5.6) and (5.8) one can see that $\tilde{i}_a = i_a$. The changed output voltage $\tilde{u}_C$ then reads

$$\begin{aligned}\tilde{u}_C &= (R_a + \Delta R_a)\tilde{i}_a + L\frac{d\tilde{i}_a}{dt} + ky \\ &= \underbrace{R_a i_a + L\frac{di_a}{dt} + ky}_{u_C} + (\Delta R_a)i_a \end{aligned} \tag{5.14}$$

Furthermore,

$$\begin{aligned}\tilde{i}_L &= i_a + C\dot{\tilde{u}}_C \\ &= i_a + C\left[\dot{u}_C + (\Delta R_a)\frac{di_a}{dt}\right] \\ &= \underbrace{i_a + C\dot{u}_C}_{i_L} + C(\Delta R_a)\frac{di_a}{dt} \end{aligned} \tag{5.15}$$

Finally, one obtains for the required input signal

$$\begin{aligned} u_{\text{req}} &= R_L\tilde{i}_L + L\frac{d\tilde{i}_L}{dt} + \tilde{u}_C \\ &= R_L\left[i_L + C(\Delta R_a)\frac{di_a}{dt}\right] + L\left[\frac{di_L}{dt} + C(\Delta R_a)\frac{d^2 i_a}{dt}^2\right] + u_C + (\Delta R_a)i_a \\ &= \underbrace{R_L i_L + L\frac{di_L}{dt} + u_C}_{mE} + \underbrace{\left[i_a + (RC)\frac{di_a}{dt} + (LC)\frac{d^2 i_a}{dt^2}\right](\Delta R_a)}_{=:\, i} \\ &= mE + \frac{1}{k}[(b\,y + J_m\dot{y}) + (RC)(b\,\dot{y} + J_m\ddot{y}) + (LC)(b\ddot{y} + J_m y^{(3)})]\Delta R_a \\ &\quad + \frac{1}{k}[T_{\text{load}} + (RC)\dot{T}_{\text{load}} + (LC)\ddot{T}_{\text{load}}](\Delta R_a) \end{aligned} \tag{5.16}$$

As a result, the required input $u_{\text{req}}(t)$ is a function of the initial input $u = mE = (1-d)E$, the desired output $y_{\text{des}}(t) = \omega(t)$ and its derivatives and of the load torque $T_{\text{load}}(t)$ and its derivatives. The time dependent modulus $\tilde{m}(t)$ reads

$$u_{\text{req}} = mE + i\Delta R_a$$
$$= \underbrace{\left(m + i\frac{\Delta R_a}{E}\right)}_{\tilde{m}(t)} E \qquad (5.17)$$

In case there is no fault ΔR_a, the required input equals the initial input.

The result (5.16) relating the required input $u_{\text{req}}(t)$ to the desired output $y_{\text{des}}(t)$ and its derivatives up to the third order obtained for this example, is not really surprising. Let

$$\dot{\boldsymbol{x}}(t) = \mathbf{A}\boldsymbol{x}(t) + \mathbf{B}\boldsymbol{u}(t) \quad \boldsymbol{x}(0) = \mathbf{0} \qquad (5.18a)$$
$$\boldsymbol{y}(t) = \mathbf{C}\boldsymbol{x}(t) + \mathbf{D}\boldsymbol{u}(t) \qquad (5.18b)$$

be a general forward LTI state space model with $\boldsymbol{x}(t) \in \mathbb{R}^n$, $\boldsymbol{u}(t) \in \mathbb{R}^m$, $\boldsymbol{y}(t) \in \mathbb{R}^m$, and constant coefficient matrices $\mathbf{A}$, $\mathbf{B}$, $\mathbf{C}$, $\mathbf{D}$ of appropriate dimensions and assume that the inverse model exists. Then by performing successive differentiation and algebraic transformations of the output $\boldsymbol{y}(t)$, classical inversion methodology has shown that the input $\boldsymbol{u}(t)$, in general, can be expressed in terms of a generalised state $\boldsymbol{z}(t)$, the output $\boldsymbol{y}(t)$ and its derivatives [25, 26]:

$$\dot{\boldsymbol{z}}(t) = \mathbf{F}\boldsymbol{z}(t) + \mathbf{G}(p)\boldsymbol{y}(t) \qquad (5.19)$$
$$\boldsymbol{u}(t) = \mathbf{H}\boldsymbol{z}(t) + \mathbf{J}(p)\boldsymbol{y}(t) \qquad (5.20)$$

where $\mathbf{F}$, $\mathbf{H}$ are constant coefficient matrices, $\mathbf{G}(p)$ and $\mathbf{J}(p)$ are polynomial matrices in the differential operator $p := d/dt$. The vector $\boldsymbol{z}$ is a generalized state vector of dimension $r \leq n$.

5.5 Passive Fault Tolerant Control by Means of an Overwhelming Controller

Given information provided by a fault diagnosis module, the challenge in active FTC is to change the parameters of a controller or to even generate a new controller law online in order to accommodate a fault. In contrast, passive FTC uses a controller of fixed structure that has been designed so that it is robust with regard to parameter uncertainties, ensures stability and the achievement of the control objectives in the presence of faults whose effects were studied and taken into account in the design phase. One possible approach to passive FTC is to design a so-called overwhelming controller implemented in software [3, 5, 24]. The scheme of a system driven by an overwhelming controller is depicted in Fig. 5.8.

An overwhelming controller is typically used for tasks such as robust trajectory control, or force control. The controller delivers a signal that causes an overwhelming of the dynamic behaviour of a system and forces it to follow a prescribed

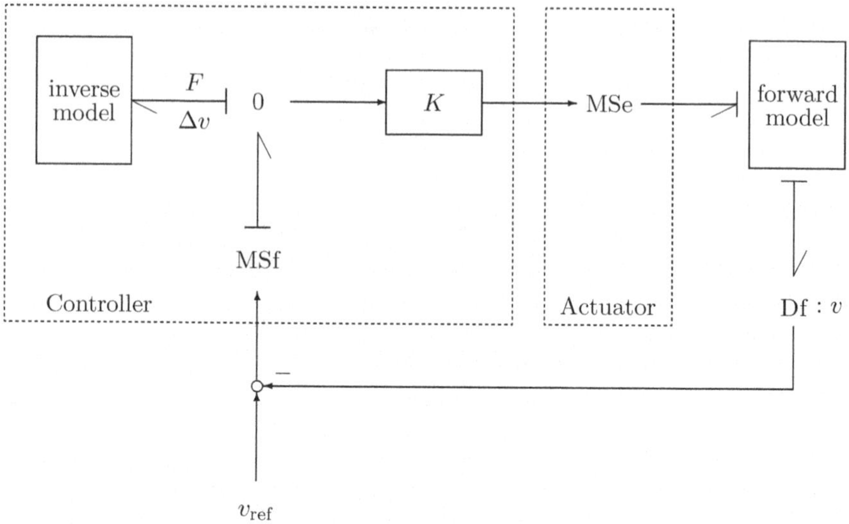

Fig. 5.8 System driven by an overwhelming controller

reference, which makes the controller suitable for passive FTC. When a parametric fault occurs that affects the dynamic behaviour of the system, the controller forces the system to follow the prescribed reference signal despite the fault.

In Fig. 5.8, the reference is chosen to be a velocity. The controller output is a force that makes sure that the system follows the reference. Accordingly, as can be seen from the block diagram, the controller uses an *inverse model* of the system and has a gain parameter K. The inverse model inside the controller is usually of reduced order. Some dynamic effects may have been neglected and the values of the inverse model parameters may differ from the ones of the forward model. The controller, however, still performs robustly if $K \gg 1$.

Bond graph methodology is well suited for the design of an overwhelming control. For the system to be controlled, a BG in preferred integral causality can be systematically developed. The forward BG model can be checked for structural invertibility. The BG of the forward model can then be turned into a bicausal BG from which the required output of the inverse model can be derived.

Let $G_{\text{inv}}(s) = F(s)/\Delta v(s)$ be the transfer function of the inverse model and $G_p(s) = v(s)/(KF(s))$ the transfer function of the forward model. From the BG in Fig. 5.8, the following equations are immediately derived:

$$G_{\text{inv}}(s)\Delta v(s) = \frac{v(s)}{KG_p(s)} \tag{5.21}$$

$$G_{\text{inv}}(s)v_{\text{ref}}(s) = \left[G_{\text{inv}}(s) + \frac{1}{KG_p(s)}\right] v(s) \tag{5.22}$$

$$G(s) := \frac{v(s)}{v_{\text{ref}}(s)} = \frac{G_{\text{inv}}(s)}{G_{\text{inv}}(s) + \dfrac{1}{K G_p(s)}} \tag{5.23}$$

$$G(s) \to 1 \quad \text{for} \quad K \to \infty$$

That is, the system output follows a prescribed reference trajectory.

5.6 Summary

Beyond FDI and failure prognosis, bond graph modelling can also support FTC. After briefly recalling some fundamental notions such as fault recovery, fault accommodation, and system reconfiguration, this chapter focuses on a BG-based input reconstruction for fault accommodation.

Section 2.7 presents an approach based on the bicausal BG concept introduced by Gawthrop [11] and elaborated for model inversion by Ngwompo [19]. Advantages of this physical bicausal bond graph model-based inversion is that *structural* invertibility can be checked on a forward system BG in preferred integral causality without deriving mathematical equations. Furthermore, after assignment of bicausality to the acausal forward BG, the equations of the inversion faulty system model can be derived.

This chapter uses the bicausal BG concept for input reconstruction. The inversion of a faulty system model is illustrated on the example of an open loop DC motor drive powered via a buck converter. It is assumed that a single fault has occurred as of some time instant and that this fault has been detected, isolated, and estimated. In Chap. 7 on applications, fault accommodation for the DC motor drive based on a bicausal BG of the faulty system is validated by two offline simulation case studies.

The chapter concludes by briefly considering the principle of an overwhelming controller as the controller is an example of a controller suited for passive FTC.

References

1. Allous, M., & Zanzouri, N. (2014). Active fault tolerant control based on bond graph approach. *Advances in Electrical Engineering*, Article ID 216153. https://doi.org/10.1155/2014/216153.
2. Alwi, H., Edwards, C., & Pin Tan, C. (2011). *Fault Detection and Fault-Tolerant Control Using Sliding Mode*. Springer. http://www.springer.com/978-0-85729-649-8.
3. Arora, R., & Bera, T. K. (2017). Trajectory tracking through overwhelming control of human vertebrae as a hybrid manipulator. *Simulation: Transactions of the Society for Modeling and Simulation International, 93*, 1–13.
4. Baïkeche, H., Marx, B., Maquin, D., & Ragot, J. (2006). On parametric and nonparametric fault detection in linear closed-loop systems. In: *Workshop on Advanced Control and Diagnosis, ACD'2006*, Nancy, CDROM.

5. Bera, T. K., Samantaray, A. K., & Karmarkar, R. (2009). Robust overwhelming control of a hydraulically driven three-degrees-of-freedom parallel manipulator through a simplified fast inverse model. *Proceedings of the IMechE, Part I: Journal of Systems and Control Engineering, 224*, 169–184.
6. Blanke, M., Frei, C., Kraus, F., Patton, R., & Staroswiecki, M. (2000). What is fault-tolerant control? Aalborg University, Department of Control Engineering. http://www.iau.dtu.dk/secretary/pdf/safeprocess_02h.pdf.
7. Blanke, M., Kinnaert, M., Lunze, J., & Staroswiecki, M. (2006). *Diagnosis and Fault-Tolerant Control*. Berlin: Springer.
8. Borutzky, W. (2015). *Bond Graph Model-Based Fault Diagnosis of Hybrid Systems*. Cham: Springer International Publishing.
9. Borutzky, W. (Ed.) (2016). *Bond Graphs for Modelling, Control and Fault Diagnosis of Engineering Systems* (2nd ed.). Springer International Publishing. https://doi.org/10.1007/978-3-319-47434-2.
10. Buchholz, J. J., & v Grünhagen, W. (2008). Inversion Impossible? Technical report, Bremen University of Applied Sciences and DLR Braunschweig.
11. Gawthrop, P. (1995). Bicausal bond graphs. In F. Cellier & J. Granda (Eds.), *ICBGM'95, International Conference on Bond Graph Modeling and Simulation* (Simulation Series, Vol. 27(1), pp. 83–88). SCS Publishing.
12. Loureiro, R. (2012). Bond Graph Model Based on Structural Diagnosability and Recoverability Analysis: Application to Intelligent Autonomous Vehicles. PhD thesis, L' Université Lille 1.
13. Lu, L. (2007). Inverse Modelling and Inverse Simulation for System Engineering and Control Applications. PhD Thesis, University of Glasgow.
14. Lunze, J., & Richter, H. (2008). Reconfigurable fault-tolerant control: A tutorial introduction. *European Journal of Control, 5*, 359–386.
15. Murray-Smith, D. (2012). Modelling and Simulation of Integrated Systems in Engineering – Issues of Methodology, Quality, Test and Application. Woodhead Publishing. Cambridge, UK.
16. Murray-Smith, D. J. (2011). Feedback methods for inverse simulation of dynamic models for engineering systems applications. *Mathematical and Computer Modelling of Dynamic Systems, 17*(5), 515–541.
17. Nacusse, M., & Junco, S. J. (2011). Passive fault tolerant control: A bond graph approach. In A. Bruzzone, G. Dauphin-Tanguy, S. Junco, & M. A. Piera (Eds.), *Proceedings of 5th International Conference on Integrated Modelling and Analysis in Applied Control and Automation (IMAACA 2011)* (pp. 75–82). Rome: DIPTEM UNIVERSITÀ DI GENOVA.
18. Ngwompo, R., Scavarda, S., & Thomasset, D. (1996). Inversion of linear time-invariant SISO systems modelled by bond graph. *Journal of the Franklin Institute, 333(B)*(2), 157–174.
19. Ngwompo, R. F., Scavarda, S., & Thomasset, D. (2001). Physical model-based inversion in control systems design using bond graph representation part 1: Theory. *Proceedings of the IMechE Part I Journal of Systems and Control Engineering*, 215(2), 95–103.
20. Ngwompo, R. F., Scavarda, S., & Thomasset, D. (2001). Physical model-based inversion in control systems design using bond graph representation part 2: Applications. *Proceedings of the IMechE Part I Journal of Systems and Control Engineering*, 215(2), 105–112.
21. Niemann, H., & Stoustrup, J. (1997). Robust fault detection in open loop vs. closed loop. In *Proceedings of the 36th IEEE Conference on Decision and Control* (Vol. 5, pp. 4496–4497).
22. Noura, H., Sauter, D., Hamelin, F., & Theilliol, D. (2000). Fault-tolerant control in dynamic systems: Application to a winding machine. *Control Systems, 20*(1), 33–49. https://doi.org/10.1109/37.823226.
23. Patton, R. J. (1997). Fault-tolerant control systems: The 1997 situation. In *Proceedings of 3rd IFAC Symposium on Fault Detection, Supervision and Safety for Technical Processes* (pp. 1033–1055). http://hull.ac.uk/control/downloads/safepr.pdf.
24. Samantaray, A., & Ould Bouamama, B. (2008). *Model-Based Process Supervision – A Bond Graph Approach* (Advances in Industrial Control). London: Springer.
25. Seraji, H. (1989). Minimal inversion, command matching and disturbance decoupling in multivariable systems. *International Journal of Control, 49*(6), 2093–2121.

26. Silverman, L. M. (1969). Inversion of multivariable linear systems. *IEEE Transactions on Automatic Control, AC-14*(3), 270–276.
27. Staroswiecki, M. (n.d.). Control systems, robotics and automation. Fault tolerant systems (Vol. XVI) In *Encyclopedia of Life Support Systems (EOLSS)*. Oxford: UNESCO Publishing-Eolss Publishers.
28. Thümmel, M., Looye, G., Kurze, M., Otter, M., & Bals, J. (2005). Nonlinear inverse models for control. In *Proceedings of 4th International Modelica Conference*, Hamburg (pp. 267–279).
29. YANG Hao, MAO Ze-Hui, & JIANG Bin. (2006). Model-based fault tolerant control for hybrid dynamic systems with sensor faults. *Acta Automatica Sinica, 32*(5), 680–685.
30. Youmin Zhang, & Jin Jiang. (2006). Issues on integration of fault diagnosis and reconfigurable control in active fault-tolerant control. In *Fault Detection, Supervision and Safety of Technical Processes* (Vol. 6, Part 1, pp. 1437–1448) P.R. China: Tsinghua University. https://doi.org/10.3182/20060829-4-CN-2909.00240.

Chapter 6
Software Support

The previous chapters cover in detail BG modelling in the context of control, fault diagnosis, and failure prognosis. This chapter gives an overview of software with a focus on free software that may be used for these tasks.

6.1 Model Development and Simulation of the Dynamic System Behaviour

The first step towards model-based control, fault diagnosis, and failure prognosis is the systematic development of a bond graph model. A widely used commercial integrated multi-formalism modelling and simulation software environment that supports the use of bond graphs among other formalisms in a user-friendly manner is *20-sim*® [5]. The software program is well suited for the development of small scale models for education purposes in academia as well as for the development of models in industrial projects. As 20-sim seamlessly enables to combine BGs with block diagrams, a BG model for a system and its control usually represented by a block diagram can be devised and tested. Another commercial software for bond graph modelling and simulation of mechatronic systems is BondSim [6, 7]. A powerful integrated modelling and simulation software package with some unique features such as the automatic generation of ARRs from a BG that was commercialised in the past is *Symbols 2000*™. Recently, the developer of the program and rights holder uploaded the software to the website of ResearchGate and made the version Symbols 6 freely available for all non-commercial academic use [25].

A graphical bond graph user interface to the sophisticated open source modelling and simulation software environment OpenModelica [21] is provided by the freely available BondLib library developed by Cellier [3, 9].

Furthermore, widely used commercial software programs designed for block diagram based modelling such as Matlab®/Simulink® or the open source software

W. Borutzky, *Bond Graph Modelling for Control, Fault Diagnosis and Failure Prognosis*, https://doi.org/10.1007/978-3-030-60967-2_6

Scilab/Xcos can also be used as bond graphs can be systematically transformed into block diagrams (Sect. 2.3).

Finally, as to the simulation of the dynamic behaviour of a system, for small to medium size BG models, equations can be manually derived from a causal BG in a systematic manner and can be directly formulated in the descriptive modelling language Modelica. Scilab/Xcos even has a Modelica compiler that supports a subset of the language so that a submodel formulated in Modelica can be used as a signal block in the block diagram interface Xcos [19]. Equations manually directly derived from a causal BG can also be coded in the script language of the open source mathematical computation programs Scilab and GNU Octave [11, 20], or in the Matlab® input language. The script language of Octave is very close to the one of Matlab so that some scripts developed for the commercial program Matlab can also be run by Octave. Some code translation from Octave to Matlab is illustrated in [22].

Commercial integrated modelling and simulation software environments such as 20-sim or BondSim as well as the free OpenModelica software and the mentioned mathematical software provide sophisticated numerical solvers for ODE system as well as for DAE systems of index one. With regard to FDI and FTC, these capabilities are certainly of importance for offline simulation studies of possible fault scenarios with predefined reference signals.

6.2 Model-Based Control

Integrated modelling and simulation software environments also support control tasks to some extent. The software 20-sim, for instance, provides a control toolbox so that transfer functions and common graphical representation in the frequency domain such as Nyquist or Bode plots can be produced. In addition, 20sim enables to export a model, a variable, or a parameter to Matlab so that further processing by means of Matlab functions can take place. However, to the best knowledge of the author, the computation of a transfer function according to Mason's loop rule by following causal paths in a BG (Sect. 2.2) is only implemented in the academic software package Archer [1] that has not been commercialised and is likely not maintained any more.

For BGs with linear elements, the commercial BG preprocessor software CAMP-G [16] can set up the matrices of a state space model which may be passed to Matlab®/Simulink® for various control tasks.

6.2.1 Observability and Controllability

In open source mathematical software such as GNU Octave and Scilab many functions beyond the transformation from a state space model to a transfer function matrix and vice versa are available for control tasks. Once the equations of a dynamic model have been derived from a causal BG, either automatically or manually, and have been formulated in linear state space form, the control package for Octave provides many functions that are usually not available in modelling and simulation software. For instance, as to the controllability and the observability of a LTI system addressed in Sect. 2.1, there are Octave functions that can be used to check a LTI model for numerical state controllability, for numerical state observability, and to compute the controllability and the observability matrices.

Given the matrices **A** and **B** of a linear state equation, then $\mathbf{O} = \texttt{obsv}(\mathbf{A}, \mathbf{C})$ provides the observability matrix **O**. The Octave command $\mathbf{Co} = \texttt{ctrb}(\mathbf{A}, \mathbf{B})$ returns the controllability matrix **Co**, and $\mathbf{W}_c = \texttt{gram}(\mathbf{A}, \mathbf{B})$ the controllability Gramian

$$\mathbf{W}_c(t) = \int_0^t e^{\mathbf{A}\tau}\mathbf{B}\mathbf{B}^T e^{\mathbf{A}^T\tau} d\tau \tag{6.1}$$

which is the solution of the ODE

$$\dot{\mathbf{W}}_c(t) = \mathbf{A}\mathbf{W}_c + \mathbf{W}_c\mathbf{A}^T + \mathbf{B}\mathbf{B}^T \tag{6.2}$$

In the case the system $\Sigma(\mathbf{A}, \mathbf{B})$ is controllable and stable, $\mathbf{W}_c(t) \to \mathbf{W}_c$ for $t \to \infty$, and $\mathbf{W}_c$ fulfils the controllability Lyapunov matrix equation

$$\mathbf{A}\mathbf{W}_c + \mathbf{W}_c\mathbf{A}^T + \mathbf{B}\mathbf{B}^T = \mathbf{0} \tag{6.3}$$

which can be solved in Octave by the call $\mathbf{W}_c = \texttt{lyap}(\mathbf{A}, \mathbf{B} * \mathbf{B}')$. In Scilab, the controllability Gramian is obtained by calling `ctr_gram(A,B)`.

Moreover, the Octave function `isdetectable()` can check whether a LTI system is detectable, i.e., whether all unstable modes are observable. The Scilab call

```
[Ao,Bo,Co] = obsvss(A,B,C)
```

returns the observable part of a linear system $\Sigma(\mathbf{A}, \mathbf{B}, \mathbf{C})$.

6.2.2 Design of a Luenberger Observer in Octave

In Sect. 3.5.1, p. 77, the Luenberger observer gain matrix has been computed by hand for a small numerical example. In the following, the steps are repeated by using Octave commands to illustrate some of the capabilities of the mathematical software with regard to control. For convenience, the equations of the LTI state space model are reproduced.

$$\dot{\boldsymbol{x}}(t) = \underbrace{\begin{bmatrix} -1 & 0 \\ 1 & -1 \end{bmatrix}}_{\mathbf{A}} \boldsymbol{x}(t) + \underbrace{\begin{bmatrix} \frac{1}{2} \\ 1 \end{bmatrix}}_{\mathbf{B}} u(t) \tag{6.4a}$$

$$y(t) = \underbrace{\begin{bmatrix} 0 & 2 \end{bmatrix}}_{\mathbf{C}} \boldsymbol{x}(t) \tag{6.4b}$$

First, the listing in Fig. 6.1 shows that the Octave symbolic package enables to compute the characteristic polynomial of the observer in symbolic form in agreement with Equation (3.78), p. 79.

Fig. 6.1 Computation of the characteristic polynomial of the observer in Octave in symbolic form

```
>>
>> A
A = (sym 2×2 matrix)

  ⎡-1  0 ⎤
  ⎢      ⎥
  ⎣1   -1⎦

>> L
L = (sym 2×1 matrix)

  ⎡__⎤
  ⎢L₁⎥
  ⎢  ⎥
  ⎢__⎥
  ⎣L₂⎦

>> C
C =

   0   2

>> A - L*C
ans = (sym 2×2 matrix)

  ⎡        ___       ⎤
  ⎢-1    -2⋅L₁       ⎥
  ⎢                  ⎥
  ⎢          ___     ⎥
  ⎣1   - 2⋅L₂  - 1   ⎦

>> det(s*eye(2) - (A - L*C))
ans = (sym)

                       ___            ___
  (s + 1)⋅(s + 2⋅L₂ + 1) + 2⋅L₁

>>
```

Fig. 6.2 Computation of the eigenvalues of **A** and of the characteristic polynomial of the plant

```
>>
>> eig(A)
ans = (sym 2×1 matrix)

  ⎡-1⎤
  ⎢  ⎥
  ⎣-1⎦

>> A = sym([-1 0; 1 -1]);
>> s = sym('s');
>> charpoly(A,s)
ans = (sym)

   2
  s  + 2·s + 1

>> poly=[1,2,1];
>> roots(poly)
ans =

  -1
  -1

>>
```

The Octave listing in Fig. 6.2 confirms that the eigenvalues of the matrix **A**, in fact, agree with the roots of the characteristic polynomial of the plant.

Given the matrices **A**, **C** of the state space model of the plant, one single Octave command is enough to compute the Luenberger observer gain matrix **L** for desired locations $-4, -4$ of the observer poles. Lines 11–38 in the listing of Fig. 6.3 just demonstrate how Octave commands can be used to verify that the result is correct. The observer matrix $\mathbf{Ao} = \mathbf{A} - \mathbf{LC}$ agrees with the manually computed one in (3.80), p. 79.

In addition, Fig. 6.4 shows how to check the controllability and the observability in Octave and how to transform the state space model into a transfer function. As can be seen, the denominator agrees with the characteristic polynomial of the matrix **A**.

6.2.3 Parameter Estimation and System Inversion on a Bicausal Bond Graph

For tasks such as parameter estimation and system inversion covered in Chap. 2, the concept of bicausal BGs is well suited. However, it appears that bicausal BGs are not supported by software apart from the free Model Transformation Tools (MTT) developed by Gawthrop in a research project [13, 14]. Likely, MTT is no longer maintained. However, equations manually derived from a bicausal BG by following

```
>>
>> A = [-1 0; 1 -1];
>> C = [0 2];
>> L = place(A',C',[-4,-4])
L =

    4.5000    3.0000

>> % Let's verify that the result is correct:

>> Ao = A - L'*C
Ao =

   -1   -9
    1   -7

>>
>> eig(Ao)
ans =

   -4.0000 + 0.0000i
   -4.0000 - 0.0000i

>>
>> s = sym('s');
>> charpoly(Ao,s)
ans = (sym)

    2
   s  + 8·s + 16

>>
>> poly=[1,8,16];
>> roots(poly)
ans =

   -4
   -4

>>
```

Fig. 6.3 Computation of the Luenberger observer gain matrix **L**

a bicausal path from a detector to a terminal element of which the parameter is to be estimated can be coded in an Octave or Scilab script. Accordingly, by following a bicausal path from an output source sensor to an input source sensor inverse system equations can be derived from a bicausal BG and implemented in a script language.

```
>>
>> A = [ -1 0; 1 -1 ]
A =

  -1   0
   1  -1

>> B = [ 1/2; 1 ]
B =

    0.50000
    1.00000

>> C = [ 0 2 ]
C =

    0   2

>>
>>[bool,ncon] = isctrb(A,B)
bool = 1    % is controllable
ncon =  2   % number of controllable states
>>
>> Co = ctrb(A,B)
Co =

    0.50000  -0.50000
    1.00000  -0.50000

>>
>> [bool,nobs] = isobsv(A,C)
bool = 1   % is observable
nobs =  2  % number of observable states
>>
>> O = obsv(A,C)
O =

    0   2
    2  -2

>>
>> [num,den] = ss2tf(A,B,C);
>>
>> G = tf(num,den)

Transfer function 'G' from input 'u1' to output ...

            2 s + 3
 y1:  -------------
        s^2 + 2 s + 1

Continuous-time model.
>>
```

Fig. 6.4 Check for controllability and observability and transformation of the state space model into a transfer function

6.3 Fault Diagnosis

6.3.1 Signal Preprocessing

One of the first steps in fault diagnosis is to filter noisy measured signals appropriately. Octave as well as Scilab and Matlab provide a comprehensive set of functions for various purposes in signal processing. The Octave signal package, for instance, contains the function `sgolayfilt()` which provides a Savitzky–Golay smoothing filter of polynomial order p and length n, where n is odd and $n > p$. The function `sgolay()` computes the filter coefficients for all Savitzky–Golay smoothing filters and can be used to get the coefficients of the mth time derivative of the polynomial approximating a time series of measured data. This is of interest for bond graph model-based FDI presented in Sect. 3.5.2, since not only measured signals but, in general, also time derivatives of measured signals are inputs into ARRs derived from a DBG. Therefore, the Savitzky–Golay filter is reviewed in Sect. 3.2.1.

6.3.2 State Estimation and Observer-Based Fault Detection

Besides the evaluation of ARRs derived from a DBG, state observers can be used to detect a fault. An abnormal change of the system state is an indication of a fault that has happened. Therefore, state estimation is of interest for FDI and for failure prognosis. For that purpose variants of the discrete Kalman filter and the particle filter are widely in use. Both filters are reviewed in Sect. 3.4.1 and in Sect. 3.4.2. In Octave, the function `kalman()` contained in the control package can be used to design a Kalman estimator for an LTI system. The input into the function is a plant model with nominal parameters. The outputs are a state space model of the Kalman filter, the gain matrix, and a solution of the Riccati equation for the state error prediction covariance matrix (3.28), p. 66.

Online recursive Bayesian state estimation for nonlinear systems by means of a time-discrete particle filter is supported by Matlab functions for the steps of the particle filter workflow [28]. A first step in the workflow is the creation of a particle filter. This can be achieved by the call

```
pf = particleFilter(StateTransitionFcn,MeasurementLikelihoodFcn)
```

which requires that a plant model and measurements are available [29], where `StateTransitionFcn()` is a function that calculates the particles at the next time step, given the state vector at a time step, while `MeasurementLikelihoodFcn()` is a function that computes the likelihood of each particle based on sensor measurements.

A pseudo code and a Matlab script of a particle filter for state estimation are given in [30]. Furthermore, a number of Matlab toolboxes for nonlinear estimation have been published, e.g., PF-Toolbox [23, 24], PFLib [4], or Nonlinear Estimation

Toolbox [17], just to name some few. The Matlab Control Systems Toolbox provides a Simulink particle filter block.

Once a behavioural BG has been developed for a real plant, a Luenberger observer can be obtained by duplicating the BG and connecting the observer BG to the system BG via detectors and modulated sources. What remains to be determined is the observer gain matrix. Their entries can be obtained from the denominator of Mason's rule by following causal paths in the behavioural BG as shown in Sect. 3.5.1.

Given a linear model $\Sigma(\mathbf{A}, \mathbf{B}, \mathbf{C}, \mathbf{D})$ for a plant with matrices obtained from a BG model, a call of the Scilab function `observer()` contained in the Computer-Aided Control System Design (CACSD) toolbox returns an observer for the observable part of the linear system. The Octave function `place(`$\mathbf{A}', \mathbf{B}', p$`)` returns the state feedback gain matrix $\mathbf{L}$ for desired observer poles $p = [p_1 \ldots p_n]$.

In [10], Ding et. al. present a Matlab toolbox for observer-based and parity space FDI methods [15]. The toolbox includes functions for the generation and the evaluation of output residuals. A collection of MATLAB functions for the analysis and solution of fault detection problems has been developed by Varga [31].

6.3.3 FDI Based on ARRs Derived from a DBG

For small to medium size models, ARRs can be derived from a DBG. A unique feature of the software Symbols 2000 is that its FDI module can generate ARRs and build a structural FSM. If component fault signatures are not unique, fault may be detected but cannot isolated by simple inspection of the structural FSM. In that case, multiple faulty parameters may be isolated by estimating all parameters of the subset of potentially faulty components with the same fault signature through nonlinear least squares optimisation of ARR residuals minimisation [2, 26]. Let $\boldsymbol{\Theta}$ be the vector of all potentially faulty parameters and $\boldsymbol{r}$ the vector of residuals of all those ARRs in which the potentially faulty parameters appear, i.e., the ARRs that contribute parameters to the unstructured part of the FSM. Then the problem is to minimise the cost function

$$f(\boldsymbol{\Theta}) = \frac{1}{2} \sum_{j=k-q}^{k} \boldsymbol{r}^T(t_j, \boldsymbol{\Theta}) \boldsymbol{r}(t_j, \boldsymbol{\Theta}) \tag{6.5}$$

by varying the parameter vector $\boldsymbol{\Theta}$. The parameters $\boldsymbol{\Theta}^*$ that make the residuals $\boldsymbol{r}(t_j, \boldsymbol{\Theta})$ close to $\mathbf{0}$ are then the ones of the faulty subsystem.

For that task, the Levenberg–Marquardt algorithm may be used. A Matlab script may be found in [12].

An alternative may be the Nelder–Mead algorithm if the numerical computation of the gradient of an objective function shall be avoided. Both algorithms are implemented in functions provided by Octave or Scilab. Octave functions such as `leasqr()` implementing the Levenberg–Marquardt algorithm are available in the nonlinear optimisation toolkit `optim`. In Scilab, the function `lsqrsolve()` is an implementation of the Levenberg–Marquardt algorithm and `fminsearch()` implements the Nelder–Mead algorithm.

6.3.4 Combined Bond Graph Model-Based Data-Driven Failure Prognosis

Bond Graph Model-Based Generation of Discrete Degradation Data

In Sect. 4.5, a hybrid bond graph model-based data-driven approach to failure prognosis has been proposed. The model-based part receives measurements at discrete sample points from a real faulty system, or simulated measurements from a model of the real faulty system and generates concurrently to the dynamic behaviour of the system discrete data for an unknown degradation behaviour in a sliding time window. The observed abnormal dynamic system behaviour is attributed to the degradation of some parameters. Estimating the values of a trending parameter at sampling time points based on a bicausal BG delivers a time series of discrete degradation data (Sect. 4.5.1).

Learning a Degradation Model

Concurrently to the dynamic behaviour of the plant or process, the degradation data in the current sliding window obtained from the model-based part are used to learn a degradation model (DM) by selecting potential functions from different classes and by performing a curve fitting which can be carried out in parallel on a multicore, multiprocessor computer. This simultaneous training of various potential degradation models is depicted in Fig. 4.15, p. 162.

Suppose that the discrete degradation data in the current window and collected in the buffer suggest an exponential decline. Then the task is to find two coefficients c_0, c_1.

$$d(t) = c_0 \exp(c_1 t) \tag{6.6}$$

Scilab provides some functions for curve fitting such as `datafit()` and `leastsq()`. Some small illustrating Scilab scripts may be found in [18]. In Octave, the functions `nlinfit()` or `lsqcurvefit()` contained in the GNU Octave optimisation package `optim` may be used. The Octave function `fsolve()` could also be used for nonlinear regression. If there are more sample points than coefficients to be determined `fsolve()` considers the over-determined set of equations as a minimisation problem.

Let $\boldsymbol{y}$ denote the observations, $\boldsymbol{\beta}$ the coefficients to be fitted, and $\mathbf{F}(\boldsymbol{\beta}, t)$ the nonlinear regression function, then `nlinfit()` solves the optimisation problem

$$\min_{\boldsymbol{\beta}} ||\boldsymbol{y}(t) - \mathbf{F}(\boldsymbol{\beta}, t)||$$

Besides the coefficients that best fit the nonlinear model function to the observed data, the function `nlinfit()` can return the solution residuals and the estimated covariance matrix of the fitted coefficients.

`[beta,R,J,covb, mse] = nlinfit(t,Y,modelfun,beta0,options)` (6.7)

where

`beta`: vector of coefficients to be fitted
`beta0`: initial guess of the coefficients
`Y`: vector of observations
`modelfun`: nonlinear regression function
`R`: vector of residuals `modelfun(beta, t) - Y`
`J`: Jacobian ∂ `modelfun`/∂ `beta`
`covb`: estimated covariance matrix of the fitted coefficients
`mse`: mean squared error

Nonlinear Regression in Octave

In the following, the use of the Octave functions `nlinfit()` and `lsqcurvefit()` is illustrated. Suppose that measurements reveal a decline of a system parameter value $p(t)$, e.g., of the capacitance of an electrolytic capacitor as of some time t_0.[1] The parameter values derived from noisy measurements may suggest a fitting by means of an exponential function. Accordingly, an exponential degradation model function $F(t)$ with three coefficients $[b_1; b_2; b_3]$ to be fitted is chosen as indicated in the listing of Octave commands in Fig. 6.5.

$$F(t) = b_1 + b_2 \exp(-b_3\, t) \tag{6.8}$$

The parameter values are generated by using actual coefficients [100; 400; 500] (line 4) and are superimposed with 10% normally distributed noise with zero mean and 0.1 standard deviation (line 9). The nonlinear regression starts with the guess beta0 = [0; 0; 0] (line 11) and is performed over the time window $[0, 2\,\text{ms}]$ for

[1]See also Sect. 7.3 where, in an offline fault scenario, the declining capacitance values $C(t)$ of a leaking electrolytic capacitor at each sampling time point following a degradation function considered as unknown are reconstructed by parameter estimation based on a DBG. The unknown coefficients of an exponential candidate function are obtained by nonlinear regression of the simulated measurements.

```
1   >>
2   >> modelfun = @(b,t)(b(1)+b(2)*exp(-b(3)*t));
3   >>
4   >> b = [100;400;500]; % actual coeficient values
5   >>
6   >> t=linspace(0,2e-3,100)';
7   >>
8   >> % noisy declining parameter values:
    >> y = modelfun(b,t) + 0.1*modelfun(b,t).*normrnd(0,0.1,100,1);
10  >>
11  >> beta0 = [0;0;0]; %guess
12  >>
13  >> [beta,R,J,covb,mse] = nlinfit(t,y,modelfun,beta0)
14  beta =
15
16       93.166
17      406.510
18      485.511
19
20  R =
21
22  J =
23
24  covb =
25
26      107.035    -99.064   199.395
27      -99.064     92.481  -182.771
28      199.395   -182.771   377.786
29
30  mse =  13.127
31  >>
32  >> betaDev = sqrt(diag(covb))'
33  betaDev =
34
35      10.3458    9.6167   19.4367
36
37  >>
38  >> beta = lsqcurvefit(modelfun,beta0,t,y)
39  beta =
40
41       93.166
42      406.510
43      485.511
44
45  >>
```

Fig. 6.5 Fitting noisy declining parameter values by means of Octave functions `nlinfit()` and `lsqcurvefit()`, respectively

which 100 generated parameter values are kept in a buffer (line 6). The two functions `nlinfit()` and `lsqcurvefit()` deliver identical values for the coefficients to be fitted that are close to the actual values. Better results can be obtained with more samples and less noise.

To save space, the residuals for each sample and the values of the Jacobian have been omitted in the listing. The values of the estimated covariance matrix of the fitted coefficients are given in lines 26–28. In Line 32, the standard deviations of the fitting coefficients `betaDev` are computed. The result is given in Line 35. In case a normal distribution can be assumed, the standard deviation values can be used for the determination of confidence intervals.

Figure 6.6 shows the results of the nonlinear regression for two overlapping time windows.

In the commercial Matlab Predictive Maintenance Toolbox™, probability distributions for the parameters of a potential degradation function may be chosen. The toolbox provides a linear and an exponential degradation model. Their coefficients can be estimated by using historical data with regard to the health of an ensemble of similar components. After that training, the fitted degradation model can be used for predicting the remaining useful life of a similar component. When new data becomes available the degradation model parameters can be updated by calling an update function before prediction of a new RUL value [27].

RUL Prediction

Once a degradation function has been found that best fits the data in the current sliding window, a RUL can be estimated by simply determining the time point of its crossing with a predefined failure alarm threshold.

Another option for forecasting is to use the data in the sliding window for determining the parameter vector of an autoregressive model. In order to solve (6.9)

$$\boldsymbol{y} = \boldsymbol{\Phi}\,\boldsymbol{\theta} + \boldsymbol{\epsilon} \tag{6.9}$$

for the unknown regression weights $\boldsymbol{\theta}$, the Octave ordinary least squares estimation function `ols()` can be used. The Octave function `pinv()` computes the Moore–Penrose pseudo-inverse of the known observation data matrix $\boldsymbol{\Phi}$. The commercial

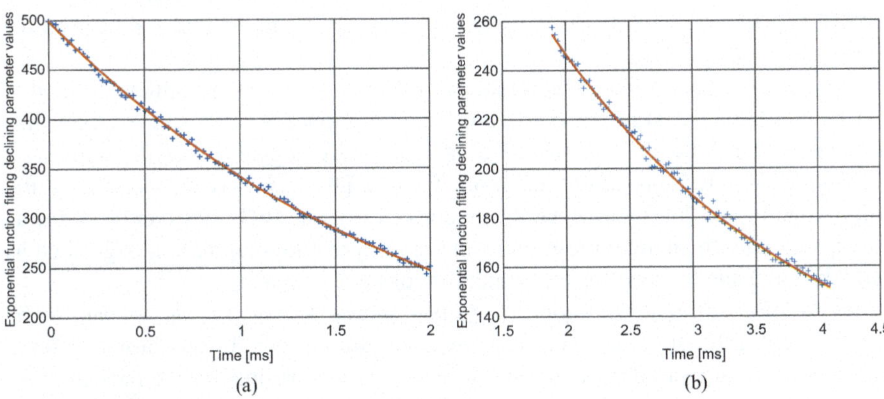

Fig. 6.6 Nonlinear regression for two overlapping windows. (**a**) Regression in window $0 \leq t \leq 2.0$ ms. (**b**) Regression in window $1.9\,\text{ms} \leq t \leq 4.1$ ms

Matlab toolbox Simulink provides a signal block for online recursive least squares estimation.

As Wileman demonstrated in [32], the autoregressive model obtained by recursive least squares minimisation (RLS) from data in the latest sliding window can be converted into a state space model and can be used in a Kalman filter for a N-step ahead prediction of the current state into the future with $N = 1, 2, 3, \ldots$ until the time instant at which a damage relevant feature reaches a failure alarm threshold.

In Sects. 3.4.1 and 3.4.2, it has been noted that the discrete Kalman filter as well as the particle filter can be used for simultaneous estimation of states and of parameters. Therefore, yet another approach to RUL prediction is to adopt a parameter degradation model with unknown parameters, to include it into the state space model, to add the unknown DM parameters to the state vector, and to perform a joint state and parameter estimation by means of a Kalman or particle filter until a monitored health indicator raises an alarm by crossing a failure level. This approach has been used by Danes et. al. in [8]. For fault detection, they use ARRs derived from a BG. The degradation model used in a particle filter may be a linear or an exponential function.

6.4 Summary

This review on software support suggests that some commercial software programs as well as open source mathematical software packages may be used for bond graph modelling and the simulation of fault scenarios. Open source mathematical software such as GNU Octave and Scilab, moreover, provide many functions for control task such as the check for controllability, or observability, signal preprocessing, state and parameter estimation. Checks of structural properties directly on a bond graph seem to be implemented only in the software Symbols 2000. Software support for bicausal BGs does not seem to be available. For linear BG models, BG preprocessor CAMP-G [16] can generate the matrices of a state space model that can be used for further purposes in Matlab.

For observer-based FDI, mathematical software provides a Kalman filter that can be used for the simulation of fault scenarios once a state space model has been derived from a BG. A survey of the literature suggests that the automatic generation of ARRs as fault indicators from a DBG is only supported by the software Symbols 2000. Once ARRs are derived from a DBG, they can be coded as a script for a mathematical program or in a descriptive modelling language such as Modelica and can be used for the simulation of fault scenarios.

As to failure prognosis, support by open source software tools appears to be scarce. State estimation for nonlinear systems until a health indicator crosses a failure level is supported by a number of library Matlab scripts for the particle filter and by the commercial dedicated Matlab Predictive Maintenance Toolbox™.

References

1. Azmani, A., & Dauphin-Tanguy, G. (1992). Archer: A program for computer aided modelling and analysis. In P. C. Breedveld & G. Dauphin-Tanguy (Eds.), *Bond Graphs for Engineers* (pp. 263–278). North-Holland: Elsevier.
2. Borutzky, W. (2015). *Bond Graph Model-based Fault Diagnosis of Hybrid Systems*. Cham: Springer International Publishing.
3. Cellier, F. E. (2020). BondLib Free library to model physical systems with bond graph. https://github.com/modelica-3rdparty/BondLib.
4. Chen, L., Lee, C., Budhiraja, A., & Mehra, R. (2007). PFLib – An object oriented MATLAB toolbox for particle filtering – art. no. 65670S. In *Proceedings of SPIE – The International Society for Optical Engineering*. https://doi.org/10.1117/12.719951.
5. Controllab Products. (2020). 20-sim Software for Modeling Complex Physics. http://www.20sim.com.
6. Damic, V. (2020). Bondsim software. https://www.bondsimulation.com/59-bond/news/130-bondsim-software.htm.
7. Damic, V., & Montgomery, J. (2016). *Mechatronics by Bond Graphs* (2nd ed.). Berlin/Heidelberg: Springer.
8. Danes, M., Ramezani, A., & Moghaddam, J. Z. (2015). Enhanced prognosis of hybrid systems with unknown mode changes. *Modares Journal of Electrical Engineering, 15*(2), 21–26.
9. de la Calle, A., Cellier, F., Yebra, L., & Dormido, S. (2013). Improvements in BondLib, the Modelica Bond Graph Library. In *Proceedings of the 8th EUROSIM Congress on Modelling and Simulation* (pp. 282–287). Cardiff: IEEE.
10. Ding, S. X., Atlas, E., Ma, Y., Jeinch, T., & Ding, E. L. (2006). An introduction to a matlab-based fdi-toolbox. *IFAC Proceedings, 39*(13), 651–656.
11. Eaton, J. W., Bateman, D., Hauberg, S., & Wehbring, R. (2020). GNU Octave – A high-level interactive language for numerical computations Edition 5 for Octave version 5.2.0 Jan 2020. https://octave.org/octave.pdf.
12. Gavin, H. P. (2019). The Levenberg-Marquardt algorithm for nonlinear least squares curve-fitting problems. Technical report, Duke University, Department of Civil and Environmental Engineering. http://people.duke.edu/~hpgavin/ce281/lm.pdf.
13. Gawthrop, P. J. (2013). Model Transformation Tools. http://mtt.sourceforge.net.
14. Gawthrop, P. J., & Bevan, G. (2003). MTT: Model Transformation Tools. http://mtt.sourceforge.net.
15. Gertler, J. (1997). Fault detection and isolation using parity relations. *Control Engineering Practice, 5*, 653–661.
16. Granda, J. (2007). CAMP-G – User's Manual. Cadsim Engineering, P. O. Box 4083, Davis, Ca 95617, https://www.bondgraph.com.
17. Steinbring, J. (2017). Nonlinear Estimation Toolbox. https://bitbucket.org/nonlinearestimation/toolbox.
18. Jasso, J. (2020). Curve Fitting for experimental data. http://www.matrixlab-examples.com/curve-fit.html.
19. Najafi, M., & Nikoukhah, R. (2006). Modeling and simulation of differential equations in Scicos. In *Proceedings of 5th International Modelica Conference 2006* (pp 177–185). The Modelica Association.
20. NN. (2020). Gnu octave. http://www.gnu.org/software/octave/.
21. OpenModelica Consortium. (2020). OpenModelica. https://www.openmodelica.org/.
22. Rawlings, J. B., Eaton, J. W., & Ekedt, J. G. (2009). A User's Guide for Translating Octave to Matlab. https://citeseerx.ist.psu.edu/viewdoc/download?doi=10.1.1.489.2734&rep=rep1&type=pdf.
23. Rosén, J. (2005). A Framework for Nonlinear Filtering in MATLAB. Master thesis, Department of Electrical Engineering, Linköpings universite. http://liu.diva-portal.org/smash/get/diva2:21103/FULLTEXT01.pdf.

24. Rosén, J. (2005). Pf toolbox. https://sourceforge.net/projects/pftoolbox/files/latest/download.
25. Samantaray, A. (2017). SYMBOLS 6 Software Installation (Windows XP to Windows 10). https://www.researchgate.net/publication/317277532_SYMBOLS_6_Software_Installation_Windows_XP_to_Windows_10.
26. Samantaray, A., & Ould Bouamama, B. (2008). Model-based process supervision – a bond graph approach. In *Advances in Industrial Control*. London: Springer.
27. The Mathwork Inc. (2020). Exponential Degradation Model. https://www.mathworks.com/help/predmaint/ref/exponentialdegradationmodel.html#mw_5913aff2-a0cf-4859-a547-8e25cfbde07e_head.
28. The Mathworks Inc. (2020). Particle Filter Workflow. https://www.mathworks.com/help/robotics/ug/particle-filter-workflow.html.
29. The Mathworks Inc. (2020). ParticleFilter – Particle filter object for online state estimation. https://www.mathworks.com/help/control/ref/particlefilter.html.
30. Tulsyan, A., Bhushan Gopaluni, R., & Khare, S. R. (2016). Particle filtering without tears: A primer for beginners. *Computers and Chemical Engineering, 95*, 130–145. http://dx.doi.org/10.1016/j.compchemeng.2016.08.015.
31. Varga, A. (2017). Fault Detection and Isolation Tools (FDITOOLS) User's Guide. CoRR. http://arxiv.org/abs/1703.08480.
32. Wileman, A. (2016). An Investigation into the Prognosis of Electromagnetic Relays. Ph.D. Thesis, Cranfield University.

Chapter 7
Applications

7.1 Introduction

With regard to multiphysics systems, bond graph methodology is particularly suited for the model-based part of an integrated hybrid approach to fault diagnosis and failure prognosis. In this chapter, bond graph-based FDI and failure prognostic is illustrated by considering small systems in various application areas.

First, a half-wave voltage doubler as a small hybrid system is modelled and analysed. The model is then used to study fault scenarios in offline simulation runs.

The subject of the following section is leakage in an electrolytic capacitor of a boost converter as semiconductors and capacitors in power electronic systems are reported to have a higher failure rate than other components [15, 16]. Leakage in hydraulic systems may also be a problem in process engineering. The third case study therefore analyses leakage from a tank due to a hole of increasing size. Moreover, a DBG of a hydraulic cylinder shows that internal cross-port leakage and external leakage from the bearings can be detected and isolated.

Another subsequent section studies increasing friction in an angular velocity controlled permanent DC motor of an electromechanical drive. In order to compensate for the increasing internal friction, the motor may overload by drawing excessively current.

Furthermore, fault accommodation using an inverse model derived from a bicausal BG is illustrated by deliberately introducing a fault in the open loop DC motor drive and by studying the effect in two offline simulation runs. The chapter on applications concludes by considering a robust overwhelming controller as an example of a controller that may be used in passive fault tolerant control.

In all case studies, the real system delivering measurements into a DBG-based residual generator for FDI is replaced by a behavioural BG model. The advantage of a behavioural model is that faults can be deliberately inserted without causing harm and their affect on the performance of the system over time can be studied by offline simulation. Simulated measurements are inserted into a DBG model for

W. Borutzky, *Bond Graph Modelling for Control, Fault Diagnosis and Failure Prognosis*, https://doi.org/10.1007/978-3-030-60967-2_7

computation of discrete values of an unknown degradation function at sampling times. A mathematical function best fitting the data is determined by regression over a sliding time window and extrapolated into the future for RUL estimation. Given a predefined failure alarm threshold, the failure time and the RUL can be estimated by simply determining the intersection of the degradation trend with the alarm level.

7.2 Half-Wave Voltage Doubler

Voltage multipliers are used in AC-DC power supplies where a DC output voltage is required that is higher than the amplitude of the AC input voltage. Figure 7.1 shows the well known Greinacher half-wave voltage doubler connected to a step-down transformer, which reduces the AC input voltages and isolates the DC output voltage from the AC primary source. Other voltage multiplier topologies such as the Delon full-wave voltage doubler are also known but are not considered in this section.

As its name suggests, the DC output voltage of the voltage doubler, u_{C_2}, is twice the amplitude E_m of a sinusoidal input minus twice the knee voltage V_g of the diodes if there is no load, or $R_L \to \infty$.

$$u_{C_2} = 2E_m - 2V_g \tag{7.1}$$

The voltage doubler is composed of a positive clamper circuit, which adds a DC component of height E_m to the sinusoidal input voltage $E(t)$. The output of the peak detector, u_{C_2}, is equal to the peak value E_m of the sinusoidal input signal $E(t)$ and carries a ripple that depends on the capacitance C_2 at the output, the resistive load R_L, and the frequency of the AC input signal.

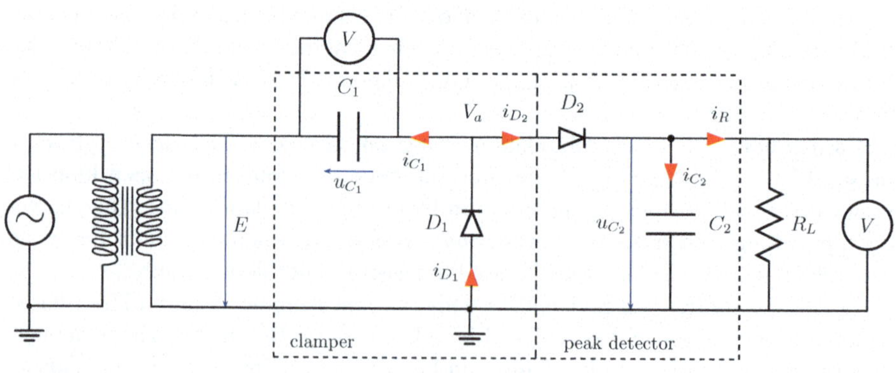

Fig. 7.1 Circuit schematic of a half-wave voltage doubler

AC-DC step-up power supplies also use a regulator circuit between the voltage multiplier and the DC load not shown in Fig. 7.1, which holds the output voltage constant independently of any input voltage fluctuations or load current variations.

The positive clamper circuit as well as the peak detector are built by means of diodes and capacitors. If the voltage multiplier is part of applications with high frequency and low load currents, small capacitors with values in the range of 20–60 nF can be used [3]. For higher load currents, the capacitor values must be selected appropriately. In [1], the authors propose design formulae which yield oversized capacitors.

For load currents much higher than the typical value of 5 mA [3], electrolytic capacitors will be needed, which, however, are critical components as they are more likely to fail than other components in power electronic circuits. Their useful life is much affected by the operating conditions. The ripple on the DC output voltage depending on the frequency of the input signal and the load current causes a current ripple depending on the equivalent series resistance (ESR) which raises the internal temperature of the capacitor and increases its failure rate. Diodes are also subject to failure even though to a much lower extent than electrolytic capacitors [20]. Failures in the components of an AC-DC power converter may affect the performance of the system driven by the power supply. In the following, the half-wave voltage doubler is modelled and some fault scenarios are considered.

7.2.1 Modelling and Analysis of the Voltage Doubler

Due to the two diodes, the voltage doubler operates in the following three modes.

Mode 1: Diode D_1 on and diode D_2 off
Mode 2: D_1 off and D_2 on
Mode 3: D_1 off and D_2 off

In Mode 1, capacitor C_1 is charged while C_2 discharges via the load resistor R_L. In Mode 2, capacitor C_1 discharges, capacitor C_2 is charged and discharges via the load resistor. In Mode 3, capacitor C_1 is not charged and C_2 discharges via the load resistor. Accordingly, a circuit can be associated with each mode. The voltage doubler model has a variable, mode-dependent structure.

Software programs such as PSpice®[10], Scilab/Xcos, or OpenModelica do have a graphical interface and can use a diode library model so that a simulation of the dynamic behaviour just requires to enter the circuit schematic in Fig. 7.1 into the graphical editor and to provide device parameters. The circuit diagram can also be transformed into a BG shown in Fig. 7.2, from which equations can be derived that hold for all three modes.

From the BG in Fig. 7.2, the following ordered set of equations can be derived:

$$V_a = E + u_{C_1} \tag{7.2}$$

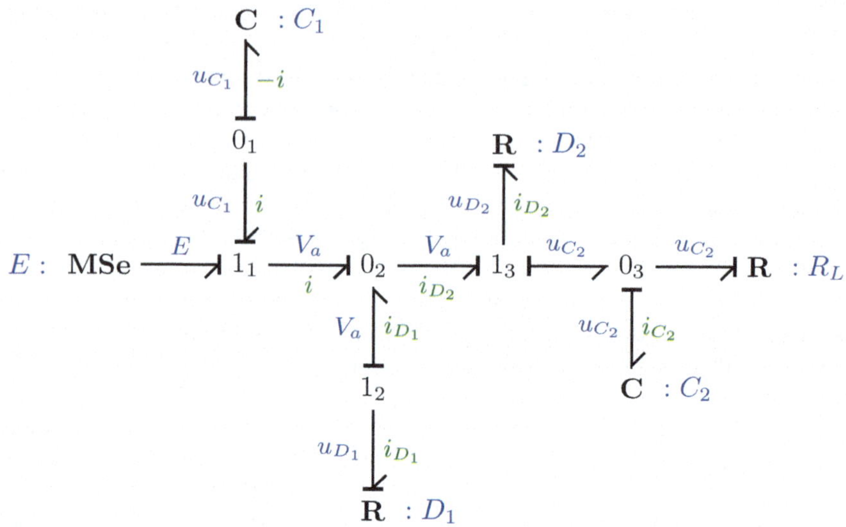

Fig. 7.2 Bond graph of the voltage doubler

$$i_{D_1} = g(-V_a) \tag{7.3}$$

$$i_{D_2} = g(V_a - u_{C_2}) \tag{7.4}$$

$$i = -i_{D_1} + i_{D_2} \tag{7.5}$$

$$C_1 \dot{u}_{C_1} = -i \tag{7.6}$$

$$C_2 \dot{u}_{C_2} = i_{D_2} - \frac{u_{C_2}}{R_L} \tag{7.7}$$

where the function $g()$ accounts for the nonlinear current versus voltage drop characteristic of the diodes.

A Switched LTI Model

One option is to use the Shockley diode equation, another one may be to replace the exponential characteristic by a piecewise linear one. In the case of the second option, the model of the voltage doubler is a switched LTI system. In each mode, the model is linear. Let the diode current i_D be zero when the voltage drop Δu across the diode is less than its knee voltage V_g and adopt an ON resistance $R_D = R_{\text{on}}$ when the diode is forward biased as depicted in Fig. 7.3a.

The diode equations then read:

$$i_{D_1} = \frac{m_1}{R_{D_1}}(-V_a - V_g) \tag{7.8}$$

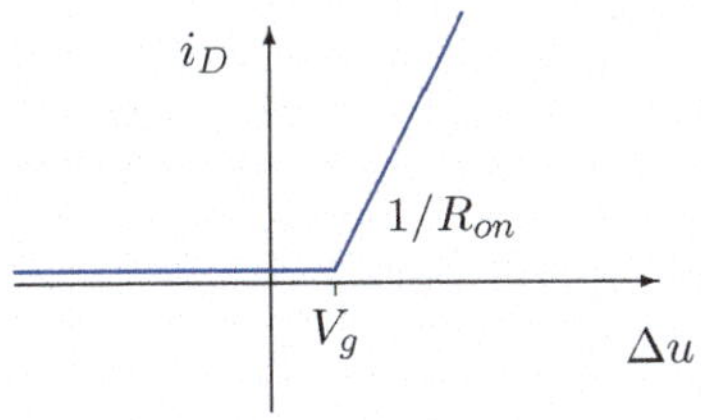

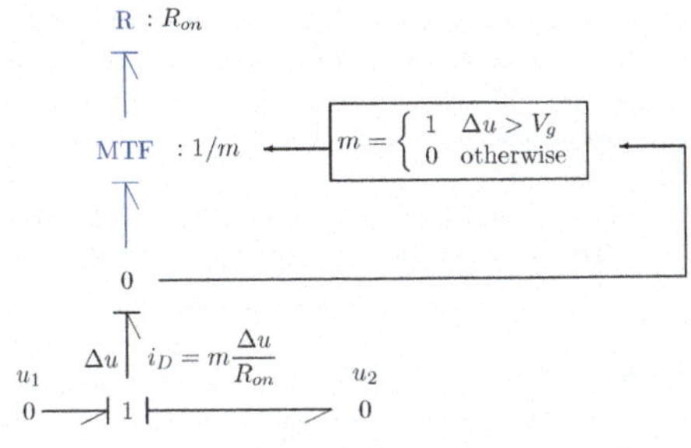

Fig. 7.3 Bond graph representation of a non-ideal switch with fixed conductance causality. (**a**) Piecewise linear approximation of a diode characteristic. (**b**) BG model of a diode with R : R_{on} in fixed conductance causality

$$i_{D_2} = \frac{m_2}{R_{D_2}}(V_a - u_{C_2} - V_g) \tag{7.9}$$

where V_g denotes the knee voltage and $m_1, m_2 \in \{0, 1\}$ depending on whether the diodes are on or off. That is, the switching from one mode to another is not a time event but an internal state event. It is not an external control signal on the gate of a MOSFET transistor that switches the transistor on and off. Accordingly, the time instances at which mode switching takes place must be detected and the numerical integration of the two explicit state equations must be restarted. In other words, a linear model associated with one mode is computed until a switching condition is met. The current time interval ends at the time point associated with the switching condition. Computation then switches to the linear model that holds for the subsequent mode by taking the state values at the switching time point as initial values for the new time interval.

Implementation of the LTI System

An easy and convenient way to achieve this for circuits with a small number of diodes is to formulate the model equations for each mode in the declarative modelling language Modelica as a condition for a subsequent section of mode-dependent equations is automatically translated into a state event that is taken into account by the ODE solver with root finding capability. More generally, an implicit hybrid DAE system with continuous states and mode-dependent discrete states may be derived from a BG that holds for all modes [7]. For the numerical solution of a single hybrid DAE system that holds for all modes, the sophisticated software OpenModelica may be used. Problems, however, occur, if the index of the DAE system changes during runtime due to a change of some discrete states.

Figure 7.4 shows a Modelica listing of the voltage doubler equations and Fig. 7.5 lists the parameters of the voltage doubler circuit.

```
// Half-wave voltage doubler circuit
// Healthy system with nominal parameters
model VD
  extends parameters;
  Real E;
  Real uC1(start = 0.0), uC2(start = 0.0), iC1, iC2, iD1, iD2;
  Boolean mode1, mode2;
  Real Va;

equation
  // AC input voltage
  E = A * sin(omega * time);
  // Voltage Doubler equations
  // D1 on and D2 off :
  mode1 = E + uC1 <= -Vg and E + uC1 - uC2 < Vg;
  // D1 off and D2 on:
  mode2 = E + uC1 > -Vg and E + uC1 - uC2 >= Vg;
  //
  if mode1 then
  // D1 on and D2 off :
    RD1*iD1 = - (E + uC1 +Vg);
    iC1 = iD1;
    iC2 = - uC2/R;
    iD2 = 0;
    C1*der(uC1) = iC1;
    C2*der(uC2) = iC2;
    Va  = E + uC1;
  elseif mode2 then
  // D1 off and D2 on:
    iD2 = 1/RD2*(E + uC1 - uC2 - Vg);
    iC1 = - iD2;
    iC2 = -iC1 - uC2/R;
    iD1 = 0;
    C1*der(uC1) = iC1;
    C2*der(uC2) = iC2;
    Va  = E + uC1;
  else
  // D1 off and D2 off :
    iC1 = 0;
    iC2 = - uC2/R;
    iD1 = 0;
    iD2 = 0;
    C1*der(uC1) = iC1;
    C2*der(uC2) = iC2;
    Va  = E + uC1;
  end if;
  //
end VD;
```

Fig. 7.4 Modelica description of the voltage doubler

```
record parameters
  // Half-wave voltage doubler circuit
  //
  constant Real pi = Modelica.Constants.pi;
  // Input Voltage
  parameter Real omega = 2 * pi * f;
  parameter Real A = 20 "Amplitude [V]";
  parameter Real f = 50 "Frequency [Hz]";
  // circuit
  parameter Real C1  = 2000e-6 "[F]";
  parameter Real C2  = 4000e-6 "[F]";
  parameter Real RD1 = 0.1 "ON resistance of diode D1 [Ohm]";
  parameter Real RD2 = 0.1 "ON resistance of diode D2 [Ohm]";
  parameter Real Vg  = 0.7 "Diode knee voltage [V]";
  parameter Real R   = 100 "Load resistance [Ohm]";
end parameters;
```

Fig. 7.5 Parameters of the voltage doubler circuit

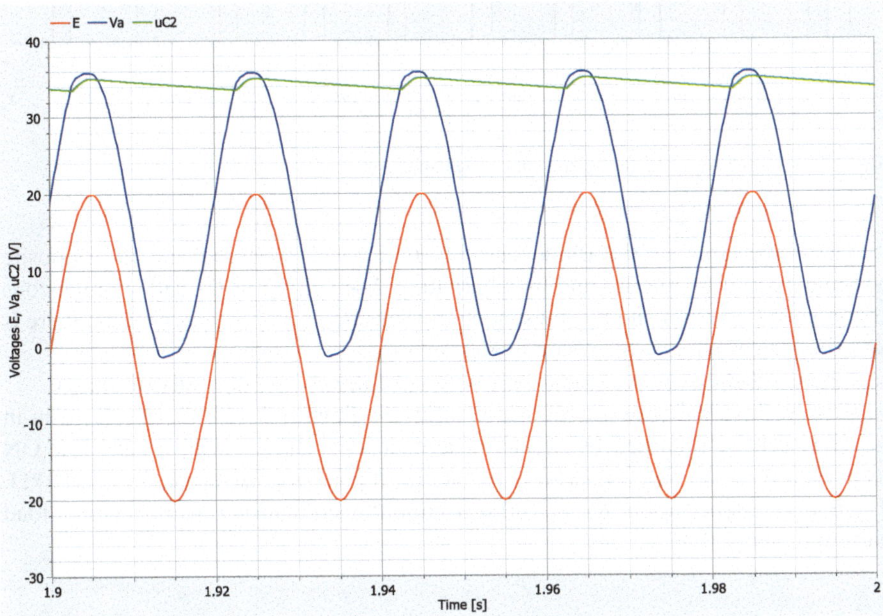

Fig. 7.6 Input voltage $E(t)$, the output of the clamper, $V_a(t)$, output of the voltage doubler $u_{C_2}(t)$

Simulation Results

Figure 7.6 depicts the typical waveforms of the input voltage $E(t)$, the output voltage of the clamper, $V_a(t)$, and the output voltage of the voltage doubler $u_{C_2}(t)$ in the steady state regime as expected.

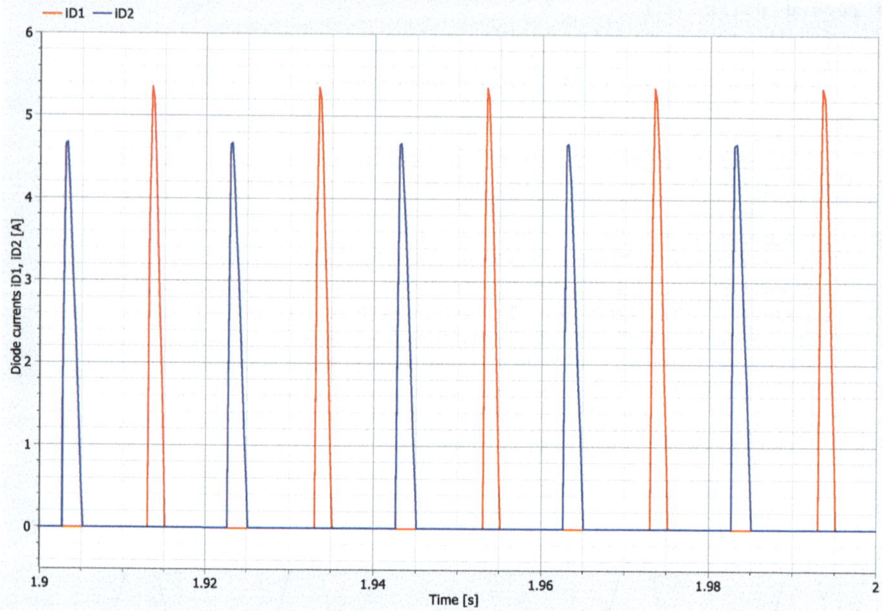

Fig. 7.7 Waveforms of the diode currents $i_{D_1}(t)$ and $i_{D_2}(t)$

The waveforms of the diode currents in Fig. 7.7 well indicate the opposite commutating of the diodes and the alternating discharging of C_1 and simultaneous charging of C_2 (blue) and the subsequent recharging of C_1. As can be seen, there is also time intervals, in which both diodes are off.

Figure 7.8 shows that capacitor C_1 does not fully charge nor fully discharge.

Finally, the ideal case often considered in the literature with the result given in Eq. (7.1) can be obtained by means of almost ideal diodes with a very low ON resistance $R_{D_1} = R_{D_2} = 0.1e - 5\,\Omega$ and a high load resistance $R_L = 10\,\text{k}\Omega$. The load resistance value $R_L = 100\,\Omega$ used in the simulation runs results in a load current of about 0.35 A.

7.2.2 *Fault Diagnosis on the Voltage Doubler*

As indicated in the circuit diagram in Fig. 7.1, it is assumed that the two capacitor voltages can be directly measured. Accordingly, from the DBG in Fig. 7.9 with detectors in inverted causality, the FSM in Table 7.1 can be obtained.

Given the two voltage detectors, all potential component faults can be detected. It is evident that a diode fault can only be detected when the diode is forward biased. As indicated in the most right column of the FSM, a faulty ON resistance of diode D_2 can be even isolated.

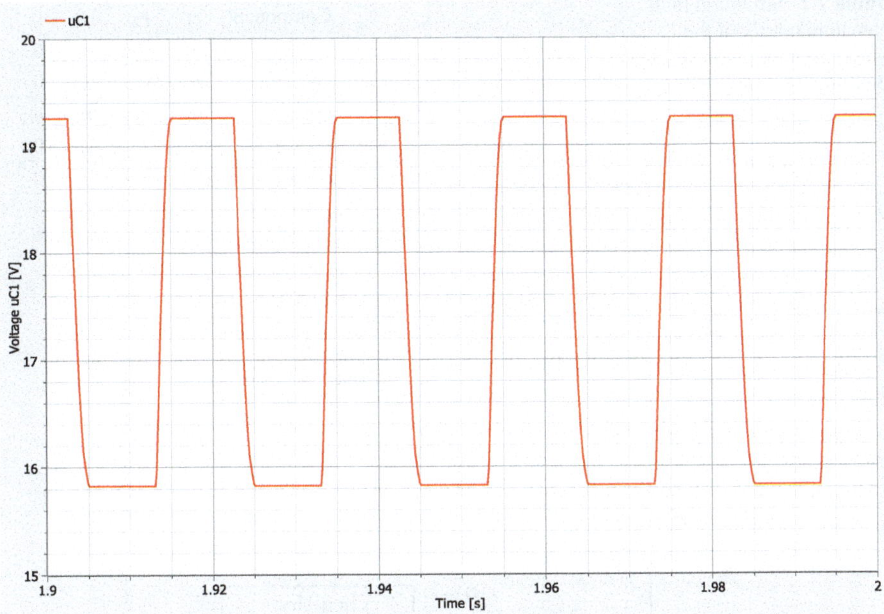

Fig. 7.8 Charge and discharge of capacitor C_1

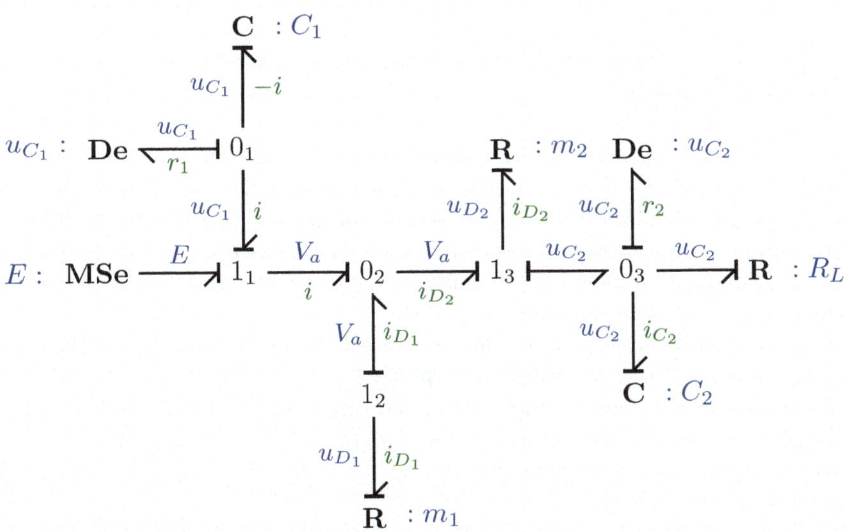

Fig. 7.9 Diagnostic bond graph of the voltage doubler with detectors for the capacitor voltages

Table 7.1 Structural fault signature matrix of the voltage doubler with voltage detectors De : u_{C_1}, De : u_{C_2}

Component	r_1	r_2	D_b	I_b
C : C_1	1	0	1	0
D_1 : R_{D_1}	m_1	0	m_1	0
D_2 : R_{D_2}	m_2	m_2	m_2	m_2
C : C_2	0	1	1	0
R : R_L	0	1	1	0

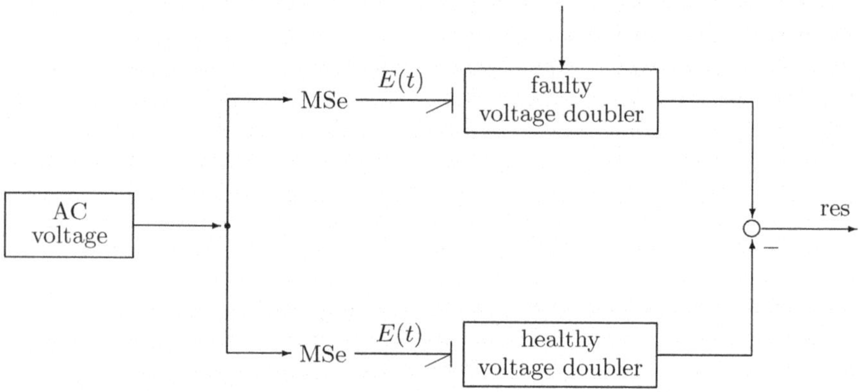

Fig. 7.10 Fault detection by means of output residuals

Fault Scenario 1: Exponential Decline of the Output Capacitance $\tilde{C}_2(t)$ as of a Time Instant t_1

FDI may be based on ARRs or on the comparison of measurements with the outputs of a model by means of an observer. In this example, the continuous states are directly measurable. Therefore, an observer is not needed. In an offline simulation, the real circuit is replaced by a model in which user defined faults can be inserted. The simulated measurements of the faulty system are compared with the outputs of the healthy system as illustrated by Fig. 7.10.

The first fault scenario assumes an exponential decline of the output capacitance C_2 in the interval [1.0, 1.6 s] from $C_2 = 4000\,\mu\text{F}$ to $\tilde{C}_2 = 4000 \cdot \exp(-2 \cdot .6)\,\mu\text{F} = 1205\,\mu\text{F}$. As can be seen from Fig. 7.11, as a result, there is a significant increase of the ripple $\Delta\tilde{V}_o$ on the faulty output voltage $\tilde{V}_o(t) = \tilde{u}_{C_2}(t)$ of the voltage doubler due to the exponential decline of $\tilde{C}_2(t)$ as of time instant $t = 1$ s as to be expected since $\Delta\tilde{V}_o \propto 1/(R_L\tilde{C}_2)$.

For simplicity, assume that in Mode 2 capacitor C_1 does not discharge and that C_2 is charged instantaneously. While C_2 discharges via the load resistor R_L, the output voltage V_o declines from a value V_o^m. The time history of this decline over the period $T = 1/f$ in the steady state regime is

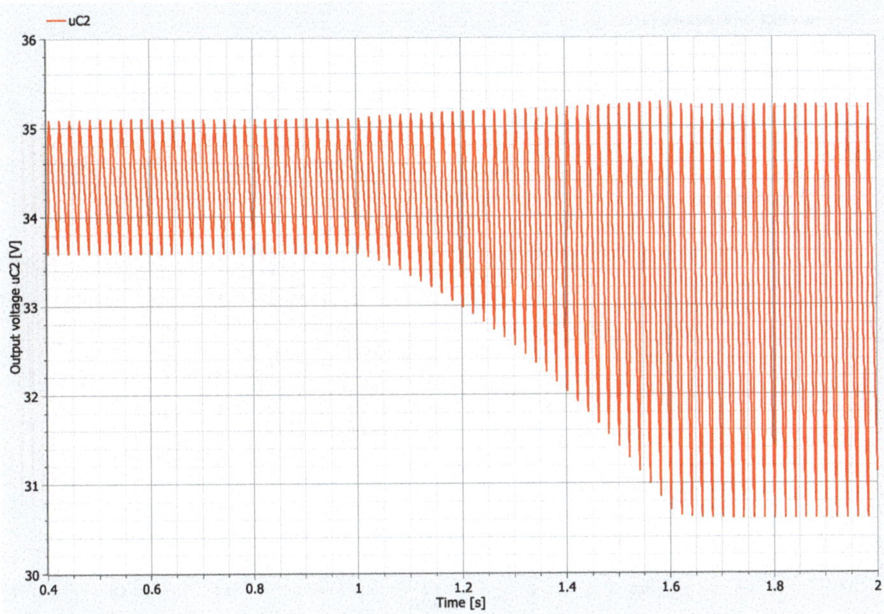

Fig. 7.11 Increase of the output voltage ripple $\Delta\tilde{V}_o$ due to the exponential decay of $\tilde{C}_2(t)$ in the time interval [1.0, 1.6 s]

$$V_o(t) = V_o^m \exp(-1/(R_L C_2 f)) \tag{7.10}$$

For the healthy voltage doubler, the ripple $\Delta V_o(t)$ then reads

$$\Delta V_o(t) = V_o^m (1 - \exp(-1/(R_L C_2 f))) \tag{7.11}$$

Due to the deliberately inserted exponential leakage, the capacitance C_2 reduces from 4000 μF to the value $\tilde{C}_2 = 1205$ μF in the interval [1.0, 1.6 s]. Accordingly, the value of the ripple $\Delta V_o(t)$ in the steady state regime is

$$\Delta V_o(t) = V_o^m (1 - \exp(-1/(R_L \tilde{C}_2 f))) \tag{7.12}$$

That is, the ripple on the output voltage of the voltage doubler increases by approximately 33%. The simulation result displayed in Fig. 7.11 agrees with this analytical estimation. A more accurate analytical estimation of the output voltage ripple that accounts for the charging and discharging of the two capacitors is given in [1].

Figure 7.12 shows the output residual $\mathrm{res}(t) = \tilde{u}_{C_2}(t) - u_{C_2}(t)$, where $u_{C_2}(t)$ is the output voltage of the healthy voltage doubler.

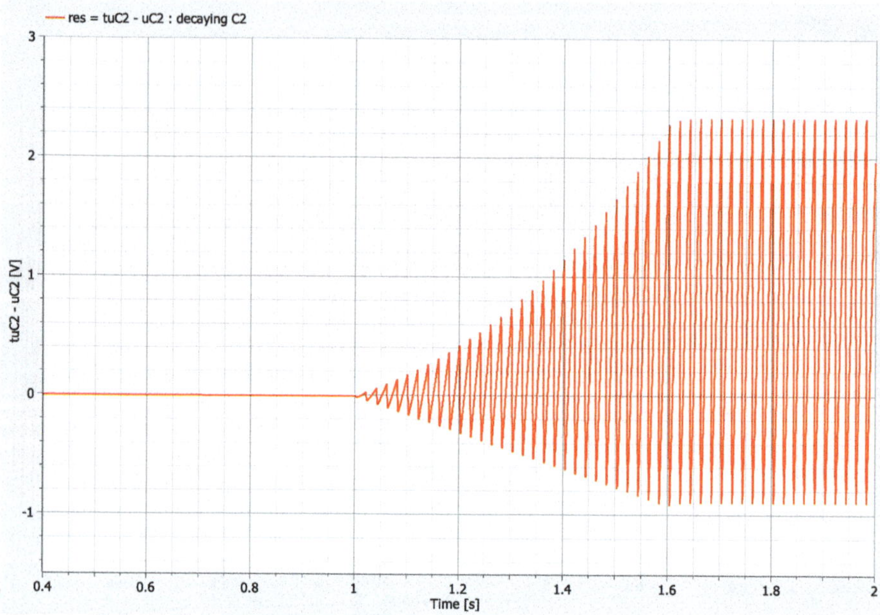

Fig. 7.12 Waveform of the output residual $\mathrm{res}(t) = \tilde{u}_{C_2}(t) - u_{C_2}(t)$

Fault Scenario 2: Open Circuit of Diode D_1

The considered power supply will not be able to deliver a DC output voltage higher than the amplitude of the AC input voltage if one of the voltage multiplier diodes fails. However, when monitoring the dynamic behaviour of a system powered by an AC-DC power source with a voltage multiplier, it may not be immediately evident that the cause for an observed anomaly is due to a diode failure in the voltage multiplier. In this second fault scenario, an abrupt open circuit failure of diode D_1 as of time instant $t_1 = .8\,s$ is inserted. In other words, diode D_1 is permanently off as of $t_1 = .8$ s. This is achieved by a slight modification of the Modelica description in Fig. 7.4.

Figure 7.13 shows the decline of the output voltage $\tilde{u}_{C_2}$ and that, in fact, there is no current through diode D_1 as of $t_1 = .8$ s. The rate of decline of $\tilde{u}_{C_2}$ depends on the values of the output capacitance C_2 and the resistive load R_L.

Figure 7.14 shows the waveform of the output residual $\mathrm{res}(t) = \tilde{u}_{C_2}(t) - u_{C_2}(t)$ according to Fig. 7.13.

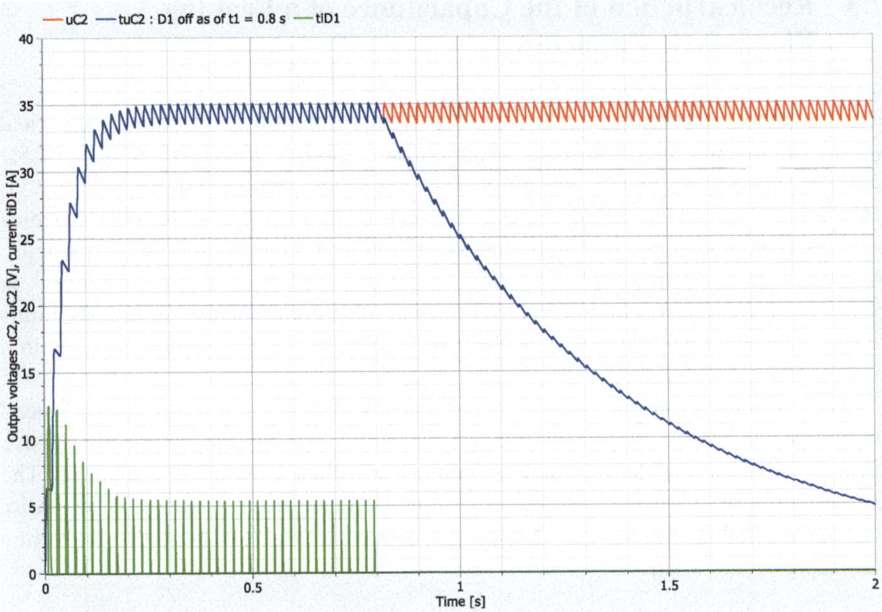

Fig. 7.13 Decline of the output voltage $u_{C_2}(t)$ due to an abrupt failure of diode D_1 as of $t_1 = .8$ s

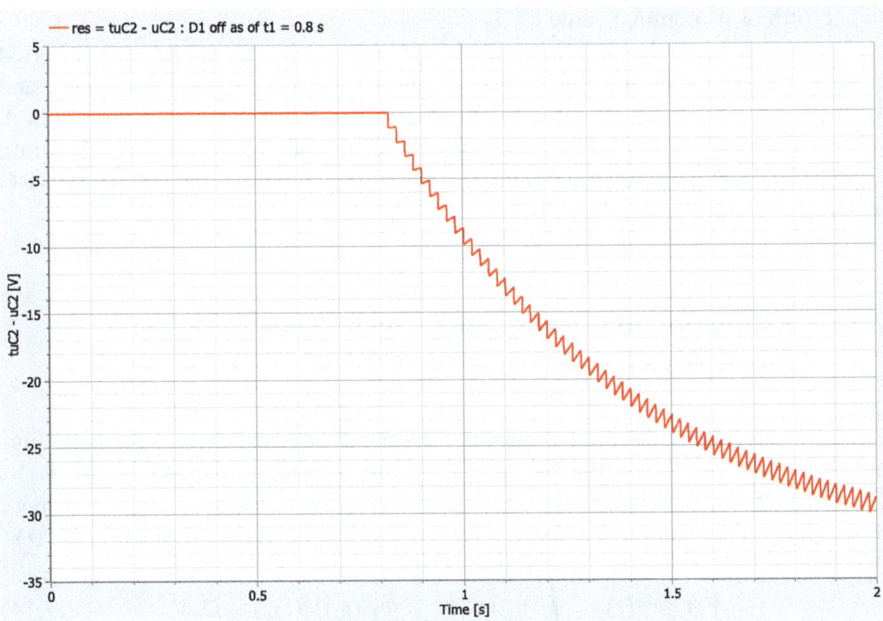

Fig. 7.14 Waveform of the output residual $\mathrm{res}(t) = \tilde{u}_{C_2}(t) - u_{C_2}(t)$

7.3 Reconstruction of the Capacitance of a Leaking Electrolytic Capacitor

In this section, the two approaches to an online determination of the numerical values of an unknown parameter degradation function presented in Sects. 4.5.1 and 4.5.2 shall be applied to the simple boost converter depicted in Fig. 4.8, p. 151. This type of circuit is often used as a component in power electronic systems such as power generation plants and is operated at high switching frequencies. According to references [15, 16], semiconductors and electrolytic capacitors in power converters have a higher failure rate than other components. As a result, faults in these components may cause performance degradation and lead to failures in a power distribution system.

In the following case study, the real circuit providing measured signals is replaced by a behavioural BG model in integral causality. The parameters of all passive element are assumed to retain their numerical value, except the capacitor. The BG model is used to study the effect of a capacitance decay. As a fault scenario, it is assumed that the decline of the capacitance $C(t)$ is exponentially with time according to the function

$$C(t) = \begin{cases} C_n & t < t_0 \\ C_n - \frac{4}{5} C_n e^{-\lambda(t-t_0)} & t \geq t_0 \end{cases} \tag{7.13}$$

That is, as of time instant t_0 the capacitance reduces exponentially with $t \to \infty$ to one-fifth of its nominal value C_n.

The objective of the offline simulation is to recover this deliberately inserted capacitance degradation from available simulated measurement data $\tilde{i}_L(t)$ and $\tilde{V}(t)$ provided by a BG model with variables averaged over the switching period. Although averaging results in some smoothing, measurement noise is taken into account by adding 1% Gaussian noise to the output signals of the behavioural BG model.

7.3.1 Estimation of the Decaying Capacitance Based on a Bicausal BG

First, the bicausal BG-based approach presented in Sect. 4.5.1 is applied. Figure 4.13 (p. 155) shows the bicausal BG with a trending capacitance $C(t) = C_n + \Phi_C(t)$. By following the causal paths from the detectors to the C element highlighted in blue and red, respectively, gives for the degradation function $\Phi_C(t)$

$$\tilde{V}(t)\Phi_C(t) = \int_{t_0}^{t} \dot{q}_C(\tau)d\tau + \Phi_C(t_0)\tilde{V}(t_{f_1}) - C_n\tilde{V} \tag{7.14}$$

(Sect. 4.5.1, Equation (4.50)).

Table 7.2 Parameters of the averaged DBG model in Fig. 4.10

Parameter	Value	Units	Meaning
E	12.0	V	Voltage supply
L	1.0	mH	Inductance
R_{L_n}	0.1	Ω	Resistance of the coil
C_n	500	μF	Nominal capacitance
R_n	5.0	Ω	Nominal load resistance
T_s	1.0	ms	Switching time period
d	0.45	–	Duty ratio
t_0	5.0	ms	Capacitance starts decline
λ	500	s^{-1}	Rate of decline

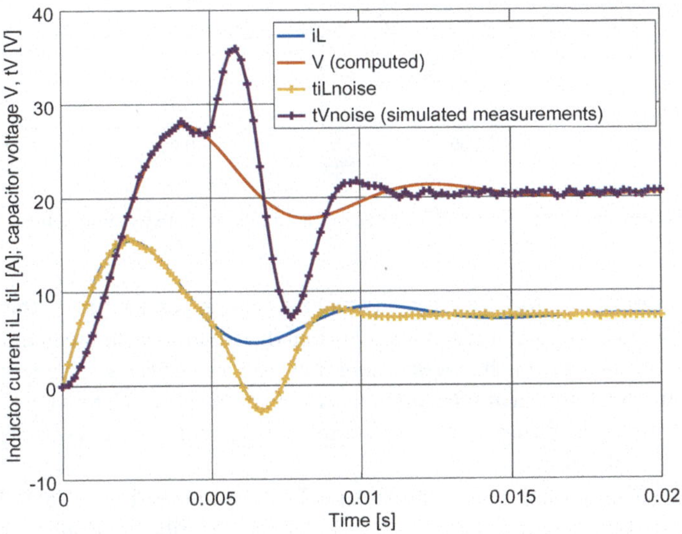

Fig. 7.15 Effect of the capacitor degradation as of $t_0 = 0.005$ s on the inductor current $\tilde{i}_L$ and the capacitor voltage $\tilde{V}$

The simulation performed by the free software GNU Octave 4.4.1 uses the parameters given in Table 7.2. The effect of the deliberately inserted capacitance degradation into the behavioural model on the inductor current $\tilde{i}_L$ and the capacitor voltage $\tilde{V}$ is displayed in Fig. 7.15 [9], in which the tilde denoting simulated measurements is substituted by the letter t prefixing the variable name.

Simulated noisy measurements are obtained by means of the Octave function `randn()`.

```
tV = tV + 0.01*tV .* randn(linspace(tV)        (7.15)
```

That is, the distribution of the generated random numbers is normal with zero mean and a variance equal to 1.

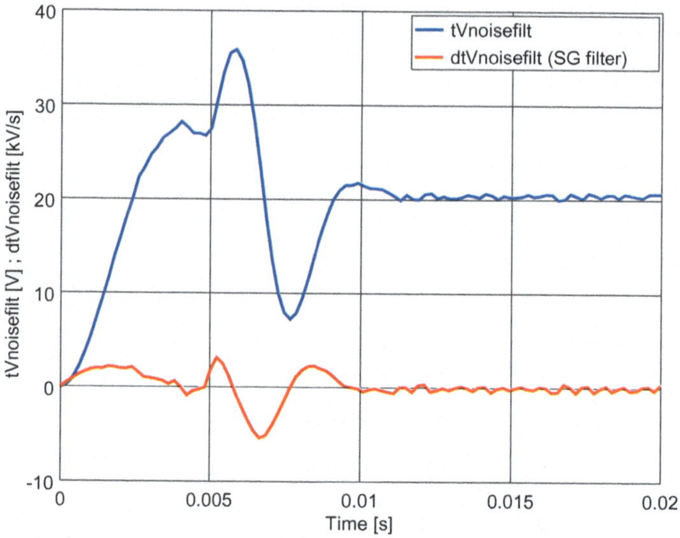

Fig. 7.16 Filtered noisy capacitor voltage tVnoisefilt and its derivative dtVnoisefilt

Other options are the function `normrnd(mu,sigma)` from the statistics package which provides samples with a normal distribution with mean `mu` and standard deviation `sigma`, or the Octave function `noise(siglen,nsigs,type)` contained in the Large Time/Frequency Analysis Toolbox (`ltfat`) [22] where `siglen` denotes the number of noise samples, `nsigs` the number of signals, and `type` the type of noise.

Figure 7.16 shows the noisy voltage signal $\tilde{V}(t)$ smoothed by a Savitzky–Golay filter (Octave function `sgolayfilt()`). As the SG filter can also provide the derivative of a smoothed measured signal (Sect. 3.2.1), the time history of $\dot{\tilde{V}}$ is also displayed for illustration. The time derivative $\dot{\tilde{V}}$ is needed in the case the load resistor R : R is deteriorating (cf. (4.44), (4.46), p. 154).

Figure 7.17 [9] displays the recovered decline $rC(t)$ of the capacitance $C(t)$. The values at sampling time points have been recovered by parameter estimation based on a bicausal BG.

The time evolution of the recovered capacitance $rC(t)$ obtained by offline simulation is sufficiently close to the decline of the capacitance $C(t)$ deliberately introduced into the behavioural model of the circuit.

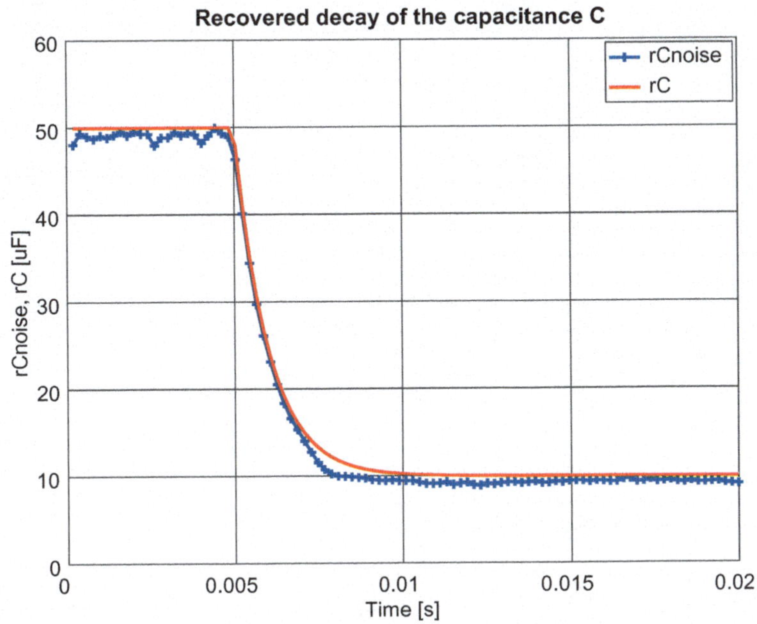

Fig. 7.17 Recovered capacitance rC(t), rCnoise(t)

7.3.2 *ARR-Based Estimation of the Capacitance Degradation Values*

The state equations of the averaged healthy system model with nominal parameters read

$$L_n \frac{di_L}{dt} = E - R_{L_n} i_L - (1-d)V \tag{7.16}$$

$$C_n \frac{dV}{dt} = (1-d)i_L - \frac{V}{R_n} \tag{7.17}$$

where d denotes the duty ratio of the signal controlling the transistor.

For the faulty system, the state equations take the form

$$L_n \frac{d\tilde{i}_L}{dt} = E - R_{L_n} \tilde{i}_L - (1-d)\tilde{V} \tag{7.18}$$

$$\frac{d}{dt}(C(t)\tilde{V}) = (1-d)\tilde{i}_L - \frac{\tilde{V}}{R_n} \tag{7.19}$$

Residual r_2^1 as an output of the first stage DBG reads

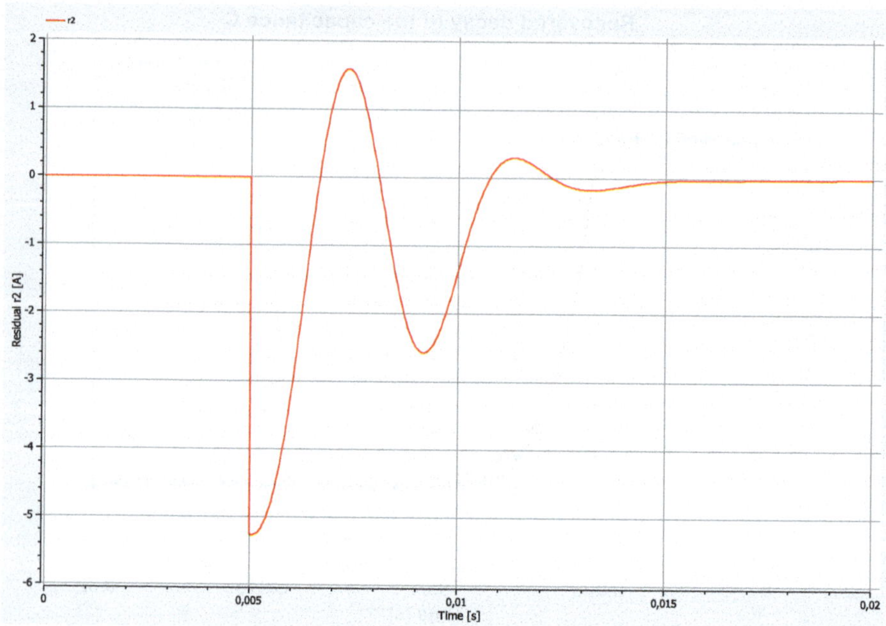

Fig. 7.18 Time evolution of residual r_2^1

$$r_2^1 = (1-d)\tilde{i}_L - C_n \frac{d\tilde{V}}{dt} - \frac{\tilde{V}}{R_n} \tag{7.20}$$

Neglecting noise on the smoothed simulated measurements, the simulation performed by means of the OpenModelica software [19] yields the time evolution of residual r_2^1 in Fig. 7.18 [8], which clearly indicates that a fault has occurred as of $t = 5$ ms.

The second stage DBG is identical with the one of the first stage except that the element $\mathrm{C} : C_n$ is replaced by the BG in Fig. 7.19.

Inserting the right faulty outputs into the second stage DBG and observing the output of the first stage DBG yields

$$\begin{aligned} r_2^2 &= 0 = (1-d)\tilde{i}_L - \frac{d}{dt}(C(t)\tilde{V}) - \frac{\tilde{V}}{R_n} \\ &= \underbrace{(1-d)\tilde{i}_L - C_n \frac{d\tilde{V}}{dt} - \frac{\tilde{V}}{R_n}}_{r_2^1} - \left(\frac{d\Phi_C}{dt}\right)\tilde{V} - \Phi_C \frac{d\tilde{V}}{dt} \end{aligned} \tag{7.21}$$

The result is a time-varying linear implicit ODE for the unknown degradation function $\Phi_C(t)$ to be determined.

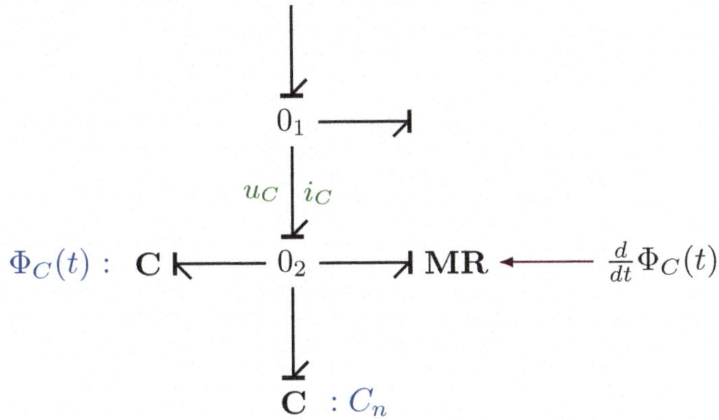

Fig. 7.19 BG model of a capacitor accounting for a degradation of its capacitance

$$\tilde{V}\frac{d\Phi_C}{dt} + \frac{d\tilde{V}}{dt}\Phi_C = r_2^1 \tag{7.22}$$

For a numerical solution, (7.22) is slightly modified in order to avoid a division by zero.

$$(\tilde{V} + \varepsilon)\frac{d\Phi_C}{dt} + \frac{d\tilde{V}}{dt}\Phi_C = r_2^1 \tag{7.23}$$

where ε is very small, i.e. $\varepsilon = 10^{-12}$.

Let $\Phi_C(t)$ be the solution of (7.23) and $rC(t) := C_n + \Phi_C(t)$ the *recovered* degradation of the capacitance. Then Fig. 7.20 shows that the time evolution of $rC(t)$ (red dashed line) matches the one of $C(t)$ with a deliberately inserted degradation (blue solid line).

Finally, Fig. 7.21 [8] confirms that, in fact, the time evolution of residual $r_2^2(t)$ as an output of the second stage DBG is actually approximately zero if the outputs of the faulty system model $\tilde{\mathbf{y}}(t)$ are input in its ARR. Numerical inaccuracies are in the order of 10^{-15} at most.

7.3.3 *RUL Prediction*

The curve fitting of the degradation data in each window w_k obtained by parameter estimation based on the DBG yields the same exponential degradation function. Let α, β, γ be the mean values of the identified coefficients of the function fitting the decline of the capacitance $C(t)$. Then the failure time t_f at which the capacitance has dropped to a critical value C_{crit}, is determined by the condition

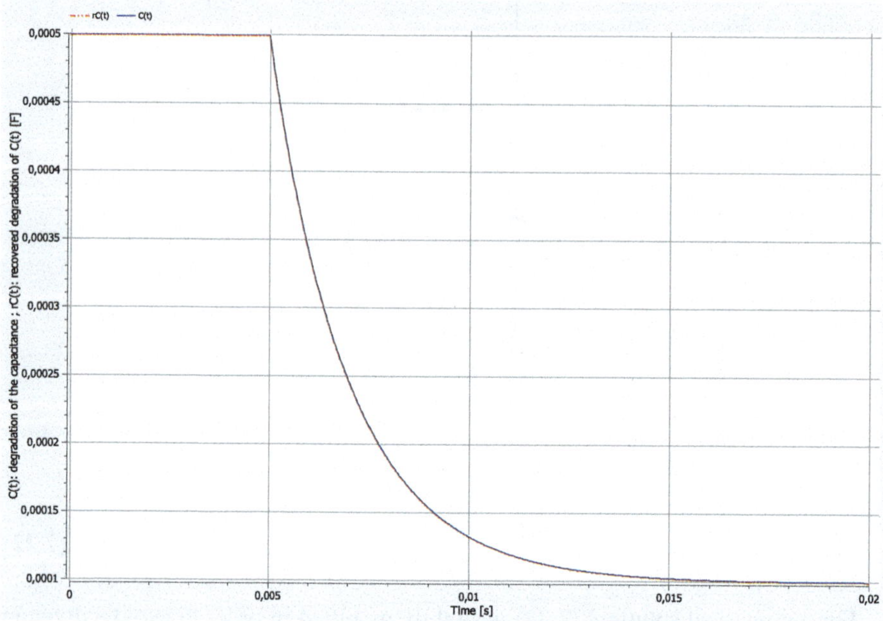

Fig. 7.20 Recovered degradation of the capacitance

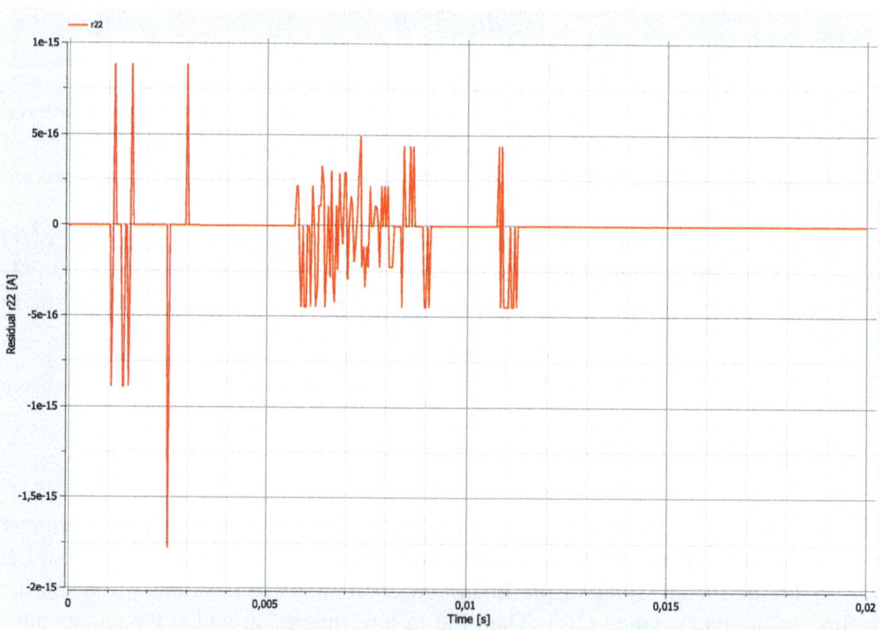

Fig. 7.21 Residual r_2^2 as an output of the second stage DBG

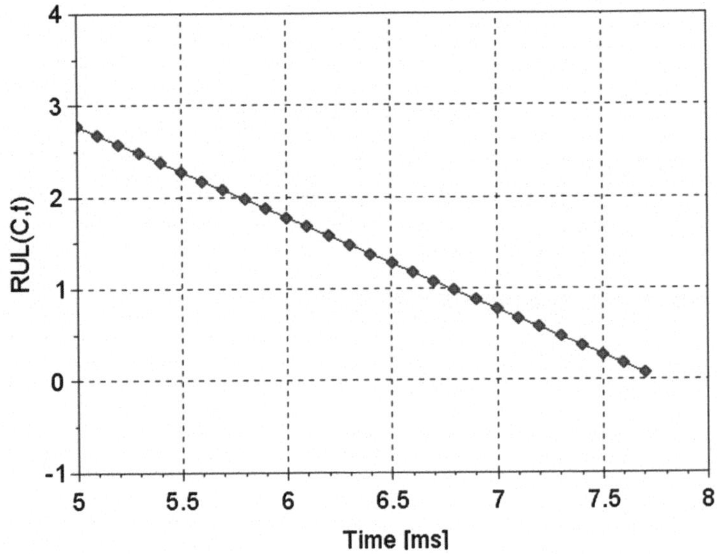

Fig. 7.22 True RUL(C,t) of the decaying capacitance C(t)

$$C(t_f) = \alpha C_n + \beta C_n e^{-\gamma(t_f - t_0)} = C_{\text{crit}} \tag{7.24}$$

Solving for t_f gives

$$t_f = t_0 - \frac{1}{\gamma}\left(-\ln\beta + ln\left(\frac{C_{\text{crit}}}{C_n} - \alpha\right)\right) \tag{7.25}$$

Equation 7.25 indicates that the failure time t_f and the $\text{RUL}(C, t) := t_f - t$ depend on the fitting parameters. The true RUL is obtained for $\alpha = 1/5$, $\beta = 4/5$, and $\gamma = 500$. These degradation parameters and the critical value $C_{\text{crit}} = 2/5C_n$ yield $t_f = 7.77$ ms and $\text{RUL}(C, t_0) = 2.77$ ms (Fig. 7.22, [9]).

7.4 External Leakage from a Closed Loop Three Tanks System

In this section, bond graph-based fault identification and failure prognostic are illustrated in a case study of a hydraulic three tanks system often considered in the literature [21] and depicted in Fig. 7.23.

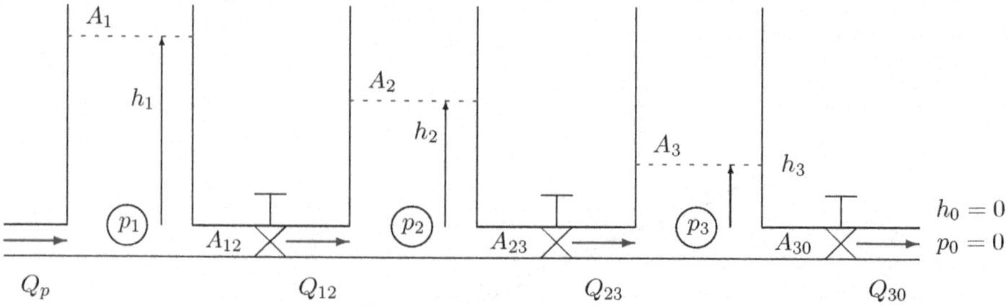

Fig. 7.23 Schematic of a hydraulic three tanks system

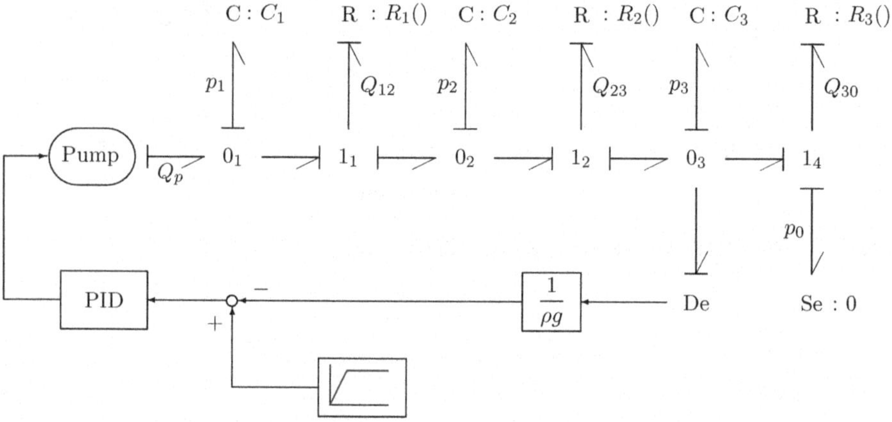

Fig. 7.24 Closed loop three tanks system

7.4.1 *Modelling and Analysis of the System*

In this example, the fluid level in the third tank, from left to right, is kept constant by means of a PID controller despite the outflow from the tank. Figure 7.24 shows a BG model and the feedback loop.

Open Loop System

The valves between the tanks are modelled as nonlinear resistors with a constitutive equation according to Bernoulli's law. The tanks are considered as hydraulic capacitors with a uniform hydrostatic pressure at their bottom.

Let A_{ij} denote the cross section of the valves between pressures p_i and p_j and let $c_{d_{ij}}$ be their discharge coefficient. The volume flow rate Q_{ij} through the valve then reads

$$Q_{ij} = c_{d_{ij}} A_{ij} \operatorname{sign}(\Delta p_{ij}) \sqrt{\frac{2}{\varrho}|\Delta p_{ij}|}$$
$$= c_{d_{ij}} A_{ij} \operatorname{sign}(\Delta h_{ij}) \sqrt{2g|\Delta h_{ij}|} \,. \tag{7.26}$$

The dynamic equations are immediately derived from the bond graph of Fig. 7.24.

$$\frac{d}{dt} h_1 = \frac{1}{A_1} (Q_p - Q_{12}) \tag{7.27}$$
$$\frac{d}{dt} h_2 = \frac{1}{A_2} (Q_{12} - Q_{23}) \tag{7.28}$$
$$\frac{d}{dt} h_3 = \frac{1}{A_3} (Q_{23} - Q_{30}) \tag{7.29}$$

An equal cross section area for all three valves implies that in steady state

$$h_1 = 3h_3 \tag{7.30}$$
$$h_2 = 2h_3 \tag{7.31}$$

and

$$Q_p = Q_{12} = Q_{23} = Q_{30} \,. \tag{7.32}$$

Adopting the values $Q_p = 1\,\text{m}^3/\text{s}$ for the flow supplied by the pump, and $A_{12} = 0.1\,\text{m}^2$ results in the steady state fluid levels in the open loop system

$$h_1 = 15.29\,\text{m}\,, \quad h_2 = 10.19\,\text{m}\,, \quad h_3 = 5.097\,\text{m} \tag{7.33}$$

as shown in Fig. 7.25.

Figure 7.26 displays the volume flow rates in the open loop system.

The parameter values used for the offline simulation of the open loop system are given in Table 7.3.

The model of the flow pump accounts for a delay in response to a step in the controller signal $u(t)$ and for a maximum outflow $Q_{p_{\max}}$ the pump can deliver. The dynamic behaviour of the pump has been approximated by a first order lag signal block. The transfer function

$$G_p(s) = \frac{k_p}{\tau s + 1} \qquad s \in \mathbb{C} \tag{7.34}$$

has been transformed into a state equation and used in an Octave script.

$$\dot{Q}_p = k_p u - Q_p \tag{7.35}$$

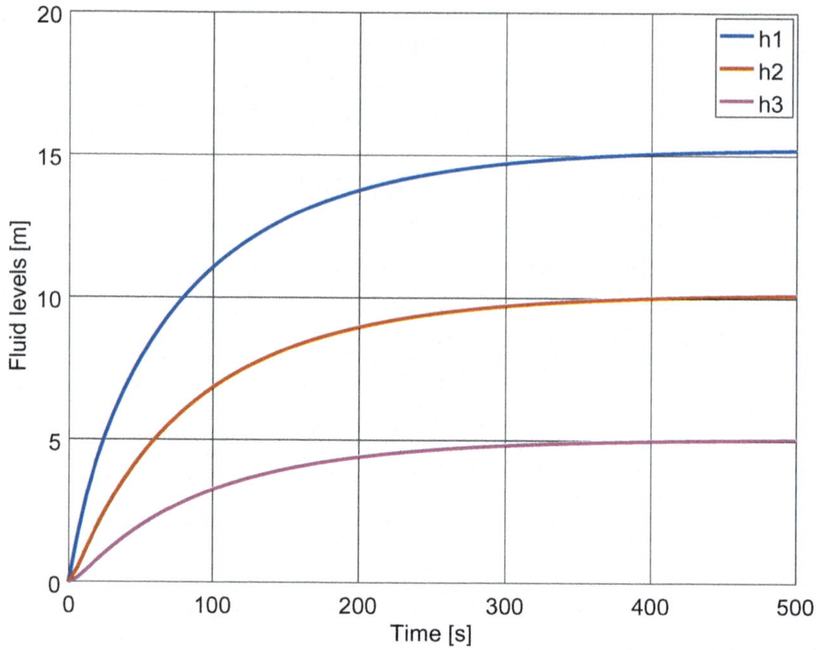

Fig. 7.25 Time evolution of the fluid levels in the open loop three tanks system

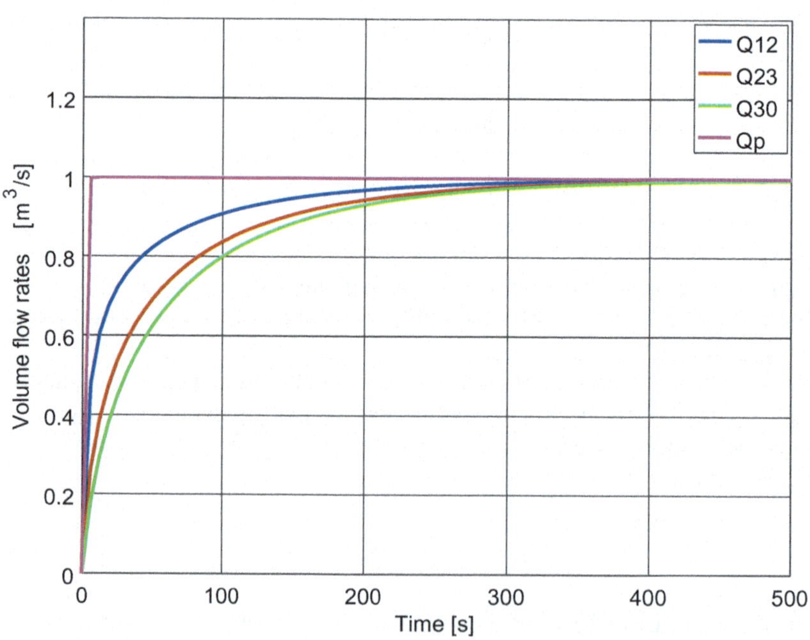

Fig. 7.26 Time evolution of the volume flow rates in the open loop three tanks system

Table 7.3 Parameters for simulation of the open loop three tank system

Parameter	Value	Units	Meaning
$A_1 = A_2 = A_3$	1.999	m^2	Cross section area of the tanks
$A_{12} = A_{23} = A_{30}$	0.1	m^2	Cross section area of the valves
ϱ	780	kg/m^3	Fluid density
Q_p	1	m^3/s	Volume flow of the supply
c_d	1	–	Discharge coefficient of the valves

Table 7.4 Parameters of the pump model

Parameter	Value	Units	Meaning
k_p	1		Proportional gain
τ	1	s	Time constant
Min	0	m^3/s	Lower bound of the pump's outflow
Max	2.5	m^3/s	Maximum outflow of the pump

Table 7.5 Parameters of the PID controller

Parameter	Value	Units	Meaning
K	5		Proportional gain
Ti	50	s	Integral time constant
Td	3	s	Derivative time constant

Numerical values of the pump parameters are given in Table 7.4.

Closed Loop System

For the simulation of the dynamic behaviour of the closed loop system, an ideal PID controller has been used.

$$u(s) = K\left[1 + \frac{1}{T_i s} + T_d s\right] e(s) \tag{7.36}$$

where $e(s) := h_{3_{\text{ref}}} - h_3(s)$. Its time domain description

$$\dot{z}(t) = \frac{K}{T_i} e(t) \tag{7.37a}$$

$$u(t) = z(t) + Ke(t) + KT_d\dot{e}(t) \tag{7.37b}$$

has been coded in an Octave script. Since $h_{3_{\text{ref}}}$ is constant $\dot{e}(t) = \dot{h}_3$. Table 7.5 gives the parameters of the PID controller.

Figure 7.27 shows the time history of the fluid levels in the tanks if all of them are initially empty and if the set point for the fluid level in the third tank is $h_{3_{\text{ref}}} = 1$ m. As can be seen, the PID controller achieves the constant reference fluid level in Tank 3 as of about 40 s. According to Equations (7.30) and (7.31), the steady state fluid levels in the other two tanks are $h_1 = 3$ m and $h_2 = 2$ m.

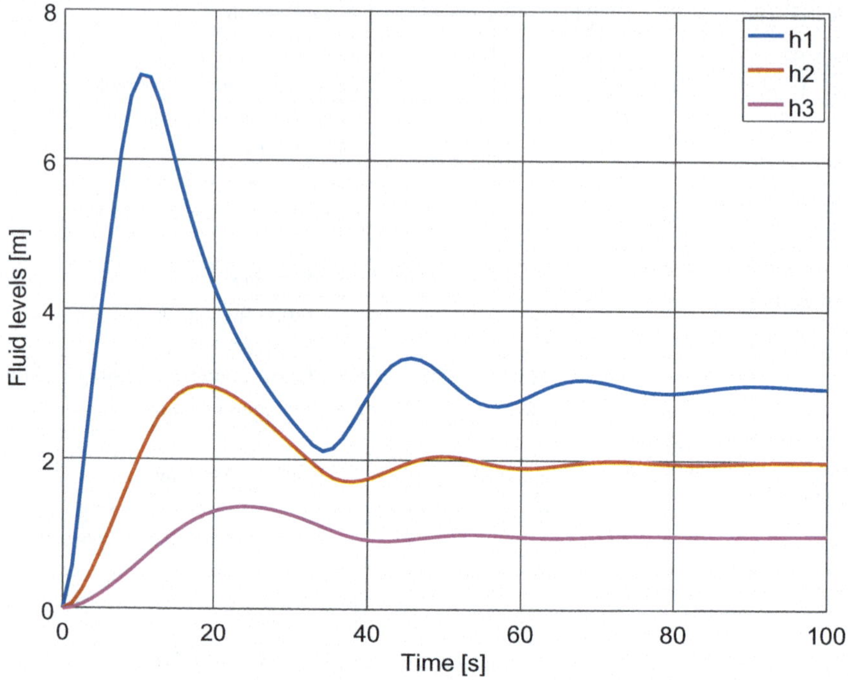

Fig. 7.27 Time evolution of the fluid levels in the closed loop three tanks system

Figure 7.28 shows the dynamics of the volume flows in the controlled system. According to (7.26), the steady state outflow of Tank 3 is

$$Q_{30} = A_{30}\sqrt{2g} = 0.1\sqrt{2 \times 9.81} = 0.4429\,\text{m}^3/\text{s} \tag{7.38}$$

Note that the volume flow rate of the pump is temporarily bounded.

The simulation results above agree with those reported in [14]. For further reading on bond graph modelling of controlled hydraulic systems, it is referred to [2, 12].

Fault Scenario: Hole of Increasing Size in the Bottom of Tank 3

For the offline study of a fault scenario, a leakage of progressing amount from Tank 3 is deliberately introduced that becomes effective as of some time instant t_1. As of time t_2, the hole does not increase in size any more (Fig. 7.29).

Figure 7.30 shows the fluid levels in the tanks of the faulty system. As can be seen, at $t_1 = 100$ s the fluid levels reach those of the healthy system. In the time interval from $t_1 = 100$ to $t_2 = 250$ s the fluid levels in the first two tanks rise, which means that the controller demands that the pump delivers more fluid into the

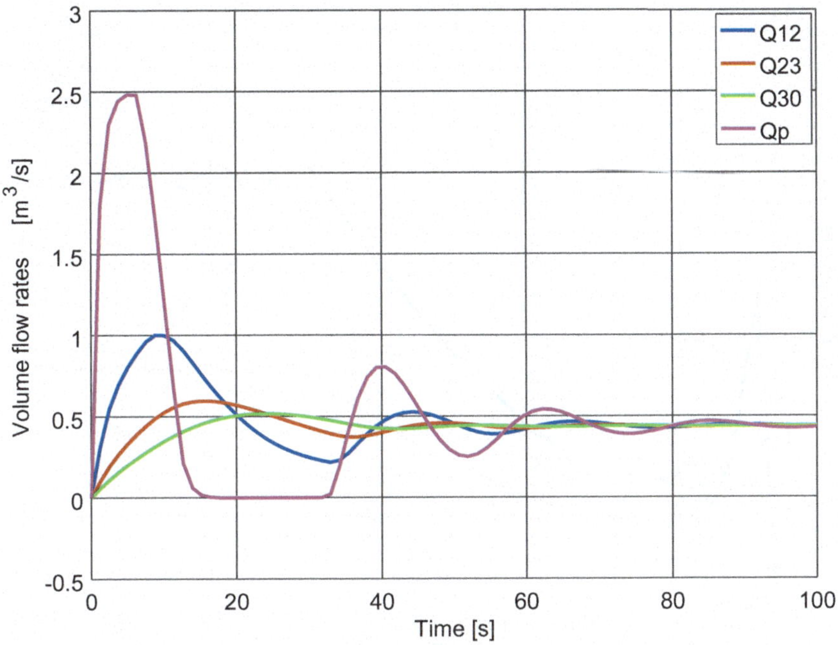

Fig. 7.28 Time evolution of the volume flow rates in the closed loop three tanks system

Fig. 7.29 Leakage volume flow rate from Tank 3

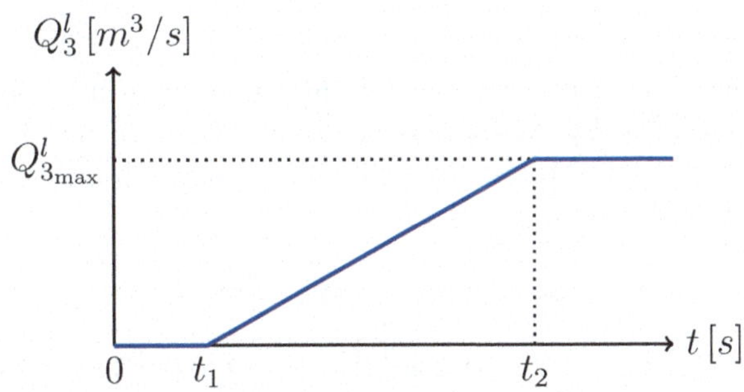

system in order to compensate for the leakage in the third tank. Despite the outflow through a hole of increasing size the controller is able to keep the level in Tank 3 at the constant reference height $h_{3_{\text{ref}}} = 1$ m at the expense of a higher energy demand. Whether the increased fluid levels in the first two tanks are acceptable depends on their size and the process.

From Fig. 7.31 it can be seen that, in fact, the pump delivers more fluid to compensate for the loss of fluid from Tank 3. According to the constant level, the steady state volume flow rate into the third tank at $t = t_1$ is $Q_{23} = 0.44\,\text{m}^3/\text{s}$.

In case the time evolutions of fluid levels and volume flow rates have not been simulated but are sensor readings from the real system they indirectly suggest that

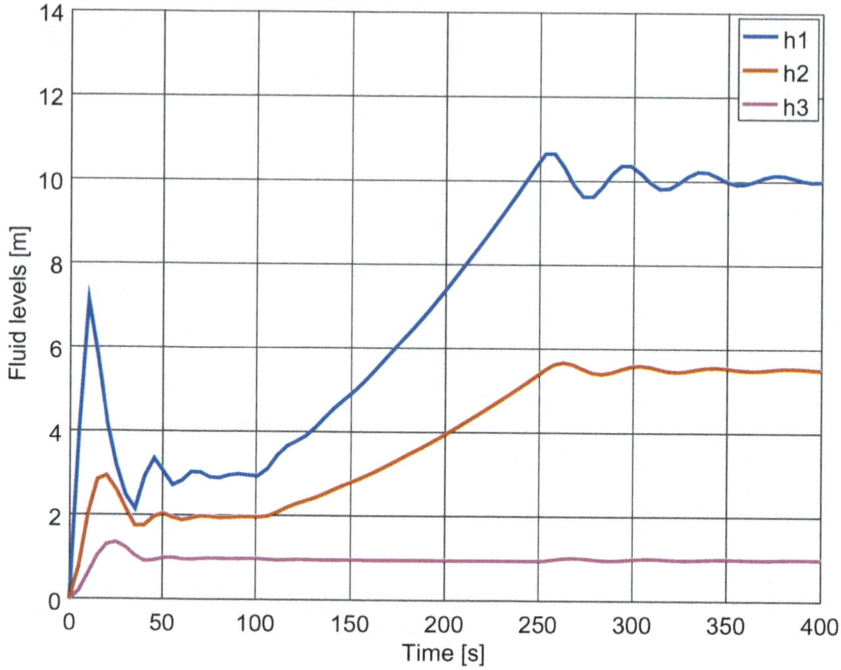

Fig. 7.30 Fluid levels in the faulty tanks system

there is a fault either in the third tank or in the outlet valve R : $R_3()$. This is confirmed by ARRs derived from a DBG of the three tanks systems and their structural FSM. If there is a fluid level sensor for each tank, then three ARRs can be derived.

$$r_1 = A_1 \dot{h}_1 - A_{12}\sqrt{2g(h_1 - h_2)} \tag{7.39}$$

$$r_2 = A_{12}\sqrt{2g(h_1 - h2)} - A_2 \dot{h}_2 - A_{23}\sqrt{2g(h_2 - h_3)} \tag{7.40}$$

$$r_3 = A_{23}\sqrt{2g(h_2 - h3)} - A_3 \dot{h}_3 - A_{30}\sqrt{2g(h_3 - h_0)} \tag{7.41}$$

The FSM in Table 7.6 shows that given three fluid level sensors, a fault in the third tank cannot be distinguished from a fault in its outlet valve which is not surprising. For fault isolation, a sensor for the flow through the outlet valve must be added. Likewise, a fluid level sensor does not enable to discriminate between internal pump leakage and leakage from the first tank.

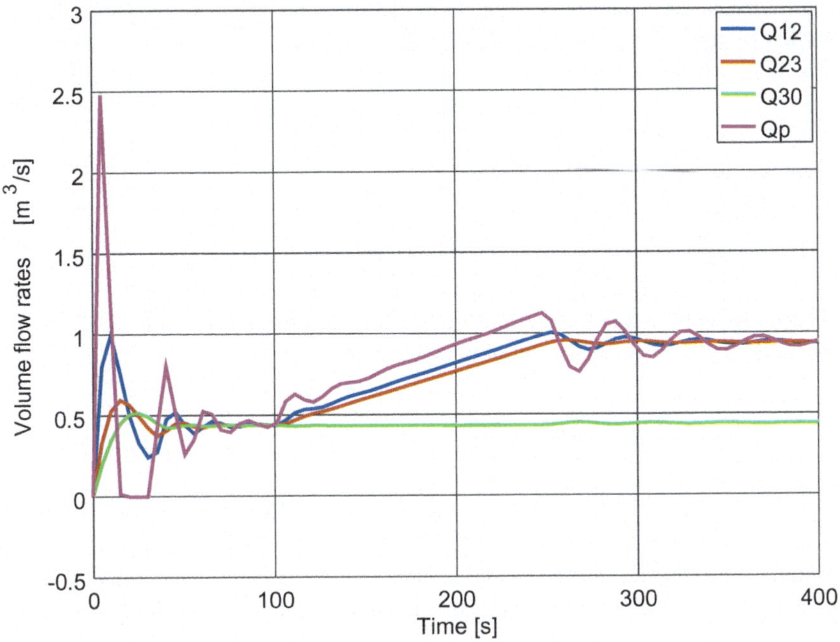

Fig. 7.31 Volume flow rates in the faulty tanks system

Table 7.6 Structural fault signature matrix of the three tanks system with level detectors De : h_1, De : h_2, and De : h_3

Component	ARR_1	ARR_2	ARR_3	D_b	I_b
Pump : Q_p	1	0	0	1	0
$Tank_1$: A_1	1	0	0	1	0
$Valve_1$: $R_1()$	1	1	0	1	1
$Tank_2$: A_2	0	1	0	1	1
$Valve_2$: $R_2()$	0	1	1	1	1
$Tank_3$: A_3	0	0	1	1	0
$Valve_3$: $R_3()$	0	0	1	1	0

Fault Estimation

Suppose that the fault can be attributed to the third tank and that the inlet valve and the outlet valve of the third tank can be confirmed to be faultless. Then it is a change in the bottom area A_3 of Tank 3 that yields a residual r_3 different from zero.

$$0 < r_3 = Q_{23} - A_3\dot{h}_3 - Q_{30} \tag{7.42}$$

As h_3 is constant for $t > 100$ s and $r_3 > 0$, the inflow Q_{23} from Tank 2 must be higher than the outflow Q_{30}. That is, r_3 equals a leakage flow $Q_l = Q_{23} - Q_{30}$. Let A_3^l denote the size of the hole in the bottom of the third tank. Then the fault can be estimated at each sampling time point. To that end, the leakage from Tank 3 is modelled by a valve that increasingly opens with time.

$$r_3(t) = Q_l(t) = A_3^l(t)\sqrt{2gh_3} \tag{7.43}$$

As a result,

$$A_3^l(t) = \frac{r_3(t)}{\sqrt{2g}} \tag{7.44}$$

The same result is obtained from the bicausal BG fragment in Fig. 7.32.
From the BG in Fig. 7.32, one obtains immediately

$$\begin{aligned} Q_l &= Q_{23} - A_3\frac{d}{dt}h_3 - Q_{30} \\ &= A_{23}\sqrt{2g(h_2 - h_3)} - A_3\frac{d}{dt}h_3 - A_{30}\sqrt{2gh_3} = A_3^l\sqrt{2gh_3} \end{aligned} \tag{7.45}$$

Clearly, online fault diagnosis and failure prognosis cannot see ahead the entire time evolution of relevant features as in an offline simulation of a fault scenario.

7.4.2 RUL Estimation

Once a fault has been detected and isolated, it is not known a priori whether an incipient fault has happened and what degradation trend its values will follow. However, as proposed in Sect. 4.5, a best fitting mathematical function can be found for the ARR evaluations obtained from measurements in a window moving forward

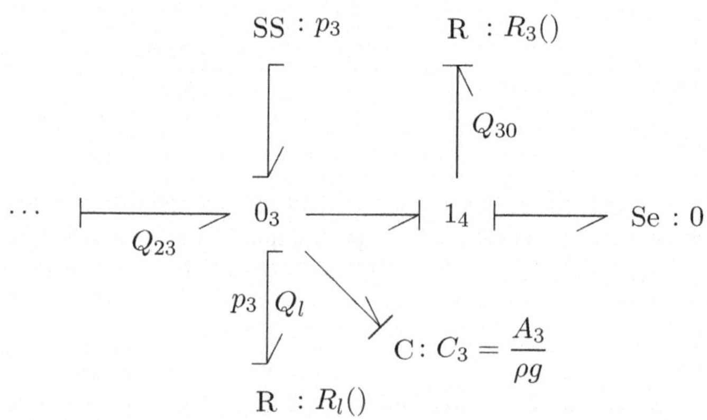

Fig. 7.32 Bicausal BG fragment of the BG in Fig. 7.24

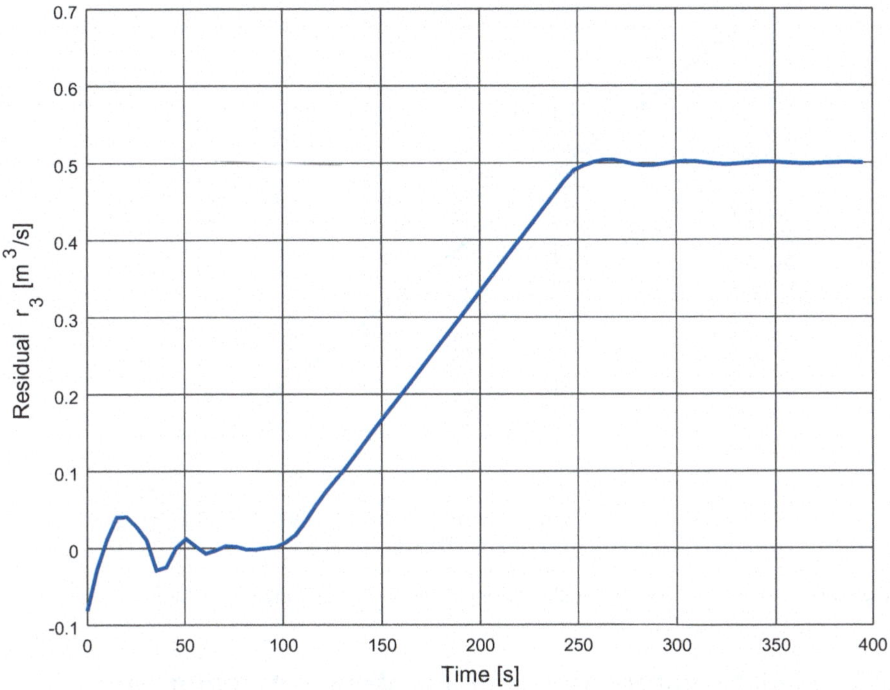

Fig. 7.33 Time evolution of residual r_3

in time. This function can be projected into the future in order to obtain a sequence of RUL estimates. In this offline simulation case study, the time evolution of residual $r_3(t)$ can be simply calculated. Figure 7.33 shows the result.

After the initially empty tanks have been filled and the controller has enforced the reference fluid level in Tank 3, residual $r_3(t)$ is equal to zero until a leakage flow starts at $t_1 = 100\,\mathrm{s}$ through a hole of a size that increases linearly with time. As of $t_2 = 250\,\mathrm{s}$ the size of the hole remains constant.

Suppose that the maximum allowable size of the hole may be 5% of the cross section area of Tank 3. That is, residual $r_3 = Q_l$ must not exceed a value of $0.443\,\mathrm{m}^3/\mathrm{s}$. This value is reached at about $t_{\mathrm{EoL}} = 234\,\mathrm{s}$. Accordingly, Fig. 7.34 shows the true RUL.

In case the value $Q_l = 0.443\,\mathrm{m}^3/\mathrm{s}$ is still admissible, (7.45) determines the fluid level in Tank 2 that would be needed to accommodate for the leakage and to ensure the set point fluid level $h_3 = 1\,\mathrm{m}$.

$$Q_l = A_{23}\sqrt{2g(h_2 - 1)} - A_{30}\sqrt{2g} \tag{7.46}$$

Inserting numerical values into (7.46) yields $h_2 = 5.0\,\mathrm{m}$ in accordance with the simulation results displayed in Fig. 7.30, i.e. $h_2(t = 234\,s) = 5.0\,\mathrm{m}$.

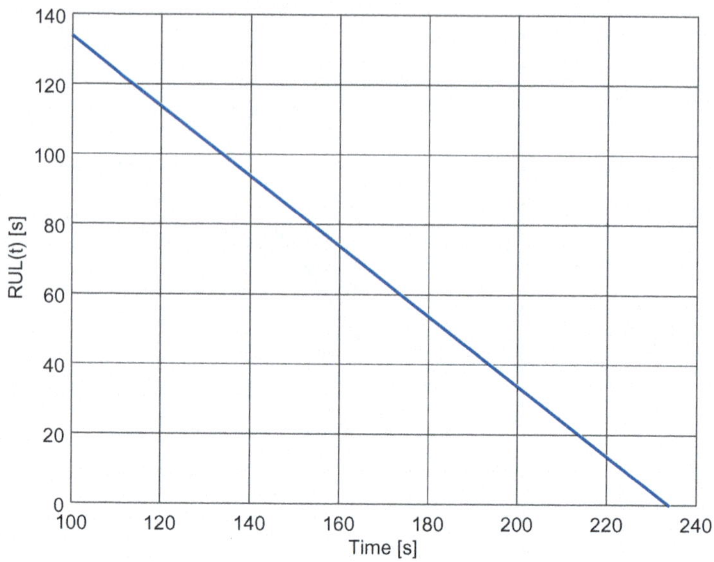

Fig. 7.34 True RUL of the three tanks system in the case of a leakage from Tank 3

7.5 Fault Signature Matrix of a Hydraulic Actuator with Leakage

Leakage is a common fault in fluid power systems. In [17], Liang An and Sepehri consider internal and external leakage from a hydraulic actuator. Internal cross-port leakage in an actuator is of concern since it reduces the actuator's power to move a load and increases the damping. In the worst case of a complete failure of the piston seals, the actuator will not be able to displace the load any more as required. External leakage into the environment may appear due to wear and tear of the bearing seals, due to abrasion and contamination, may pollute the environment, result in a sluggish dynamic behaviour, and reduce the efficiency of a closed loop system. In [17], the authors use sensors for the chamber pressures and the displacement of the rod and an extended Kalman filter (Sect. 3.4.1) based on a mathematical model of the healthy system for leakage fault detection and perform fault scenarios on a test rig by opening bleeding valves. Results indicate the inserted faults and enable to identify the type of leakage fault.

In the following, a diagnostic BG model of a double-acting cylinder with a double-ended piston rod is developed that accounts for internal cross-port as well as external leakage between the rod and the cylinder. From the DBG, a FSM is derived which shows that the easy to measure chamber pressures and the displacement of the actuator rod are sufficient to detect and to isolate internal and external leakage. Figure 7.35 shows a hydraulic circuit diagram. The actuator is powered by a variable pump via a three-way servovalve.

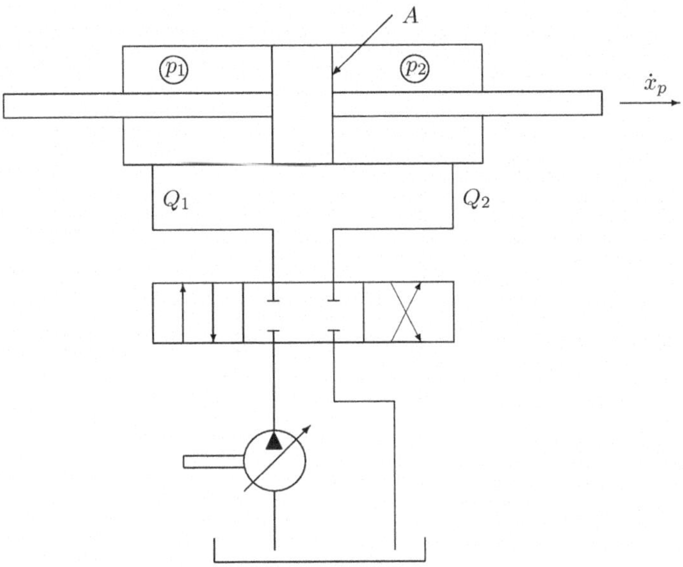

Fig. 7.35 Hydraulic circuit diagram

In [5, 6], the author presents an energetically correct BG model of the viscous fluid flow and the laminar flow across a cylinder piston and the forces they cause on the piston. To this BG model of a hydraulic double-acting cylinder, linear resistors are added that account for external leakage from the chambers following a suggestion of Merritt [18], and detectors in inverted causality delivering measurements into the DBG depicted in Fig. 7.36.

The DBG in Fig. 7.36 shows that the internal leakage Q_{il} is caused by the laminar fluid flow Q_{HP} due to the difference of chamber pressures and a volume flow rate Q_N entrained by the motion of the piston. The fluid flow through the clearance exerts the shear force F_N and the force F_{HP} on the piston. The latter one lowers the friction effect caused by the viscous flow due the motion of the piston.

Stick-slip friction on the seals of a hydraulic actuator is significant and is taken into account by the nonlinear resistor $\mathrm{R} : R_f()$. The linear resistors $\mathrm{R} : R_{el_1}$ and $\mathrm{R} : R_{el_2}$ model external leakage from the chambers. The volume flow rates at the ports of the servovalve controlled by the displacement of its spool are Q_1 and Q_2, respectively. The electrical command input into the servovalve is $u(t)$.

From the DBG in Fig. 7.36 with three detectors, the following three ARRs can be easily derived:

$$0_1 : \mathrm{ARR}_1 : r_1 = Q_1 - C_1 \dot{p}_1 - Q_{el_1} - Q \tag{7.47}$$

$$0_2 : \mathrm{ARR}_2 : r_2 = Q - C_2 \dot{p}_2 - Q_{el_2} - Q_2 \tag{7.48}$$

$$1_2 : \mathrm{ARR}_3 : r_3 = A_p(p_1 - p_2) + F_{HP} - F_N - F_c - F_{\text{load}} - m_p \ddot{x}_p \tag{7.49}$$

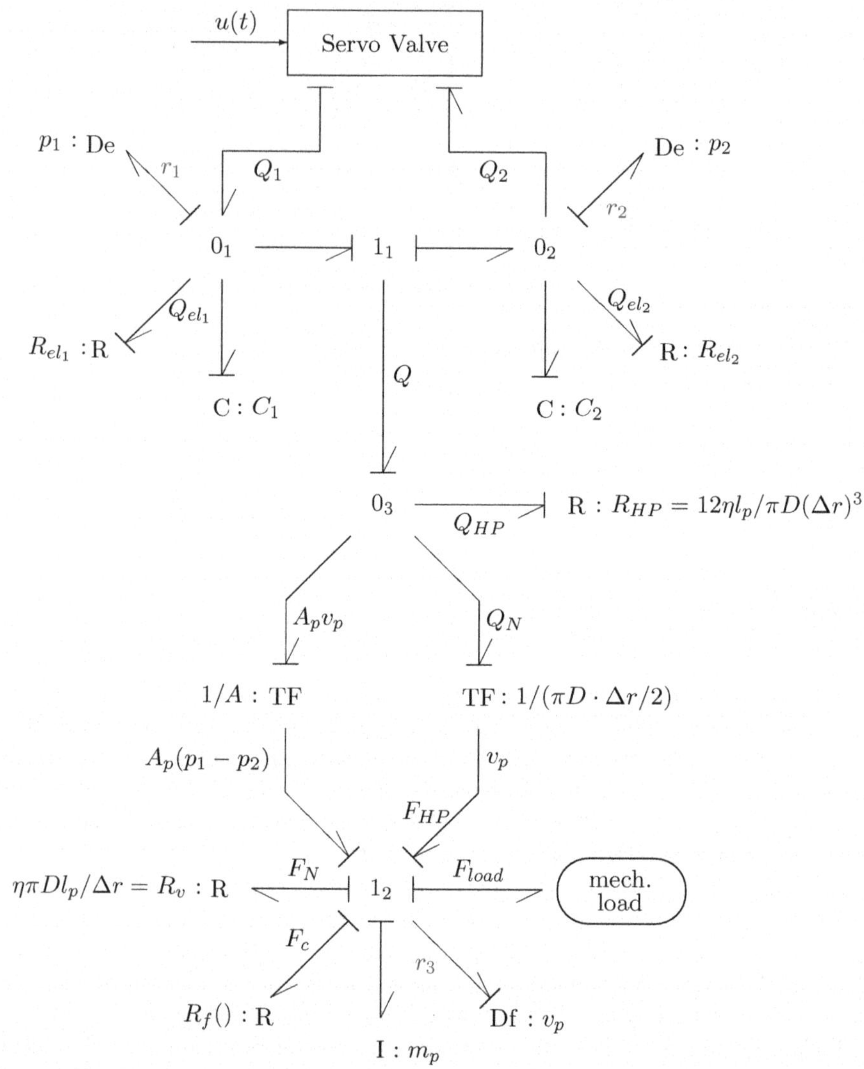

Fig. 7.36 DBG of a double-acting cylinder with double-ended piston rod accounting for internal and external leakage

where

$$Q_{HP} = \frac{\pi D (\Delta r)^3}{12 \eta l_p} (p_1 - p_2) \tag{7.50}$$

$$Q_N = \pi D \frac{\Delta r}{2} v_p \tag{7.51}$$

Table 7.7 Structural fault signature matrix of the hydraulic cylinder with detectors De : p_1, De : p_2 and Df : v_p

	$\mathrm{ARR}_1 : r_1$	$\mathrm{ARR}_2 : r_2$	$\mathrm{ARR}_3 : r_3$	D_b	I_b
$Q_{el_1} : R_{el_1}$	1	0	0	1	1
$Q_{el_2} : R_{el_2}$	0	1	0	1	1
$Q_{il} \ : \Delta r$	0	1	1	1	1

$$Q_{il} = Q_N + Q_{HP} \tag{7.52}$$

$$Q = Q_{il} + A_p \dot{x}_p \tag{7.53}$$

$$F_{HP} = \pi D \frac{\Delta r}{2} (p_1 - p_2) \tag{7.54}$$

$$F_N = \eta \frac{\pi D l_p}{\Delta r} v_p \tag{7.55}$$

$$Q_{el_i} = \frac{1}{R_{el_i}} p_i \quad i = 1, 2 \tag{7.56}$$

and D denotes the piston's diameter, l_p its length, Δr the radial clearance, and η the fluid's viscosity.

The FSM in Table 7.7 captures the structure of the ARRs.

As can be seen, external leakage from each chamber and internal cross-port leakage can be detected and isolated given that the two pressure chambers and the piston displacement are measured.

Suppose that there is no external leakage from the chambers but internal cross-port leakage and that the size of radial clearance Δr shall be estimated. To that end, Fig. 7.37 depicts a part of the DBG in Fig. 7.36 with a bicausal path from detector De : p_1 to resistor R : R_{HP} highlighted in blue colour.

Following the bicausal path one obtains an equation that determines the radial clearance Δr.

$$Q_{HP} = Q - A_p v_p - Q_N \tag{7.57}$$

$$\frac{\pi D (\Delta r)^3}{12 \eta l_p} (p_1 - p_2) = Q_1 - C_1 \dot{p}_1 - A_p \dot{x}_p - \pi D \frac{\Delta r}{2} \dot{x}_p \tag{7.58}$$

As a second fault scenario it is assumed that there is no internal leakage but external leakage from the left chamber. Figure 7.38 shows a part of the DBG with a bicausal path from detector De : p_1 to the leakage resistor R : R_{el_1}, from which an equation estimating its resistance R_{el_1} can be derived.

$$Q_{el_1} = Q_1 - C_1 \dot{p}_1 - Q \tag{7.59}$$

$$\frac{1}{R_{el_1}} p_1 = Q_1 - C_1 \dot{p}_1 - (Q_2 + C_2 \dot{p}_2) \tag{7.60}$$

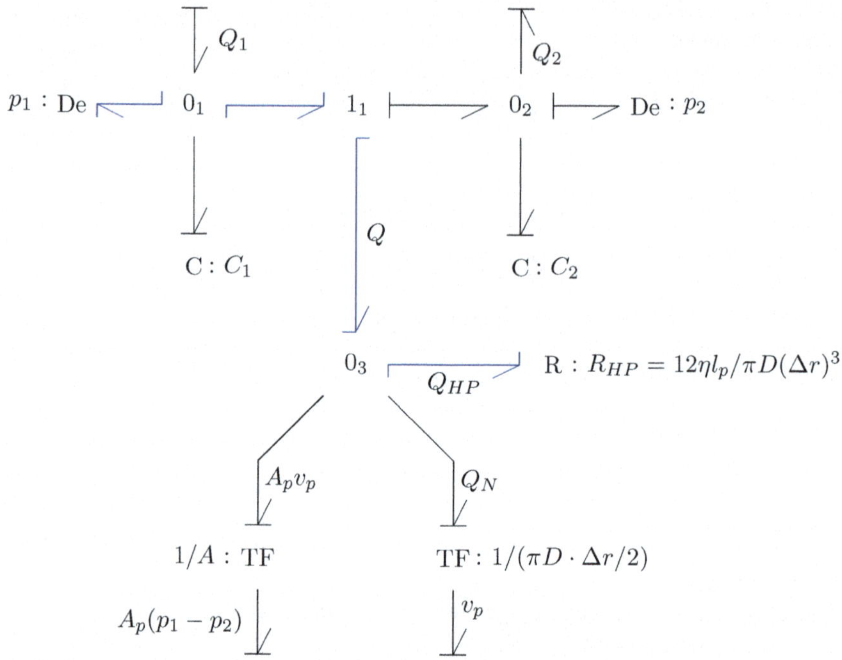

Fig. 7.37 Part of the DBG in Fig. 7.36 with a bicausal path from De : p_1 to R : R_{HP}

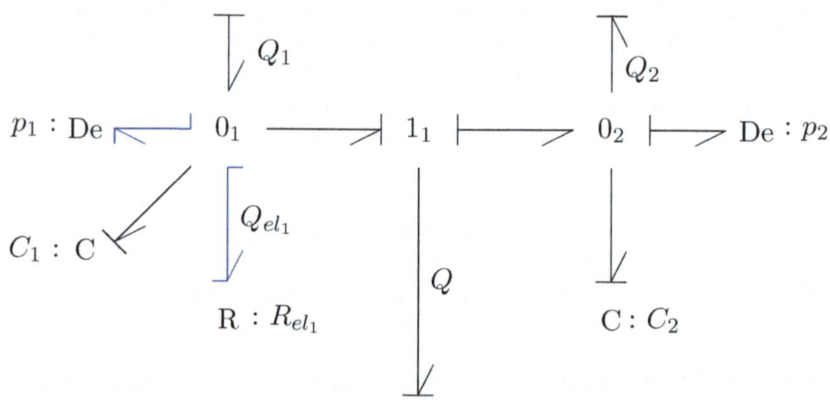

Fig. 7.38 Part of the DBG in Fig. 7.36 with a bicausal path from De : p_1 to R : R_{el_1}

Similarly, the coefficient for the external leakage from the right chamber can be estimated.

7.6 Internal Friction in a Permanent Magnet DC Motor

DC motors are widely used as a basic component for high precision positioning in industrial machinery, e.g. in semiconductor manufacturing systems as well as in robotics. A humanoid robotic hand, for instance, may need up to 15 separately controlled small DC motors for precise independent positioning of the fingers. A prevailing problem for precise motion control, however, is friction, and accurate friction values are generally not known in advance. Accordingly, various publications on DC motor parameter identification and compensation of friction by feedforward control may be found in the literature [4, 11, 23].

In general, DC motors are quite reliable and require little maintenance. Misalignment and shaft imbalances can cause bearing failures and excessive current draw can result in a motor overload.

The following offline simulation study considers a permanent magnet DC motor driven by a buck converter. The angular velocity of the load is controlled in a feedback loop as depicted in Fig. 7.39. As of some time instant, a linear increase of friction inside the motor is deliberately introduced. The objective is to detect this fault and to predict the RUL.

7.6.1 *Modelling of the DC Motor Drive*

The circuit schematic of the buck converter transforms immediately into the BG in Fig. 7.40 in which the transistor Q and the diode D are modelled as ideal switches Sw.

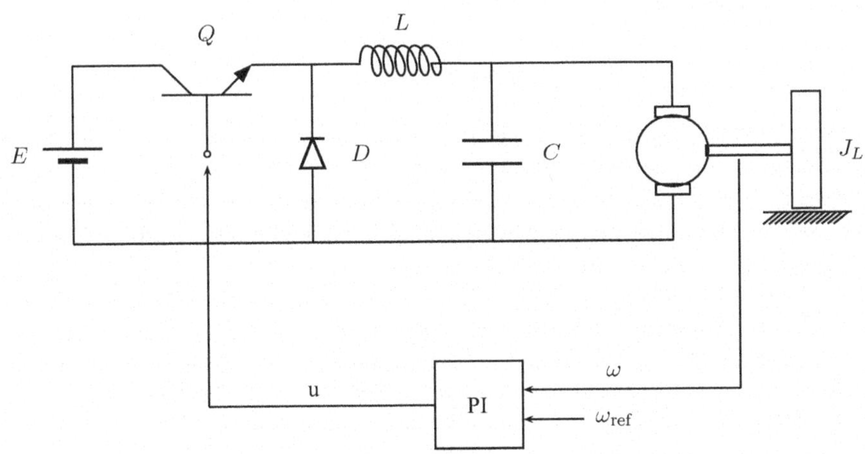

Fig. 7.39 Schematic of a closed loop DC motor drive

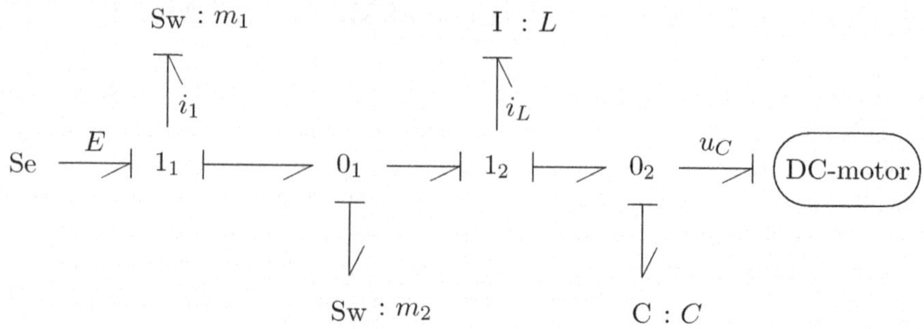

Fig. 7.40 BG of the buck converter circuit

The BG of the buck converter can be simplified if the switches are considered ideal with linear *implicit* constitutive equations.

$$\begin{aligned}0 &= m_1 u_1 + \bar{m}_1 i_1 \\ &= m_1(E - u_2) + \bar{m}_1 i_1 \end{aligned} \tag{7.61}$$

$$\begin{aligned}0 &= m_2 u_2 + \bar{m}_2 i_2 \\ &= m_2 u_2 + \bar{m}_2(i_1 - i_L) \end{aligned} \tag{7.62}$$

Elimination of switch variable i_1 yields

$$(\bar{m}_1 m_2 + m_1 \bar{m}_2) u_2 = m_1 \bar{m}_2 E + \bar{m}_1 \bar{m}_2 i_L \tag{7.63}$$

In this circuit, the two switches commutate oppositely, i.e. $\bar{m}_2 = m_1$. Therefore, (7.63) simplifies to

$$u_2 = m_1 \bar{m}_2 E = m_1 E \tag{7.64}$$

The switching signal $m_1(t)$ controlling the transistor Q can be replaced by its duty ratio d turning the buck converter model into a model with variables averaged over the duty cycle. The resulting model is used in the BG of the DC motor drive (Fig. 7.41) and captures the free wheel mode and the load mode. In closed loop, the duty ratio is controlled by the controller signal.

The BG of the DC motor accounts for the rotor inertia I : J_m and friction in the motor R : R_m. The shaft is modelled by a torsion spring of high stiffness. If the shaft is considered rigid, then inertia and friction of the motor can be combined with the inertia of the load I : J and the external friction R : R as the DBG in Fig. 7.42 shows. The load rotates against a time-varying torque $T(t)$.

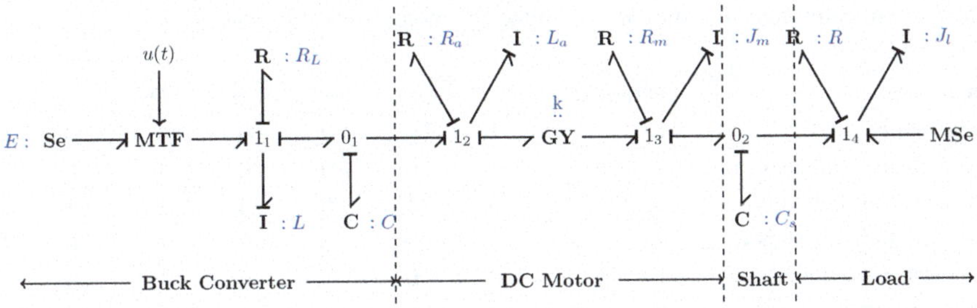

Fig. 7.41 BG of the open loop DC motor drive

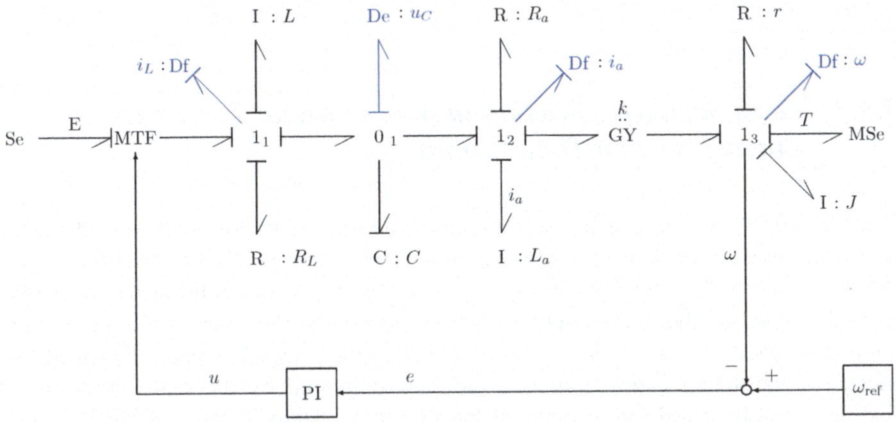

Fig. 7.42 DBG of the DC motor drive with four sensors

7.6.2 *Fault Detection*

From the DBG with four detectors, the following ARRs can be derived:

$$1_1: \quad r_1 = uE - R_L i_L - L\frac{di_L}{dt} - u_C \tag{7.65}$$

$$0_1: \quad r_2 = i_L - C\dot{u}_C - i_a \tag{7.66}$$

$$1_2: \quad r_3 = u_C - R_a i_a - L_a\frac{di_a}{dt} - k\omega \tag{7.67}$$

$$1_3: \quad r_4 = ki_a - r\omega - J\dot{\omega} - T \tag{7.68}$$

The structural FSM in Table 7.8 indicates which component parameters affect which residual. The controller and the sensors are assumed to be faultless.

Table 7.8 Structural fault signature matrix of the DC motor drive with sensors Df : i_L, De : u_C. De : i_a and De : ω

Component	Parameter	r_1	r_2	r_3	r_4	D_b	I_b
Power supply	E	1	0	0	0	1	0
Buck converter inductance	L	1	0	0	0	1	0
Buck converter capacitor	C	0	1	0	0	1	1
Motor armature resistance	R_{L_a}	1	0	1	0	1	0
Motor armature inductance	L_a	0	0	1	0	1	0
Motor constant	k	0	0	1	1	1	0
Moment of inertia	J	0	0	0	1	1	0
Friction	r	0	0	0	1	1	0
Load moment	T	0	0	0	1	1	0

7.6.3 Fault Scenario: Friction in the DC Motor Increases Linearly as of a Time Instant

If there is friction of increasing magnitude inside the DC motor, then an evaluation of ARR_4 yields a residual $r_4 \neq 0$ as of some time instant t_4. That is, friction inside the motor can be detected but cannot be isolated by inspection of the structural FSM as r_4 depends on other parameters as well. Suppose that the onset of the increasing friction moment $M_C(t) = b(t)\, r_C \text{sign}(\tilde{\omega})$ inside the motor has been identified by means of parameter estimation. Let variables delivered by the faulty system be distinguished by a tilde from those of the faultless system. If the effective friction value $M_C(t)$ were known, then residual $r_4'(t)$ would vanish.

$$\begin{aligned} r_4'(t) &= 0 = k\,\tilde{i}_a - r\,\tilde{\omega} - J\,\dot{\tilde{\omega}} - T - M_C(t) \\ &= \underbrace{k\,\tilde{i}_a - r\,\tilde{\omega} - J\,\dot{\tilde{\omega}} - T}_{r_4(t)\,\neq\, 0} - M_C(t) \end{aligned} \tag{7.69}$$

That is, in this example, residual $r_4(t)$ simply equals the unknown friction inside the motor.

The offline simulation of this fault scenario with the open source software Scilab, uses the parameter values in Table 7.9.

Figure 7.43 shows the waveforms of the reference signal $\omega_{\text{ref}}(t)$ and of the external torque $T(t)$ acting on the load I : J.

The linearly increasing friction $M_C(t)$ inside the motor deliberately inserted into the model is

$$M_C(t) = \begin{cases} 0 & t < t_4 \\ (t - t_4)\, r_C \text{sign}(\tilde{\omega}) & t > t_4 \end{cases} \tag{7.70}$$

where $t_4 = 3.5$ s and $r_C = 0.1$ Nm/s.

Table 7.9 Parameters of the DC motor drive [13]

Parameter	Value	Units	Meaning
E	12.0	V	Voltage supply
L	20	mH	Inductance
C	400	μF	Capacitance
L_a	2.6	mH	Armature inductance
R_a	2.0	Ω	Armature resistance
k	0.046	Vs/rad	Motor constant
J	$7.0 \cdot 10^{-5}$	kgm^2	Moment of inertia
r_l	$8.0 \cdot 10^{-4}$	Nms/rad	Coefficient for the friction caused by the load
T_0	0.05	Nm	Load moment for $0 \leq t < 1$ s
k_p	0.0072	s/rad	PI controller's proportional gain
T_i	0.015	s	Integral time constant
ω_{ref}	50	rad/s	Reference speed at $t = 0$ s

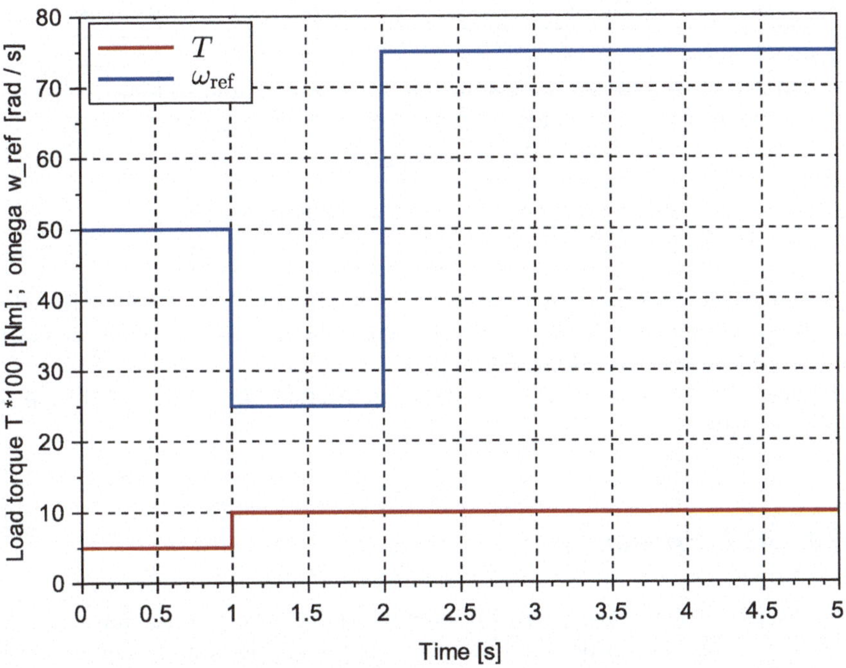

Fig. 7.43 Waveforms of the reference angular velocity $\omega_{\text{ref}}(t)$ and of the external torque $T(t)$ acting on the load I : J

As can be seen from Fig. 7.44, the controller can cope with the increasing internal friction as of t_4 so that the waveform of the angular velocity $\tilde{\omega}(t)$ of the faulty system is close to the speed $\omega(t)$ of the healthy one.

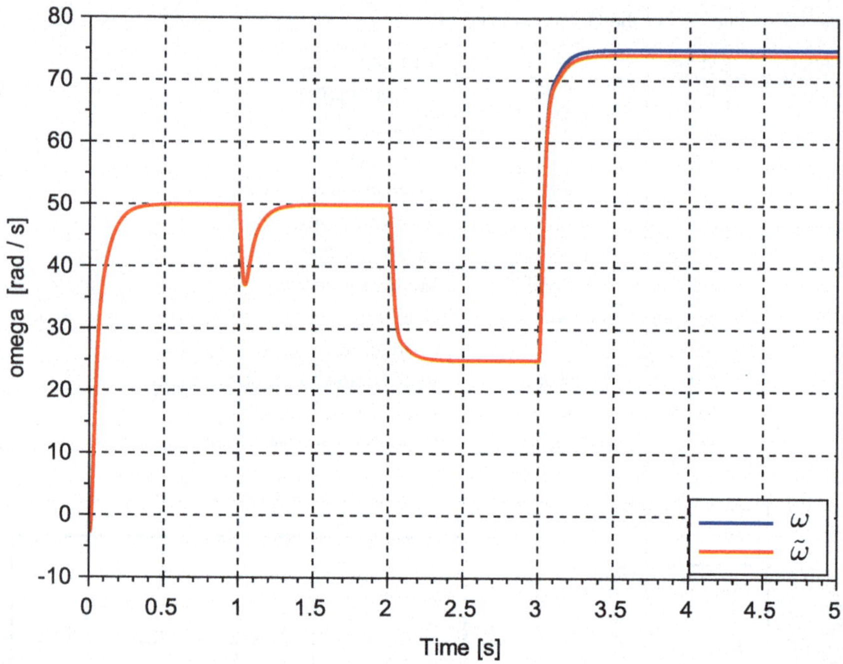

Fig. 7.44 Angular velocities ω_{ref} of the faulty motor and of the healthy motor

To compensate for the increasing internal friction as of $t_4 = 3.5$ s and to follow the reference angular velocity $\omega_{\text{ref}}(t)$, the controlled motor draws an increasing current $\tilde{i}_a(t)$ as depicted in Fig. 7.45.

Figure 7.46 shows the time evolution of the friction $M_C(t)$ recovered from the residual of ARR$_4$.

7.6.4 RUL Estimation

Once discrete values of the internal friction $M_C(t)$ have been fitted and the incline $r_C = 0.1$ of the regression line has been determined, the time to failure can be estimated. In the case of friction inside the motor, the sum of moments reads for $t > t_4$:

$$0 = k\,\tilde{i}_a - r_l\,\tilde{\omega} - J\,\dot{\tilde{\omega}} - T - (t - t_4)\,r_C \tag{7.71}$$

If there is no internal friction, the sum of moments yields

$$0 = k\,i_a - r_l\,\omega - J\,\dot{\omega} - T \tag{7.72}$$

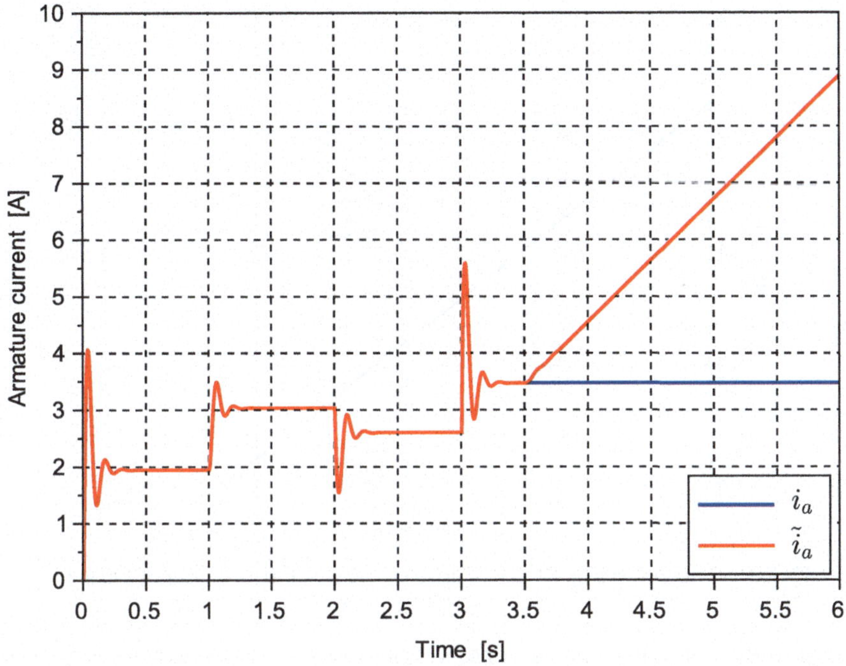

Fig. 7.45 Armature current

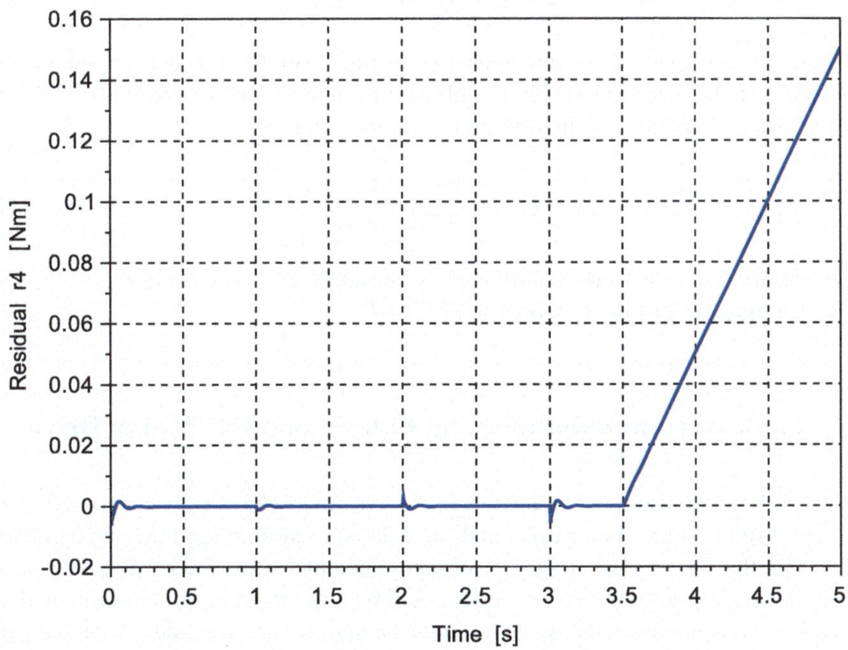

Fig. 7.46 Recovered friction $M_C(t)$ inside the motor

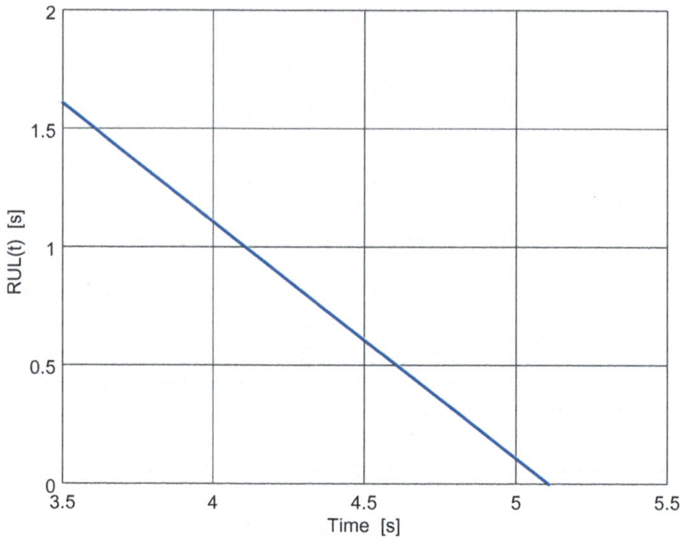

Fig. 7.47 RUL of the closed loop DC motor drive in the case of friction inside the motor

Due to the controller one can observe that $\tilde{\omega} \approx \omega \approx 75\,\text{rad/s}$ for $t > 3.5\,\text{s}$. Hence,

$$0 = k(\tilde{i}_a - i_a) - (t - t_4)\, r_C \tag{7.73}$$

An excessive current draw due to friction inside the motor may damage the motor. Suppose that the armature current is allowed to increase by a maximum of 3.5 A. As a result, one obtains for the time to failure the estimate

$$t_{\text{EoL}} - t_4 = \frac{0.046 \cdot 3.5}{0.1} = 1.6\,\text{s} \tag{7.74}$$

in accordance with the time evolution of the armature current $\tilde{i}_a(t)$ in Fig. 7.45. The RUL as a function of time is shown in Fig. 7.47.

7.7 Fault Accommodation in an Open Loop DC Motor Drive

In Sect. 5.4, the equations of an inverse faulty system model have been derived from a bicausal BG of a DC motor drive and formulated as an implicit DAE system. Due to the fact that in a bicausal BG all storage elements receive derivative causality, all unknown states of the inverse model could be eliminated in this example and an analytical expression, (5.16), p. 188, could be set up for the system input $u_{\text{req}}(t)$ required for a desired angular velocity ω_{des} of the motor.

7.7.1 *Fault Scenario 1: Increase in the Motor Armature Resistance*

In the following, the constructed control law is used in an offline simulation study to compensate an abrupt increase of the armature resistance so that the desired angular velocity of the faulty motor equals the speed of the healthy motor. As a reference, first, the steady state values of the angular velocity of the healthy and the faulty motor are determined analytically.

Analytical Determination of Steady State Values

The implicit DAE system (5.13), p. 187, yields for ω_{des} in steady state

$$\left(1 + \frac{k^2}{b} \frac{1}{R_{\text{on}} + R_L + R_a}\right) \omega_{\text{des}} = \frac{k}{b} \frac{1}{R_{\text{on}} + R_L + R_a} (1 - d)E - \frac{1}{b} T_{\text{load}} \tag{7.75}$$

and for the armature current i_a in the healthy motor

$$i_a = \frac{1}{k}(b\,\omega - T) \tag{7.76}$$

Given the parameter values in Table 7.10, the following steady state values are obtained analytically $\omega_{\text{des}} = 41.9$ rad/s and $i_a = 1.851$ A.

Table 7.10 Component parameters of the DC motor drive in Fig. 5.4

Parameter	Value	Units	Meaning
E	12.0	V	Voltage supply
L	20	mH	Inductance
R_L	0.1	Ω	Resistance of the coil
R_{on}	0.1	Ω	ON resistance (switch, diode)
d	0.5	–	Duty ratio
C	40	μF	Capacitance
L_a	2.6	mH	Armature inductance
R_a	2.0	Ω	Armature resistance
ΔR_a	2.0	Ω	
k	0.046	Vs/rad	Motor constant
J_m	$7.0 \cdot 10^{-4}$	kgm^2	Moment of inertia
b	$8.4 \cdot 10^{-4}$	Nms/rad	Friction coefficient
T_{load}	0.05	Nm	Load moment
R_l	0.2	Ω	Leakage from capacitor (fault scenario 2)

In the case that the armature resistance abruptly doubles as of some time instant, i.e. $\tilde{R}_a = 2R_a$, the steady state motor velocity drops to the value $\tilde{\omega} = 11.7$ rad/s if the fault is not accommodated by a change in the command input.

Simulation of the Recovery from the Fault

The offline simulation performed with the open source software Scilab assumes that the armature resistance doubles at $t_1 = 1.5$ s. As of that time instant, the constant modulus m of the MTF element in the averaged BG model of the buck converter i.e. the signal controlling the switching of the transistor Q, is replaced by the time dependent modulus $\tilde{m}(t)$, (5.17), p. 189, to enforce that the angular velocity $\tilde{\omega}(t)$ follows one of the healthy motor despite the faulty armature resistance $\tilde{R}_a$.

Figure 7.48 depicts the time history of the desired angular velocity, $\omega_{\text{des}}(t)$, the faulty velocity $\tilde{\omega}(t)$, and the accommodated faulty velocity $\omega_{\text{acc}}(t)$.

As can be seen, the simulation confirms the analytically computed steady state values. Furthermore, the reconstructed input $u_{\text{req}}(t) = \tilde{m}(t)E$ into the buck converter can accommodate the increase of the armature resistance as of $t_1 = 1.5$ s so that the angular velocity of the motor is able to recover and to follow the desired velocity $\omega_{\text{des}}(t)$ of the healthy motor. At $t_1 = 1.5$ s, there is a short small drop of

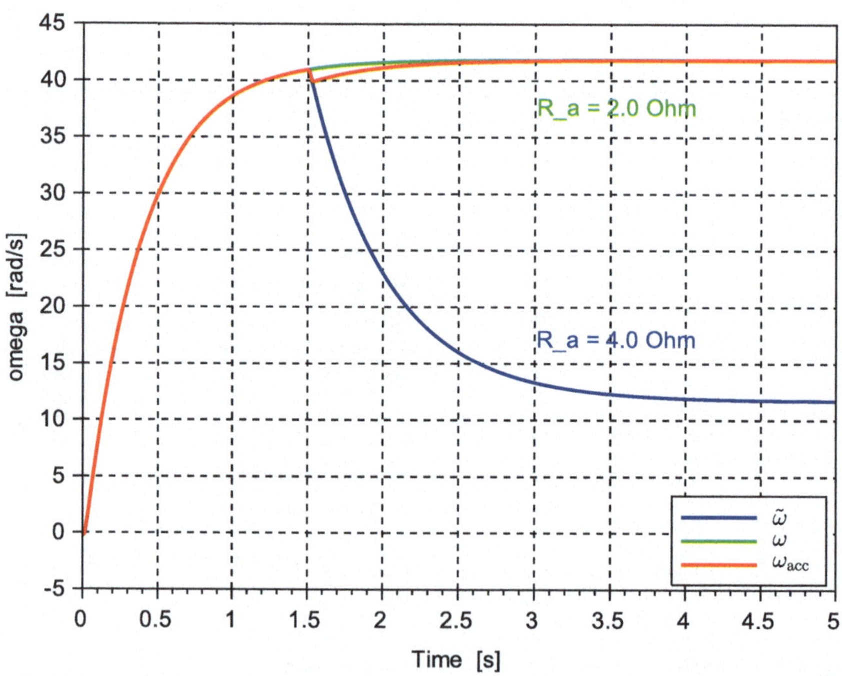

Fig. 7.48 Time histories of the desired, the faulty, and the accommodated faulty angular velocity

the velocity. This drop reflects that immediately after the occurrence of the fault, the motor is still controlled by the unchanged input signal until fault accommodation becomes effective.

7.7.2 Fault Scenario 2: Leakage in the Buck Converter Capacitor

The second fault scenario reconsiders the buck converter DC motor drive system but now assumes that there is an abrupt and persistent leakage from the capacitor as of some time instant t_2. As a result of this leakage, the voltage across the capacitor C driving the motor will drop and so will the angular velocity ω. Given a constant voltage supply E of the buck converter, the duty cycle of the signal controlling the transistor Q must be changed to keep up the angular velocity of the healthy system considered the desired output y_{des}.

The leakage may be captured by a small resistance R_s in parallel to the capacitance that becomes effective as of some time t_2. In the bicausal BG in Fig. 7.49, the added resistor R : R_s is highlighted in blue.

Determination of a New System Input

For simplicity, let $T_{\text{load}} = 0$. Equations can be derived from the bicausal BG in Fig. 7.49 and may be formulated again as an implicit DAE system for the inverse faulty system model.

$$0 = u_{\text{req}} - R\tilde{i}_L - L\frac{d\tilde{i}_L}{dt} - \tilde{u}_C \tag{7.77}$$

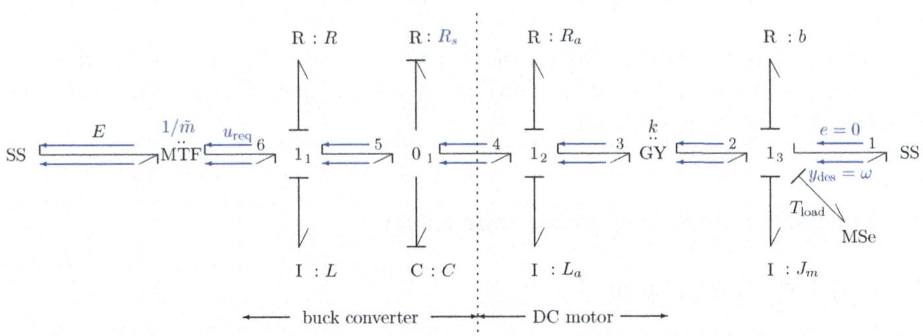

Fig. 7.49 Bicausal BG of the buck converter DC motor system accounting for a leakage of the capacitor by means of the resistor R : R_s

$$0 = \tilde{i}_L - C\dot{\tilde{u}}_C - \frac{\tilde{u}_C}{R_s} - \tilde{i}_a \tag{7.78}$$

$$0 = \tilde{u}_C - R_a\tilde{i}_a - L_a\frac{d\tilde{i}_a}{dt} - ky \tag{7.79}$$

$$0 = k\tilde{i}_a - b\omega - J_m\dot{\omega} \tag{7.80}$$

$$y = \omega \tag{7.81}$$

where $u_{\text{req}}(t) = \tilde{m}(t)E$ is to be determined, $y(t)$ is known (measured), and $R := R_L + R_{\text{on}}$.

As in the previous case, there is no need to solve the implicit DAE system numerically in order to obtain values for the reconstructed input required for accommodation of the fault. Instead, an analytical expression for $u_{\text{req}}(t)$ can be found. As all storage elements in the bicausal BG in Fig. 7.49 are in derivative causality, the input $u_{\text{req}}(t)$ is expected to depend only on the output y and its time derivatives as in the first fault scenario.

Let i_a denote the armature current and u_C the capacitor voltage in the healthy system obtained by solving the forward model. From (7.80) one can conclude that $\tilde{i}_a = i_a$ and $\tilde{u}_C = u_C$. Substitution of equations yields for the required input

$$\begin{aligned} u_{\text{req}} &= \tilde{m}E \\ &= R(C\dot{u}_C + \frac{u_C}{R_s} + i_a) + L\frac{d}{dt}(C\dot{u}_C + \frac{u_C}{R_s} + i_a) + u_C \\ &= mE + \frac{R}{R_s}u_c + \frac{L}{R_s}\dot{u}_C \end{aligned} \tag{7.82}$$

and

$$u_C = R_a\frac{1}{k}(by + J_m\dot{y})\; L_a\frac{1}{k}(b\dot{y} + J_m\ddot{y}) \;+\; ky \tag{7.83}$$

That is, $u_{\text{req}}(t)$ only depends on the output $y = \omega$ and its time derivatives in accordance with the derivative causality of all storage elements in the bicausal BG of the faulty system in Fig. 7.49.

Analytical Determination of Steady State Values

In steady state, (7.82) reduces to

$$u_{\text{req}} = \tilde{m}E = R\left[\frac{1}{R_s}(R_a\frac{b}{k} + k) + \frac{b}{k}\right]\omega + \;(R_a\frac{b}{k} + k)\omega \tag{7.84}$$

As long as the leakage in the capacitor has not occurred, $R_s \rightarrow \infty$ and $\tilde{m}E = mE$. This gives an equation the steady state value of the desired angular velocity must fulfil.

$$u = mE = \left[(R + R_a)\frac{b}{k} + k\right]\omega \tag{7.85}$$

If the fault is not accommodated by a changed input $\tilde{m}E$, then the angular velocity reaches a faulty steady state value $\tilde{\omega}$ that is given by the equation

$$u = mE = R\left[\frac{1}{R_s}(R_a\frac{b}{k} + k) + \frac{b}{k}\right]\tilde{\omega} + (R_a\frac{b}{k} + k)\tilde{\omega} \tag{7.86}$$

The parameters in Table 7.10 give the steady state values $\omega = 69.63$ rad/s and $\tilde{\omega} = 35.57$ rad/s.

Simulation of the Recovery from the Fault

Simulation runs have been performed by means of the open source software Scilab and have used the parameters given in Table 7.10.

Figure 7.50 shows the time evolution of the armature current $\tilde{i}_a(t)$ and the output voltage $\tilde{u}_C(t)$ of the buck converter in the case that the capacitor leakage is not compensated by a reconstructed input.

The steady state values obtained by simulation agree with the analytically computed ones for the healthy system $i_a = 1.27$ A, $u_C = 5.75$ V and for the faulty not accommodated system $\tilde{i}_a = 0.65$ A, $\tilde{u}_C = 3.0$ V.

Figure 7.51 depicts the time history of the angular velocity of the healthy motor, $\omega(t)$, the faulty velocity $\tilde{\omega}(t)$, and the accommodated faulty velocity $\omega_{\text{acc}}(t)$. Again, the simulation confirms the analytically computed steady state values.

As of $t_1 = 1.5$ s, the small resistance of $R_s = 0.2\,\Omega$ in parallel to the capacitor becomes effective. As a result of the capacitor leakage modelled this way, the capacitor voltage drops sharply and so does the angular velocity. If this fault is not compensated, the faulty steady state value is roughly half of the desired one.

It is assumed that detection and isolation of this fault takes about 0.02 s so that fault accommodation can start at $t_2 = 1.52$ s. As can be seen, after the leakage of the capacitor has happened, the reconstructed input $u_{\text{req}}(t)$, in fact, forces the angular velocity $\omega_{\text{acc}}(t)$ to follow the desired velocity $\omega_{\text{des}}(t)$ despite the capacitor leakage. The dynamics of the recovery depends on how much the capacitor voltage has dropped and on the parameters of the systems. Given the parameters in Table 7.10, it takes about 1.5 s to recover from this sharp drop of the angular velocity.

Finally, Fig. 7.52a confirms that the armature current $i_{a_{\text{acc}}}(t)$ in the accommodated system, apart from a peak at $t = 1.5$ s caused by the abrupt leakage of the capacitor, in fact, remains unchanged.

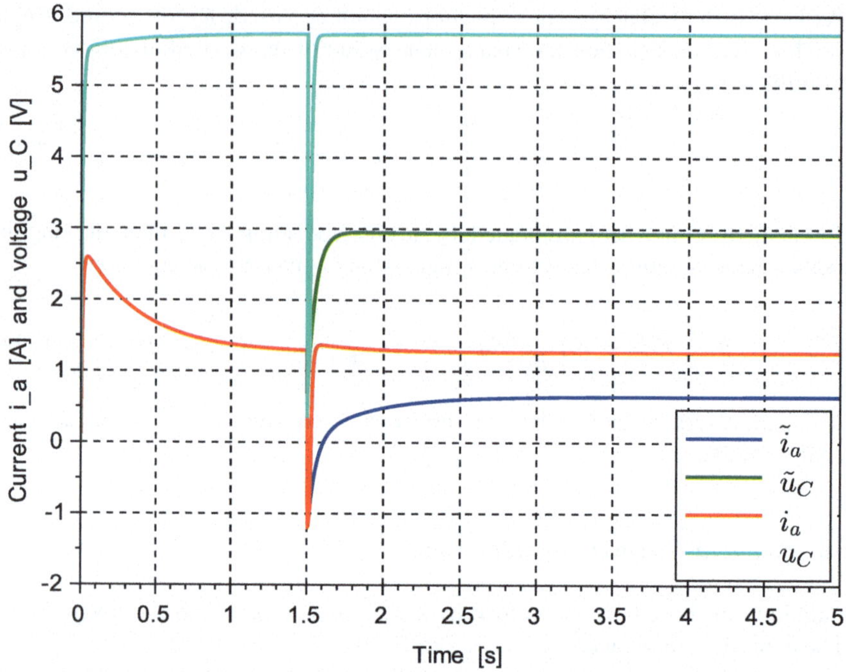

Fig. 7.50 Time evolution of the armature current and the output voltage of the buck converter in the case of no capacitor leakage accommodation

Figure 7.52b indicates that the voltage $u_{C_{\text{acc}}}(t)$ in the accommodated system does not drop to the steady state value of 3.0 V but is forced in a very short time to regain the value of the healthy system so that the motor continues operating with the desired angular velocity despite the leakage of the capacitor. As a result, the inductor current $i_{L_{\text{acc}}}(t)$ increases significantly (Fig. 7.52c). The steady state value rises from 1.27 to 30.07 A.

7.8 Robust Overwhelming Control of a Mechanical Oscillator

The operating principle of passive FTC by means of an overwhelming controller considered in Sect. 5.5 (Fig. 5.8) is illustrated by applying it to the simple classical one-dimensional mechanical spring-mass oscillator depicted in Fig. 7.53.

Figure 7.54 shows a BG of the mechanical oscillator connected to a BG of the overwhelming controller according to Fig. 5.8, p. 190.

In Fig. 7.54, parameters and variables in the inverse model are marked with the subscript ‘c’, while those of the oscillator carry an index ‘p’. As can be seen from the BG, the inertia element I : m_c in the inverse plant model used by the controller

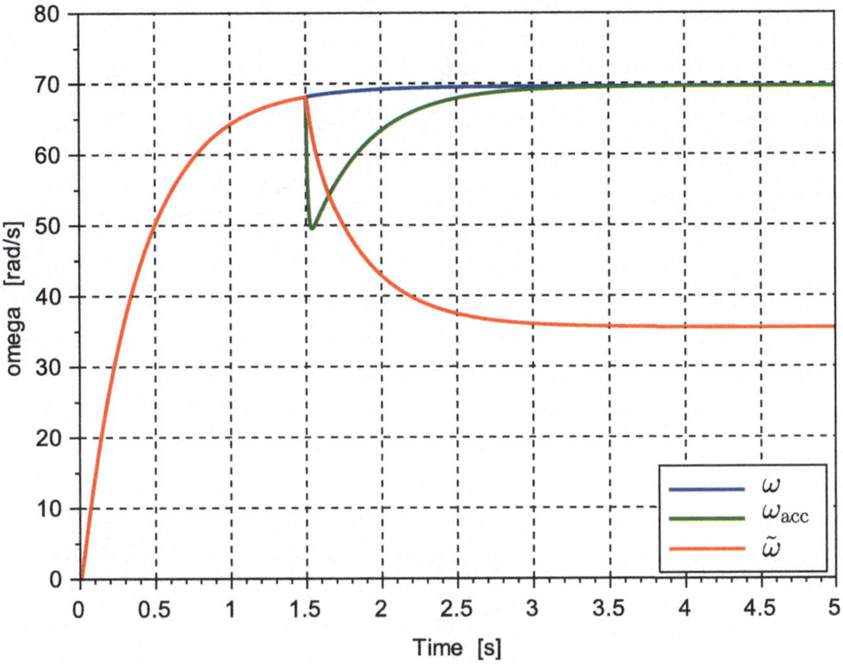

Fig. 7.51 Time histories of the desired velocity, ω, the faulty velocity, $\tilde{\omega}$, and the accommodated faulty angular velocity, ω_{acc}, in the case of a capacitor leakage

has received derivative causality. Only the storage element C : C_c of the spring remains in integral causality. That is, the inverse model is of reduced order. From an inspection of the causal BG it can be concluded that the transfer function of the inverse model has only one pole and is improper. The degree of its denominator polynomial is less than the degree of the numerator polynomial. Let $k_c := 1/C_c$. Then

$$G_{\text{inv}}(s) = \frac{F(s)}{\Delta v(s)} = \frac{m_c s^2 + r_c s + k_c}{s} = \frac{N_c(s)}{s} \tag{7.87}$$

Although the transfer function of the inverse model is improper, the transfer function $G(s)$ of the controlled oscillator is biproper. The degree of its numerator polynomial equals the degree of its denominator polynomial.

$$G(s) := \frac{v(s)}{v_{\text{ref}}(s)} = \frac{G_{\text{inv}}(s)}{G_{\text{inv}}(s) + \dfrac{1}{K G_p(s)}} \tag{7.88}$$

$$= \frac{K N_c(s)}{K N_c(s) + D_p(s)} \tag{7.89}$$

Fig. 7.52 Currents and voltages in the accommodated system. (**a**) Time evolution of the currents i_a and $i_{a_{acc}}$. (**b**) Time evolution of the voltages u_C and $u_{C_{acc}}$. (**c**) Time evolution of the currents i_L and $i_{L_{acc}}$

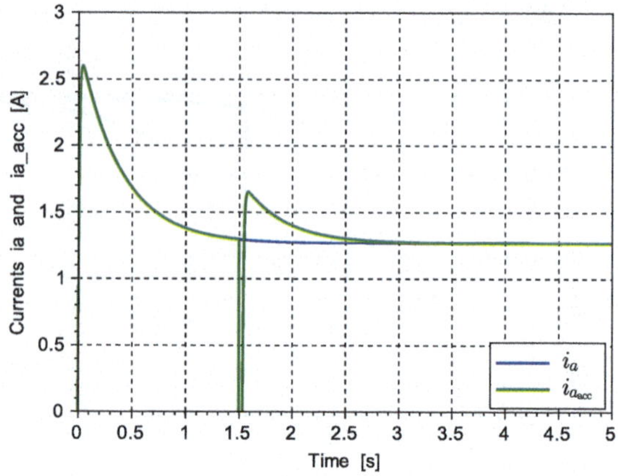

(a)

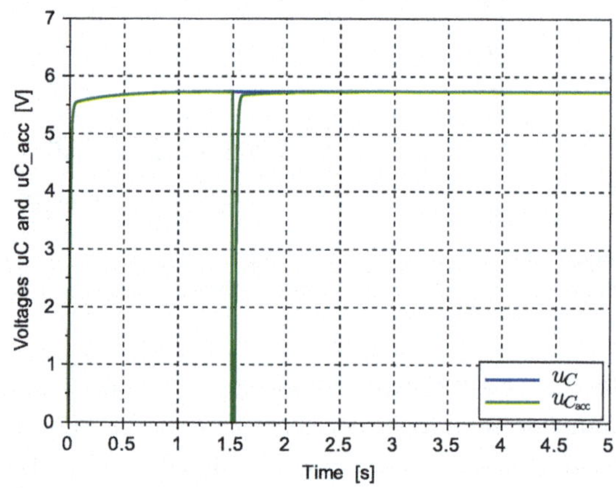

(b)

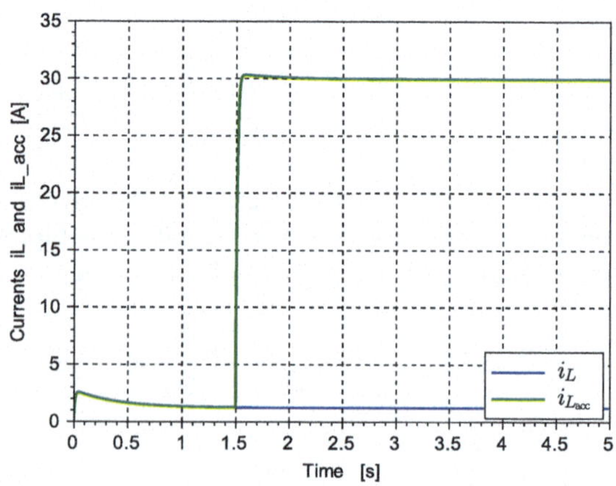

(c)

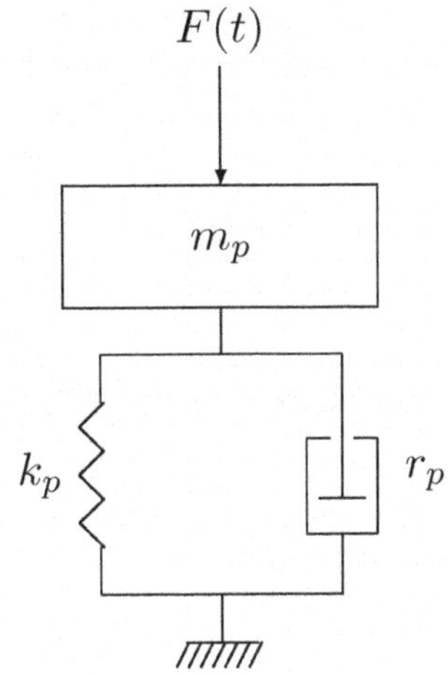

Fig. 7.53 Classical spring-mass-oscillator

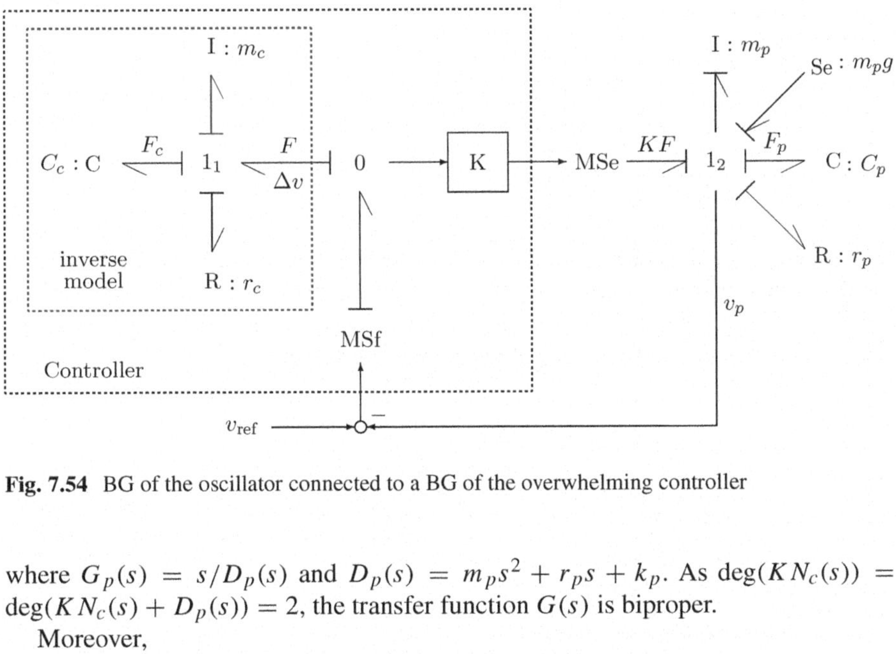

Fig. 7.54 BG of the oscillator connected to a BG of the overwhelming controller

where $G_p(s) = s/D_p(s)$ and $D_p(s) = m_p s^2 + r_p s + k_p$. As $\deg(K N_c(s)) = \deg(K N_c(s) + D_p(s)) = 2$, the transfer function $G(s)$ is biproper.

Moreover,

$$\lim_{K \to \infty} G(s) \to 1$$

even when the parameters of the inverse model inside the controller differ from the ones of the forward model, i.e. if $N_c(s) \neq D_p(s)$, or $G_{\text{inv}}(s) \neq G_p^{-1}(s)$. That is, the overwhelming controller with the transfer function $G_c(s) = K G_{\text{inv}}(s)$ makes sure that the oscillator velocity $v(t)$ follows the prescribed trajectory $v_{\text{ref}}(t)$ provided its gain K is sufficiently high.

If the parameters used by the inverse model inside the overwhelming controller would match exactly the ones of the plant, which is quite unlikely in practice, then the transfer function of a controlled plant would be

$$G(s) = \frac{K}{1+K} \tag{7.90}$$

Simulation of the Closed Loop Oscillator

Equations for an offline time domain simulation are immediately read out from the BG in Fig. 7.54.

$$\dot{v}_p = \frac{1}{m_p}[KF + m_p g - F_p - r_p v_p] \tag{7.91}$$

$$\dot{F}_p = \frac{1}{C_p} v_p \tag{7.92}$$

$$F = m_c \Delta\dot{v} + F_c + r_c \Delta v \tag{7.93}$$

$$\dot{F}_c = \frac{1}{C_c} \Delta v \tag{7.94}$$

The output of the inverse model, F, can be eliminated from the DAE system for the four unknowns v_p, F_p, F, F_c such that an explicit set of ODEs for the states, v_p, F_p, F_c is obtained that can be easily implemented as an Octave script and numerically solved by means of the `lsode()` solver.

$$\begin{aligned}\dot{v}_p = \frac{1}{M}[&-(r_p + Kr_c)v_p - F_p + KF_c + Kr_c v_{\text{ref}} \\ &+ Km_c \dot{v}_{\text{ref}} + m_p g]\end{aligned} \tag{7.95}$$

$$\dot{F}_p = \frac{1}{C_p} v_p \tag{7.96}$$

$$\dot{F}_c = \frac{1}{C_c}(v_p - v_{\text{ref}}) \tag{7.97}$$

where $M := m_p + Km_c$.

For test purposes, an undamped sinusoidal waveform $v_{\text{ref}}(t) = \sin(\omega t)$ is adopted as the prescribed trajectory. The objective of the simulation is to show that the controller forces the oscillator to follow the reference $v_{\text{ref}}(t)$. The simulation uses the parameter in Table 7.11.

Table 7.11 Parameters of the overwhelming controller and the oscillator

Parameter	Value	Units	Meaning
m_p	20.0	kg	Mass of the oscillator
m_c	1.0	kg	Mass of the oscillator in the controller
C_p	1e-5	m/N	Flexibility of the spring
C_c	1e-3	m/N	Flexibility of the spring in the controller
r_p	10.0	Ns/m	Friction coefficient
r_c	1.0	Ns/m	Friction coefficient in the controller
K	10^{+6}	–	Controller gain
ω	$2\pi/50$	rad/s	Angular frequency of the prescribed trajectory
$v_p(0)$	1.0	m/s	Initial value of the oscillator velocity
$F_p(0)$	10.0	N	Initial value of the oscillator spring force

Figure 7.55b clearly displays that the overwhelming controller with a gain $K = 10^6$ forces the oscillator velocity v_p to follow the prescribed trajectory $v_{\text{ref}}(t)$, which is an undamped sinusoidal waveform, although the oscillator includes a damper and the parameters the controller uses for the inverse model differ from the actual parameters of the plant. In addition, the initial values of the oscillator states are different from zero, while the initial values of the spring force in the inverse model and of the reference velocity are equal to zero.

It can be analytically proven that the error $\epsilon(t) = \Delta v(t)$ tends to zero for $t \to \infty$.

$$\epsilon(t \to \infty) = \lim_{s \to 0} s\epsilon(s) = \lim_{s \to 0} s\,(1 - G(s))\, v_{\text{ref}}(s) \tag{7.98}$$

$$= \lim_{s \to 0} s \cdot \frac{D_p(s)}{KN_c(s) + D_p(s)} \cdot \frac{\omega}{s^2 + \omega^2} = 0 \tag{7.99}$$

where $D_p(0) = k_p \neq 0$ and $N_c(0) = K_c \neq 0$.

This clearly demonstrates that the overwhelming controller is robust with regard to parameter and initial value uncertainties, respectively. Hence, the controller can accommodate parametric faults given its gain is sufficiently high. However, a high controller gain may drive an actuator into saturation. Moreover, a measured noisy error signal into the controller is differentiated due to the storage elements in derivative causality in the inverse model used by the overwhelming controller which affects its output so that the noise on the error signal should be reduced by a low pass filter before the signal is used by the controller.

7.9 Summary

Leakage in electrolytic capacitors and hydraulic tanks as well as internal friction in DC motors considered in the case studies in this chapter are typical faults that may

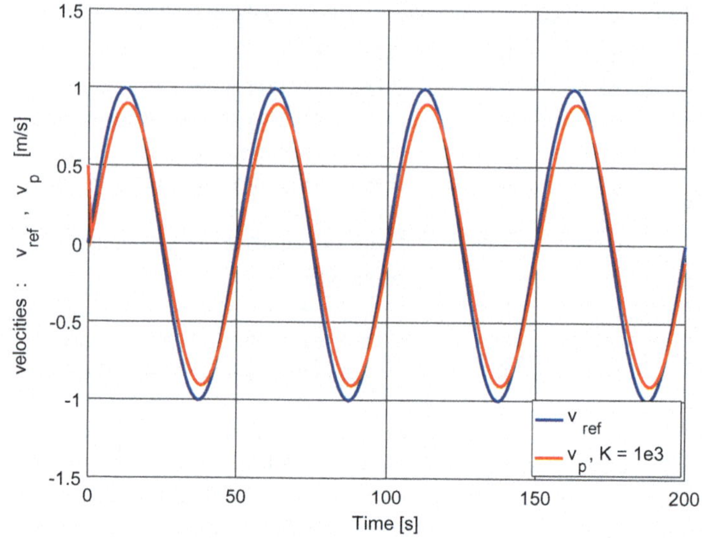

(a)

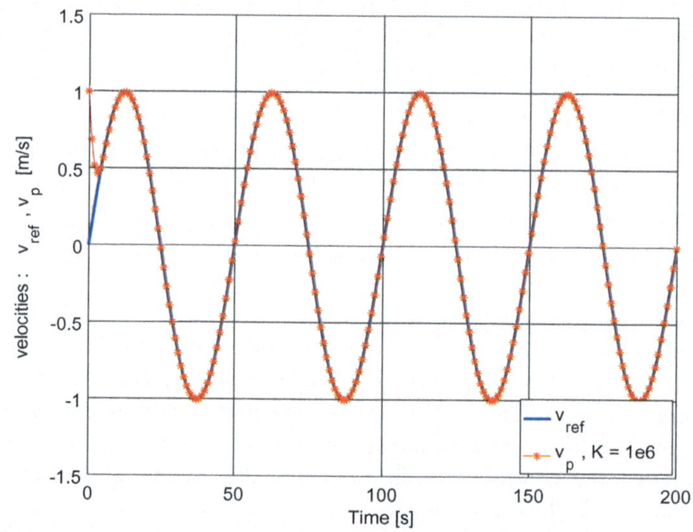

(b)

Fig. 7.55 Prescribed velocity v_{ref} and velocity v_p of the oscillator forced by the overwhelming controller

significantly affect the performance of a system and ultimately may cause a failure if these faults are not addressed in good time.

First, a behavioural BG model of the considered small example systems is developed in which deliberately a single fault, leakage, or friction, is inserted. The faulty system model delivers simulated measurements into a DBG, from which a time series of discrete data for the degradation model is obtained. A RUL is computed by intersection of the degradation trend with a predefined failure alarm threshold. A value for the latter one has been chosen on the basis of physical assumptions. For instance, for the leakage from a tank it was assumed that the area of the hole must not be more than 5% of the area of the tank bottom.

The considered systems show that the inserted fault can be detected with the sensors in use but cannot be isolated by simple inspection of the FSM capturing the structure of the ARRs derived from the DBG. Therefore, subsequent failure prognosis assumes that the onset of the incipient fault was isolated, i.e. by parameter estimation.

For two example systems it was assumed that noise was smoothed before 'measurements' were used in the computation of degradation data. In a more realistic offline simulation, output signals of the behavioural model could be superimposed by a white noise signal. The corrupted signals are then to be fed through an appropriate low pass filter.

Furthermore, for simplicity, the offline simulation neglected that estimated degradation data obtained from real measurements are random. This has not been accounted for so that only the true RUL could be predicted and no performance metrics could be applied. The considered small example systems well demonstrate that ARRs derived from a DBG can not only be used for FDI but also for RUL estimation.

As to FTC, the bicausal BG-based input reconstruction for fault accommodation illustrated on the example of a DC motor drive in Sect. 5.4 is validated by two offline simulation case studies, in which a single fault hypothesis is assumed. In both cases, the reconstructed input into the buck converter enables the system to recover from the fault so that the angular velocity of the healthy motor can be followed with some small time delay despite the presence of a fault. Simulation results obtained by the open source software Scilab confirm analytically computed steady values of variables.

The presentation of bicausal BG model-based inversion has been confined to LTI forward models. In the case of nonlinear forward models an explicit analytical expression for the new required system input in response to an abrupt fault is not guaranteed. In that case, numerical values of the input to be determined may be obtained by inverse simulation. That is, desired outputs are obtained by computing a forward model of the healthy system. Values of the reconstructed input into the faulty system needed at the next time instant are obtained by computing the implicit DAE system of the inverse model simultaneously. To that end, freely available sophisticated software for solving nonlinear DAE systems such as OpenModelica can be used. For complex systems with fast dynamics, the time needed for computing a reconstructed input after a fault has occurred may however become an issue. This time adds to the time needed to detect and to isolate a fault and increases the time delay until the reconstructed input becomes effective.

Another subject of further research may be fault accommodation for hybrid systems with various modes of operation. Their dynamic behaviour in one mode may be quite different from that in another mode.

Finally, the small example of a mechanical oscillator governed by an overwhelming controller illustrates that BG methodology can also serve passive FTC. The inverse plant model used by the controller can be directly obtained from the BG of the plant by just changing causalities. The invertibility of the forward model can be directly checked on its BG. From the interconnection of the controller and the plant model, the equations needed for simulation can be either manually or automatically derived. The simulation run shows that the overwhelming controller, in fact, forces the plant output to follow a prescribed trajectory although the parameters of the inverse model and the initial values of its state significantly differ from those of the plant. That is, the controller performs robustly and is suited for accommodating faults given its gain is sufficiently high and is not affected by the saturation of the actuator.

References

1. Amaral, A.M.R. & Cardoso, A.J.M. (2016). Voltage doubler for AC-DC step-up linear power supplies: Design, modelling and simulation. *Acta Electronica et Informatica, 16*(4), 3–10. https://doi.org/10.15546/aeei-2016-0025.
2. Barnard, B. (1973). Predicting the Dynamic Response of a Hydraulic System Using Power Bond Graphs. Master's Thesis, Monash University, Melbourne.
3. Beck, J. (2002). Using Rectifiers in Voltage Multiplier Circuits. Application Note, Document Number 88842, Vishay Semiconductors.
4. Bjekic, M., Antic, S., & Milovanovic, A. (2011). Permanent magnet DC motor friction measurement and analysis of friction's impact. *International Review of Electrical Engineering (IREE), 6*(5).
5. Borutzky, W. (1993). An energetically consistent bond graph model of a double acting hydraulic cylinder. In A. Pavé (Ed.), *Modelling and Simulation 1993, Proceedings of the 1993 European Simulation Multiconference* (pp. 203–207). Lyon: SCS Publishing.
6. Borutzky, W. (2010). *Bond Graph Methodology – Development and Analysis of Multidisciplinary Dynamic System Models*. London: Springer. ISBN:978-1-84882-881-0.
7. Borutzky, W. (2017). A new approach to the derivation of a single set of implicit state equations from a fixed causality bond graph of a hybrid model. In A. Bruzzone, G. Dauphin-Tanguy, S. Junco, & F. Longo (Eds.), *Proceedings of the 10th International Conference on Integrated Modelling and Analysis in Applied Control and Automation (IMAACA 2017), part of the I3M Multi conference*, Barcelona.
8. Borutzky, W. (2018). Determination of a function for a degradation process by means of two diagnostic bond graphs. *IFAC PapersOnLine, 51–24*, 636–642. Full text access.
9. Borutzky, W. (2020). A hybrid bond graph model-based – data driven method for failure prognostic. *Procedia Manufacturing, 42*, 188–192, Open-access article.
10. Cadence Design Systems, Inc. (2020). PSpice Technologies for Academic. https://www.pspice.com/academic-user.
11. Dinesh Chinthaka, M.K.C., Harsha, A.M., & Abeykoon, S. (2015). Friction compensation of DC motors for precise motion control using disturbance observer. *Ecti Transactions on Computer and Information Technology, 9*(1), 74–82.

12. Dransfield, P. (1981). *Hydraulic Control Systems – Design and Analysis of Their Dynamics*. New York: Springer.
13. Ganesh Kumar, S. & Hosimin Thilagar, S. (2011). Comparative study of proportional integral and passivity based control for buck converter. In *Proceedings of 1st International Conference on Simulation and Modeling Methodologies, Technologies and Applications (SIMULTECH-2011)* (pp. 309–314). https://doi.org/10.5220/0003577703090314.
14. Hoffmann, J. (1998). *MATLAB und Simulink – Beispielorientierte Einführung in die Simulation dynamischer Systeme*. Addison-Wesley.
15. Kulkarni, C., Biswas, G., & Koutsoukon, X. (2009). A prognosis case study for electrolytic capacitor degradation in DC-DC converters. In *Proceedings of Annual Conference of the Prognostics and Health Management Society*, San Diego.
16. Kulkarni, C., Biswas, G., Koutsoukos, X., Goebel, K., & Celaya, J. (2010). Physics of failure models for capacitor degradation in DC-DC converters. In *The Maintenance and Reliability Conference*, MARCON, Knoxville.
17. Liang, A.N. & Sepehri, N. (2005). Hydraulic actuator leakage fault detection using extended Kalman filter. *International Journal of Fluid Power, 6*(1), 41–51.
18. Merritt, H. (1967). *Hydraulic Control Systems*. Hoboken: Wiley.
19. OpenModelica Consortium. (2020). OpenModelica. https://www.openmodelica.org/.
20. Pang, H.M. & Bryan, P.M.H. (2010). A life prediction scheme for electrolytic capacitors in power converters without current sensor. In *2010 Twenty-Fifth Annual IEEE Applied Power Electronics Conference and Exposition (APEC)* (pp. 973–979). IEEE, Palm Springs. https://doi.org/10.1109/APEC.2010.5433384.
21. Sobhani, M. & Poshtan, J. (2012). Fault detection and isolation using unknown input observers with structured residual generation. *International Journal of Instrumentation and Control Systems (IJICS), 2*(2). https://doi.org/10.5121/ijics.2012.2201.
22. Soendergaard, P.L. (2018). ltat the large time-frequency analysis toolbox. https://octave.sourceforge.io/ltfat/index.html.
23. Tjahjowidodo, T., Al-Bender, F., & Van Brussel, H. (2005). Friction identification and compensation in a DC motor. In *IFAC 16th Triennial World Congress*, Prague

Chapter 8
Conclusions

The focus of this book has been on contributions of bond graph methodology to model-based control, fault diagnosis, failure prognosis and fault tolerant control. Model-based fault diagnosis has become a quite mature discipline over the past decades. Accordingly, fault detection and isolation in continuous time as well as in hybrid systems has been a subject in a number of bond graph related publications. Due to its importance in various fields, especially in condition based maintenance (CBM) and integrated systems health management (ISHM), failure prognosis has received increasing attention in industry and is an ongoing research topic. However, few publications have related bond graph methodology to failure prognosis so far. One aim of this book is to make a contribution by presenting a combined bond graph model-based, data-based approach to failure prognosis.

Bond graph methodology supports an intuitive graphical, systematic step-by-step development of models for multidomain systems based on the universal concept of energy exchange between subsystem and component ports respectively and on conservation principles for physical quantities such as mass, momentum, and charge. Modellers can focus on an insight into the physical phenomena taking place in a system to be designed rather than on details of a purely mathematical model and on constraints the software in use is imposing. The unique step-by-step approach of bond graph modelling from a schematic towards a mathematical model that can be automatically generated from the bond graph has been well presented in various books.

It is the computational structure superimposed on an acausal bond graph, i.e., the assignment of computational causality and the powerful concept of causal paths that makes bond graph methodology useful for model-based control, fault detection and isolation, failure prognosis, and fault tolerant control beyond the development of models for the simulation of the dynamic behaviour of multidisciplinary systems.

W. Borutzky, *Bond Graph Modelling for Control, Fault Diagnosis and Failure Prognosis*, https://doi.org/10.1007/978-3-030-60967-2_8

Model-Based Control

As to model-based control, *structural* observability and *structural* controllability are a prerequisite for numerical controllability and for the design of a controller and can be directly checked on a causal BG without the need to set up the matrices of a LTI state space model used by the numerical criteria of Kalman or Hautus.

For the controllable and observable part of a system, transfer functions can be directly derived from a bond graph according to Mason's loop rule by following causal paths (Sect. 2.2.2). This derivation can be manually carried out for small bond graph models. For large bond graph models composed of several subsystem models, it may be too cumbersome and error-prone. An alternative is then to have software generate the matrices of a state space model from a bond graph and have mathematical software such as GNU Octave, Scilab, or Matlab® to convert the state space model into a transfer function matrix.

The denominator of a transfer function is the characteristic polynomial of a model. This can be used for a determination of the gain matrix entries of a Luenberger observer directly from a causal BG as illustrated in Sect. 3.5.1.

Furthermore, asymptotic stability can be checked on a causal BG according to Lyapunov's second method without the need for state space equations (Sect. 2.8). Once the matrices of an LTI model have been derived from a causal BG, mathematical software can be used for stability analysis.

Fault Detection and Isolation

In regards to fault diagnosis, bicausality (Sect. 2.4) as an extension of the original concept of BG causality has proven useful for parameter estimation and for the creation of inverse models. The determination of the size of an isolated parametric fault is a prerequisite for fault accommodation. Moreover, online parameter estimation can provide data for the approximation of a parametric degradation trend for which a physics-based degradation model is not available and difficult to develop but needed for estimating future deteriorated system states starting from its current state.

Diagnostic BGs (Sect. 3.5.2) well support fault detection and isolation (FDI). By following causal paths, it can be determined which physical component parameter affects which analytical redundancy relation (ARR) without having to derive ARRs in analytical form from a causal BG. The result can be expressed in a structural fault signature matrix (FSM). An inspection of this FSM reveals which faults are detectable with a given set of sensors and which of them can be even isolated. In case several components have the same fault signature, least squares parameter estimation on a subset of ARRs then is one option to isolate faults (Sect. 3.5.4). Section 3.8 proposes a graphical approach to a sensor placement that aims at increasing the number of isolated faults.

The numerical evaluation of ARRs derived from a DBG provides fault indicators. As to the detection of faults, it is important that the model is robust with parameter uncertainties to avoid misdetections. Parameter uncertainties can be taken into account by incremental BGs or BGs in LFT form. Diagnostic bond graphs as a residual generator for FDI use measurements and, in general, also need time derivatives of measurements as storage elements are in derivative causality to get rid of unknown initial conditions. Therefore, Sect. 3.2.1 revisits the Savitzky–Golay filter as a suitable low pass filter for signal preprocessing which can not only smooth measured signals but can also provide time derivatives and is available in widely used mathematical software packages. Section 3.5.3 briefly considers a way to avoid the differentiation of measurements.

Failure Prognosis

The hybrid bond graph model-based, data-based approach to failure prognosis presented in Sect. 4.5 uses parameter estimation on a DBG to obtain discrete degradation values at sample points in a sliding window of fixed length. Concurrently to the monitoring of an engineering system or process, the parameters of several function candidates can be fitted in parallel on one and the same data set. The best fitting function in respect of a criterion such as the RSME can then be projected from the current sliding window into the future to determine its crossing with a fault alarm threshold in order to determine the remaining useful life (RUL), which is the core of failure prognostic. As this regression of data provided by parameter estimation on the DBG is repeatedly performed while the window is moving forward, the approach is suited for hybrid systems in which the degradation behaviour may change from one mode of operation to a subsequent one such that a change of the mathematical function with coefficients to be determined is needed.

An advantage of the DBG-based parameter estimation for a finite number of time instances in a sliding window of fixed length is that a physical model of the degradation process is not needed for the failure prognosis which can be difficult to develop. Instead, the parameters of a mathematical model approximating the degradation trend in the current sliding window are determined online. Also no historical data or a preceding training is necessary.

Alternatively, a state space model derived from a BG with parameters estimates updated at every sample time point on a DBG may be used by a Kalman or particle filter for advancing the estimate of the deteriorating state until an alarm threshold is reached and appropriate action is required. Particle filters have become quite popular in purely mathematical approaches to failure prognosis. Bond graph methodology can contribute the state space model with time-varying parameters to be used by the filter. If one has an indication what type of mathematical function might capture the degradation, one may just use the BG model and incorporate the estimation of the fitting parameters into an extended Kalman or particle filter that simultaneously estimates the system state and the curve fitting parameters.

Bayesian-based filters enable to account for various uncertainties such as model or parameter uncertainties, disturbances, or unforeseen load changes on the system by considering the states of a system as random variables with an expectation value and a pdf at each sample time point and to propagate the uncertain state values so that the RUL is also a random variable. Often, a confidence interval is of more interest than a single value for a RUL. This can also be achieved by assuming pdfs for the parameters of the regression function approximating the deteriorating data in the sliding window which is supported by the Matlab Predictive Toolbox™. State estimation by Bayesian-based filters has been addressed by revisiting Kalman and particle filters to some extent in Sects. 3.4.1 and 3.4.2, respectively. Section 4.6 briefly considers various uncertainties associated with failure prognosis.

In case ARRs cannot be derived offline in symbolic closed form from a DBG because some unknown variables cannot be eliminated in the constraints between known inputs and measurements due to nonlinear constitutive element equations, in the worst case, it may become necessary to evaluate online the entire DBG model instead of some ARRs in order to deliver fault indicators into a fault diagnosis software module. The generation of ARR residuals can be performed concurrently to the online monitoring. Fault indicator values are needed at time sampling points. However, if the numerical computation of a DAE system derived from a DBG is needed to provide these values, some time delay will result. A subject of further research may be how much delay between the onset of an incipient fault, its clear detection, and the start of failure prognosis is admissible under which constraints and, if necessary, how the delay can be reduced.

Fault Tolerant Control

Fault accommodation needs an inverse model for the reconstruction of the command input into a system after a fault has been isolated and its size has been estimated. Bicausal bond graphs are well suited for the determination of the inverse model needed in active FTC as illustrated in Sect. 5.4. An inverse model may also be needed in passive FTC. This is shown in Sect. 7.8 by considering an overwhelming controller as an example used to force the velocity of a classical mechanical mass-spring oscillator to follow a prescribed undamped velocity trajectory.

Some Subjects of Further Work

In this book, small application systems and mainly continuous time models have been considered for illustration. As the regression of estimated deteriorating data provided by the diagnostic BG and its projection into the future is based on a moving window, the combined bond graph-based, data-based approach to failure prognosis presented in Sect. 4.5 is also applicable to hybrid systems in which the deteriorating behaviour may be mode dependent. An a priori choice of a degradation function with parameters to be fitted is not necessary. When the estimated data indicate a significant change in the degradation rate, a function candidate from another more appropriate class may be chosen for the current window. By this way, degradation rate changes due to an abrupt fault that may be caused by an accumulated deterioration, e.g., a total blocking of a valve or the clogging of a filter, can be captured.

Incremental bond graphs and BGs in LFT form enable to set up adaptive fault thresholds for parametric uncertainties. As long as the variations of estimated parameter values are within these bounds, a system is considered to be in a healthy state. Misdetections of faults are thus avoided. However, once the estimated value of a parameter crosses a fault threshold, values of the parameter are to be collected for a number sampling points to decide with certainty whether the onset of an incipient fault had happened. This delay in fault detection postpones the start of the failure prognosis. How much time delay is admissible depends on the dynamic system behaviour and on the degradation process.

Multiple parametric faults that cannot be isolated by simple inspection of a FSM derived from a diagnostic BG may be isolated and estimated by least squares ARR residual minimisation as outlined in Sect. 3.5.4. However, in the case of multiple concurrent faults, one fault may temporarily reduce the effect of another one. Also, a hybrid system part may switch off shortly after an incipient fault had started in that part. That is, the fault cannot be detected until the system part is reactivated again. This may be the case when a trending parameter in an ARR is multiplied by a discrete switch state. Both cases have an impact on failure prognosis.Also, after an isolated fault is accommodated by a reconstruction of the command input, estimations of future states must continue to estimate the RUL of the new faulty system. As to incipient faults, so far, not many works have been reported that consider multiple incipient fault prognosis [2].

If it is not possible to perform a run-to-failure experiment in order to determine the actual end of life time of a system with a deteriorating dynamic behaviour, the remaining useful life as of some current time point till the system's end of life cannot be computed. That is, a failure threshold and a failure zone according to a performance criterion have to be chosen. A failure threshold does not necessarily indicate a total system failure [1]. The crossing of a fault indicator or of a feature with this threshold defines a time instant beyond which the system is not useable any more. Thus, the RUL depends on the choice of the failure threshold which in turn is specific for the monitored system and depends on the requirements the system dynamics must meet and on engineering experience. Possible criteria are that the system must remain stable despite a parametric degradation, that the dynamic system response must reach at least a certain percentage of the steady value the healthy system would reach, or that the liquid in a hydraulic system must not exceed a certain temperature and that the pressure remains below a certain limit. The simulation of a boost converter with a leaking capacitor in Sect. 7.3.3 assumes that the failure threshold is reached when the decaying capacitance has lost $3/5$ of its initial value.

Hybrid systems with various modes of operation may require to set a failure threshold that depends on the current system's mode of operation and also accounts for the impact changing environment conditions may have on a system so that a failure threshold to be defined is not a time independent constant with some uncertainty but may be a piecewise constant function of time.

A bond graph model of a system developed on the basis of a good understanding of the physical phenomena can be used for offline simulations of various fault scenarios to study the impact of deliberately injected parametric degradations on the system behaviour as illustrated by the application examples in Chap. 7, or to analyse the system stability with regard to different values of a trending parameter.

References

1. Saxena, A., Celaya, J., Saha, B., Saha, S., & Goebel, K. (2010). Metrics for offline evaluation of prognostic performance. *International Journal of Prognostics and Health Management, 1*(4), 2153–2648, Open-access Article.
2. Yang, Y., Bin, R., Huang, Z., Qin, Y., & Yi, J. (2014). Multiple incipient faults prognosis for CCBII braking system. In *The 26th Chinese Control and Decision Conference (2014 CCDC)* (pp. 789–794). https://doi.org/10.1109/CCDC.2014.6852272.

Appendix A
Some Definitions

A.1 Fault Diagnosis

A number of key terms have been used throughout this book. This appendix provides a list of definitions that are in accordance with the outcome of a standardisation effort of the IFAC SAFEPROCESS Technical Committee. The definitions have been taken from [1, 2].

Definition A.1 (Fault) A fault is an unpermitted deviation of at least one characteristic property or parameter of the system from an acceptable, usual, or standard condition.

Definition A.2 (Failure) A failure is a permanent interruption of a system's ability to perform a required function under specified operating conditions.

Definition A.3 (Malfunction) A malfunction is an intermittent irregularity in the fulfilment of a system's desired function.

Definition A.4 (Symptom) A symptom is a change of an observable quantity from normal behaviour.

Definition A.5 (Fault detection) Fault detection: Determination of the faults present in a system and the time of detection.

Definition A.6 (Fault isolation) Fault isolation: Determination of the kind, location, and time of detection of a fault.

Definition A.7 (Fault identification) Fault identification means the determination of the size- and time-variant behaviour of a fault.

Definition A.8 (Fault diagnosis) Fault diagnosis encompasses the determination of kind, size, location, and time of detection of a fault by evaluating symptoms.

W. Borutzky, *Bond Graph Modelling for Control, Fault Diagnosis and Failure Prognosis*, https://doi.org/10.1007/978-3-030-60967-2

Definition A.9 (Diagnostic model) A set of static or dynamic relations which link specific input variables – the symptoms - to specific output variables – the faults.

Definition A.10 (Analytical redundancy) Use of two, not necessarily identical ways to determine a quantity where one way uses a mathematical process model in analytical form.

Definition A.11 (Disturbance) A disturbance is an unknown and uncontrollable system input.

Definition A.12 (Residual) Fault indicator based on deviations between measurements and model equation based calculations.

A.2 Failure Prognostic

Definition A.13 (Prognostic) Prognostic is defined as the estimation of time to failure and risk for one or more existing and future failure modes (ISO 13381-1, 2004)

Definition A.14 (Remaining Useful Life (RUL)) The RUL can be defined as the length of the time span from the current life time instant t_c (age of the system) to the end of the useful life instant t_{EoL}.

The following definitions have been from [3].

Definition A.15 (Failure Threshold (FT)) Failure Threshold – a limit on damage level beyond which a UUT (Unit Under Test) is not usable. FT does not necessarily indicate complete failure of the system but a conservative estimate beyond which risk of complete failure exceeds tolerance limits.

Definition A.16 (Run-to-Failure (RtF)) refers to a scenario where a system has been allowed to fail and corresponding observation data are collected for later analysis.

Definition A.17 (End-of-Life (EoL)) time instant when a prediction crosses a FT. This is determined through RtF experiments for a specific UUT.

Definition A.18 (End-of-Prediction (EoP)) time index for the last prediction before EoL is reached. This is a conceptual time index that depends on the frequency of prediction and assumes that predictions are updated until EoL is reached.

Definition A.19 (End-of-Useful-Predictions (EoUP)) time index beyond which it is futile to update a RUL prediction because no corrective action is possible in the time available before EoL.

References

1. Blanke, M., Kinnaert, M., Lunze, J., & Staroswiecki, M. (2006). *Diagnosis and Fault-Tolerant Control*. Berlin: Springer.
2. Isermann, R. (2006). *Fault-Diagnosis Systems An Introduction from Fault Detection to Fault Tolerance*. Berlin/Heidelberg: Springer.
3. Saxena, A., Celaya, J., Saha, B., Saha, S., & Goebel, K. (2010). Metrics for offline evaluation of prognostic performance. *International Journal of Prognostics and Health Management, 1*(4), 2153–2648, Open-access Article.

Appendix B
Short Introduction into Bond Graph Modelling

Bond graphs were devised by Professor Henry Paynter[1] at Massachusetts Institute of Technology (MIT), Cambridge, Massachusetts, U.S.A. as early as 1959 [21]. The concept was elaborated into a formal physical modelling methodology for multidisciplinary systems by his former Ph.D. students Professor Dean Karnopp and Professor Donald Margolis (University of California at Davis, California) and Professor Ronald Rosenberg (Michigan State University, East Lansing, Michigan). Since then this modelling approach has spread all over the world and is used in academia as well as in industry by many individuals. Software programs supporting bond graph modelling as well as a number of textbooks in various languages have emerged. The very first bond graph program for bond graph modelling and simulation was ENPORT™ developed by R. Rosenberg. Other programs are, for instance, 20sim [10], SYMBOLS [22], or the bond graph preprocessor CAMP-G [16] for Matlab®/Simulink® [25]. Some of the textbooks written in English besides the well known textbook of Karnopp, Margolis, and Rosenberg [18] are referenced in a list not meant to be exhaustive at the end of this appendix [4–6, 9, 11, 12, 15, 19, 20, 23]. There are also some short introductions to bond graph modelling available [3, 8, 14]. In the following, some fundamentals of bond graph modelling are recalled so that the use of bond graph modelling for FDI and prognosis in this book can be followed more easily.

B.1 Basic Concepts

Bond graph-based physical systems modelling starts from considering the exchange of energy between real or conceptual subsystems, components or elements of a multidisciplinary engineering system in which energy may be present in different

[1] 1923–2002.

W. Borutzky, *Bond Graph Modelling for Control, Fault Diagnosis and Failure Prognosis*, https://doi.org/10.1007/978-3-030-60967-2

forms and conversions from one form into another may take place. Like any other graph, bond graphs consist of nodes and edges. Nodes may represent models of subsystems, system components, or elements. In the nodes, energy may be temporarily stored, transformed from one form into another, especially irreversibly into heat, or power may be distributed to other nodes. In bond graphs, the latter ones have got the so-called *power ports* where energy may enter or leave a node. Of course, nodes may have more than one power port. They are called *multiports*. The edges of a bond graph are connections between the power ports of different nodes. They are called power bonds, or just *bonds*. They represent the transfer of power between ports and may be associated with physical links between real systems such as a shaft between a motor and a mechanical load, or a hydraulic line between hydraulic components if it is assumed that energy is neither stored nor dissipated into heat in a physical link. Otherwise, a physical model is to be developed for the physical link.

B.1.1 Power Variables and Energy Variables

Bond graph methodology assumes that the current amount of power $P(t)$ at time instant $t > 0$ transferred between two power ports can be expressed equally in all energy domains as the product of two power conjugated variables called *effort*, e, and *flow*, f, respectively.

$$P(t) = e(t) \cdot f(t) \tag{B.1}$$

Thus, the transfer of power between two subsystem ports can be depicted as indicated in Fig. B.1. The *half arrow* attached to the bond denotes the *reference* direction of the energy flow.

As a bond in a bond graph represents the energy exchange between two power ports of different nodes, all edges of a bond graph carry two power variables. Bonds may be annotated by the names of these power variables. It is a convention to place the name of an effort above a horizontal bond and the flow below the bond (cf. Fig. B.1). For vertical bonds the convention is to place the effort to the left and the flow to the right of the bond. For inclinations different from a multiple of 90°, a more sophisticated convention is to consider that variable as flow that is on the same side as the half arrow.

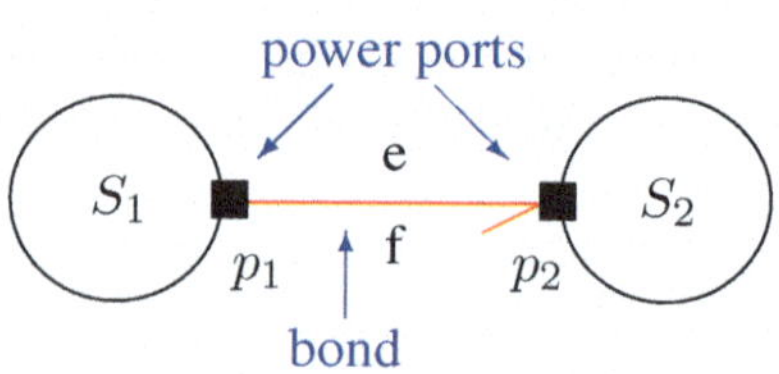

Fig. B.1 Power bond connecting two power ports of two subsystem models

Effort and flow can be identified as force and linear velocity in translational mechanical engineering. In electrical engineering, the product of the voltage drop across the two pins of an electrical port and the joint current in both pins is the instantaneous amount of electrical power at this port. Correspondingly, in the thermal domain, these variables are the absolute temperature and the rate of change of the entropy. Table B.1 lists the effort and flow variables in the various energy domains.

The second and the third column of Table B.1 list the effort and flow variables in the various energy domains. The variables in the fourth column of Table B.1 are the time integral of the efforts and the variables in the fifth column are the time integral of the flows. They are called *energy variables* because they quantise the amount of energy in the energy storage elements of a model.

$$p(t) := p(t_0) + \int_{t_0}^{t} e(\tau)d\tau \tag{B.2}$$

$$q(t) := q(t_0) + \int_{t_0}^{t} f(\tau)d\tau \tag{B.3}$$

where $p(t_0)$, $q(t_0)$ are initial values.

Table B.1 Power and energy variables in various energy domains

Energy domain	Effort e	Flow f	Generalised momentum p	Generalised displacement q
Translational mechanics	Force F [N]	Velocity v [m/s]	Momentum p [N s]	Displacement x [m]
Rotational mechanics	Angular moment M [Nm]	Angular velocity ω [rad/s]	Angular momentum p_ω [Nms]	Angle θ [rad]
Electrical domain	Voltage u [V]	Current i [A]	Linkage flux λ [Vs]	Charge q [As]
Hydraulic domain	Total pressure p [N/m^2]	Volume flow rate Q [m^3/s]	Pressure momentum p_p [N/m^2 s]	Volume V_c [m^3]
Thermodynamic domain	Temperature T [K]	Entropy flow rate $\dot{S}$ [J/K/s]	–	Entropy S [J/K]
Chemical domain	Chemical potential μ [J/mole]	Molar flow $\dot{N}$ [mole/s]	–	Molar mass N [mole]

B.1.2 Analogies

The power variables effort and flow play an equal role with respect to each other. They are just characterised by the fact that they are a factor in the power product. Given a pair of power variables, it is a matter of preference which of them is chosen as effort and which consequently serves as the flow. This gives rise to two possible *analogies*. One choice could be to relate a mechanical force, or a moment to an electrical voltage drop. Consequently, velocities, or angular velocities, correspond to electrical currents. This analogy has been widely used for a long time. Therefore, occasionally, it is referred to as the *classical* analogy. The other possible relation of a velocity with a voltage drop also makes perfectly sense from the point of *measurements*. It has been introduced by Firestone around 1933 and is called *mobility* analogy. If two modelling approaches just differ with respect to the analogy that is used, the resulting bond graph models will look different as well as the mathematical models derived from the graphs. However, as to the numerical evaluation of the mathematical models, simulation results should be the same.

- Classical Analogy:

$$\begin{array}{|ll|} \hline \text{Force} & \triangleq \text{Voltage} \\ \text{Velocity} & \triangleq \text{Current} \\ \hline \end{array}$$

- Mobility Analogy:

$$\begin{array}{|ll|} \hline \text{Velocity} & \triangleq \text{Voltage} \\ \text{Force} & \triangleq \text{Current} \\ \hline \end{array}$$

B.1.3 Hierarchical Bond Graph Models

As other graphical modelling formalisms, bond graph methodology supports a hierarchical modelling approach. Bond graph models may be developed in a hierarchical combined top-down and bottom-up approach by using component models or elements from model libraries. For each hierarchy level, the structure of the model may be represented by a bond graph. While in iconic diagrams, or in electrical or hydraulic networks, application-specific icons are used for the nodes of the graph, in bond graphs, nodes are presented by words enclosed by an optional ellipsis. For that reason, the notion of a *word bond graph* is common in the process of a bond graph-based model decomposition approach.

At the lowest hierarchy level, bond graph nodes represent basic energetic processes, that is, the delivery or storage of energy, the irreversible transformation of energy into heat, or the power conservative distribution of power. For these fundamental energetic processes, specific type names for nodes are used. For instance, the storage of electrical energy in a capacitor or the storage of potential energy in a mechanical spring is represented by a node of type C. Of course, as with other graphical representations, user defined nodes may be introduced, e.g. a node labelled `DisplacementPump` representing a mathematical model of a hydraulic displacement pump.

As bond graph modelling starts from considering the energy exchange between system components and since this exchange is associated with physical quantities such as momentum, mass, electrical charge, or entropy, bond graph models should comply with physical conservation laws. In contrast, signal processing blocks in block diagrams may represent any functional relation between signals.

B.2 Bond Graph Elements

At the lowest hierarchy level, bond graph modelling uses a set of nine basic conceptual elements for representation of the fundamental energetic processes. They can be grouped into five categories.

B.2.1 Supply and Absorption of Energy

The supply of energy into a system is modelled by source elements. The absorption of energy flowing out of a system into its environment can be represented by sinks, which can be considered negative sources. As a power port has two variables, two kinds of sources exist. Sources may impose either an effort or a flow onto a system. For instance, a battery serving as a constant voltage source can be modelled by an effort source, while a hydraulic pump providing a constant volume flow rate can be modelled as a flow source. In bond graphs, sources and sinks, respectively, are denoted by the character S (Source). The type is naturally indicated by adding either the characters e or f, respectively, (Se or Sf). Sources may have more than one power port. Moreover, as there exist, e.g. stabilised voltage sources, or controlled hydraulic pressure pumps, sources may also have a signal port for feedback control. At a signal port, the amount of power is negligible small. In this case, the node type identifier is prefixed by the character M standing for *modulated* source (MSe, or MSf). (Conventionally, the signal port is located on the M side of the element.)

B.2.2 Energy Storage

The generalised momentum and the generalised displacement are not just the integral of an effort and a flow, respectively, but have a physical meaning in most energy domains as indicated in Table B.1. There is no generalised momentum in thermodynamics. In chemical engineering and in hydraulics, the use of a pressure momentum is rather uncommon. Except in the thermodynamic and in the chemical domain, the generalised momentum and the generalised displacement, respectively, can be related to a power variable resulting in the constitutive relation of a 1-port storage element. That is, one of the two power variables of the port is the rate of change of a so-called *conserved*, or stored quantity, also called a *state*, while the other power variable is an equilibrium determining variable. As either the effort or the flow can be the rate of change of the conserved quantity, two types of energy stores can be distinguished.

In a C energy storage element, the flow f is integrated and the resulting generalised displacement q is related to the conjugate effort of the port.

$$q(t) = q(t_0) + \int_{t_0}^{t} f(\tau)d\tau \tag{B.4a}$$

$$q(t) = \Phi_C(e(t)) \tag{B.4b}$$

where Φ_C is a one-to-one function $\Phi_C : \mathbb{R} \to \mathbb{R}$ that has a unique single valued inverse Φ_C^{-1}

For the second type of energy store, the I energy storage element, the role of effort and flow is just interchanged. The effort is integrated and the resulting generalised momentum is related to the conjugate flow.

$$p(t) = q(t_0) + \int_{t_0}^{t} e(\tau)d\tau \tag{B.5a}$$

$$p(t) = \Phi_I(f(t)) \tag{B.5b}$$

where Φ_I is a one-to-one function $\Phi_I : \mathbb{R} \to \mathbb{R}$ that has a unique single valued inverse Φ_I^{-1}.

In this sense, both storage types are dual to each other. For instance, an electrical capacitor or a mechanical spring can be modelled as a C-type energy store, while a rigid body storing kinetic energy or a coil storing magnetic energy can be modelled by an I type storage element. Energy stores can be multiport elements. Note that, in contrast to sources, modulation of storage elements would violate the principle of energy conservation.

B.2.3 Irreversible Transformation of Energy into Heat

The irreversible transformation of energy into heat, e.g. in an electrical resistor, or due to friction in mechanical and hydraulic systems, is often modelled as a loss of *free* energy. In bond graphs, it is represented by an R element (resistive element). If the production of entropy is taken into account, the two-port RS element introduced by Thoma [26] is used (cf. Fig. B.2). The character S (Source) indicates the thermal port and expresses the entropy production.
The RS element conserves power. That is,

$$e \cdot f = T \cdot \dot{S} \tag{B.6}$$

where T denotes the temperature and S the entropy.

While the relation between the power variables of the non-thermal port may be linear, for the thermal port it is always nonlinear. If a linear constitutive equation

$$e(t) = R \cdot f(t) \tag{B.7}$$

with a resistance R is assumed for the non-thermal port, then

$$\dot{S}(t) = \frac{R \cdot f^2(t)}{T(t)} \tag{B.8}$$

According to the second law of thermodynamics, entropy production must be positive. Consequently, the graph of the constitutive relation must be within the first and third quadrant.

Like sources, resistive elements may be modulated. For instance, variable hydraulic orifices in spool valves controlled by the displacement of the spool may be modelled by displacement controlled R elements.

B.2.4 Reversible Transformation of Energy

In this kind of process, entropy is neither stored nor produced. Consequently, it is power conservative. There are two types of bond graph elements representing this kind of process. They are denoted by the acronyms TF and GY, respectively. In the simplest case they are 2-port elements. Power conservation means that the instantaneous power at one port equals the instantaneous power at the other port. An

Fig. B.2 Irreversible transformation into heat [26]

element of type TF relates the efforts at the ports and separately relates the flows, while an element of type GY relates the effort of one port to the flow of the other port and vice versa. In the constitutive relations of both elements, a variable is multiplied by either a constant or by a function of time. In the second case, the elements must have a signal port in addition to the power ports. This is pointed out by prefixing their acronyms with the character M (MTF, MGY). Physical components that may be modelled by a TF element are electrical transformers, mechanical gear boxes, or hydraulic displacement pumps. The GY element may be a simple model for the transformation of electrical energy into mechanical energy in an electrical motor. Transformers as well as gyrators can be multiport elements.

B.2.5 Power Conservative Distribution of Energy

Modelling of energy flows in a system means that energy is supplied by sources and conveyed and distributed between the conceptual elements of the model. As energy storage and irreversible transformation of energy have already been taken into account by energy storage elements and by resistors, distribution of energy between elements can be considered power conservative. There are two types of multiport interconnection elements, called *0-junction* and *1-junction*. The two of them distribute power and have linear constitutive relations. For a 0-junction, the efforts of all power ports are the same and all power conjugated flows sum up to zero taking into account their signs. The sign is determined by the orientation of the half arrow for the energy reference direction. If a half arrow is oriented towards a port, the flow is taken positive, otherwise the flow has a negative sign. The 1-junction plays the dual role. That is, the flows of all bonds incident the node are the same and all conjugate efforts sum up to zero by taking into account their signs. A 0-junction corresponds to an interconnection node in a network. Such a node has an effort (voltage, pressure) and according to the generalisation of Kirchhoff's current law, all flows in the branches incident to the node sum up to zero. In contrast, a 1-junction in a bond graph has no node as a counterpart in networks. The constitutive equation of a 1-junction relating all efforts is embodied implicitly in networks as it corresponds to the generalisation of Kirchhoff's voltage law.

According to the orientations of the half arrows in Fig. B.3, the constitutive equations of the 0-junction read

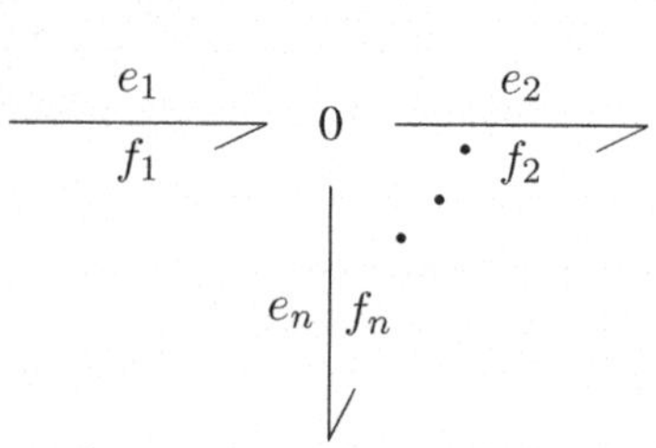

Fig. B.3 0-junction with *n* ports

$$e_1 = e_2 = \ldots = e_n \tag{B.9a}$$

$$0 = f_1 - f_2 - \ldots - f_n \tag{B.9b}$$

Definition B.1 (Junction structure) The subgraph of a bond graph that contains only nodes of type 0, 1, (M)TF, (M)GY is called a *junction structure*. □

B.3 Systematic Construction of Acausal Bond Graphs

One of the advantages of the bond graph approach is that the topological connectivity of components in a system can guide the systematic construction of a bond graph. Two procedures can be formulated, one for the construction of bond graphs for mechanical subsystems and one for the construction of bond graphs for subsystems in energy domains other than the mechanical domain (non-mechanical subsystems).

B.3.1 Mechanical Subsystems (Translation and Fixed-Axis Rotation)

1. Identify distinct inertial velocities and angular velocities; represent them by a 1-junction. Zero absolute velocities may be represented by a 0-junction also.
2. Insert C- and R-ports via a 0-junction between a proper pair of two 1-junctions. (The 0-junction represents a difference of velocities and, at the same time, a force or a moment. A spring or a dashpot reacts to a difference of velocities at its terminals.)
3. Insert TF- and GY-elements between appropriate pairs of 1-junctions representing either absolute or relative velocities.
   (A TF element relates a velocity at one port to a velocity at another port. A GY element relates a velocity at one port to a force or moment at another port. Velocities may be either inputs at all ports of a GY element or they all must be outputs. Consequently, TF- and GY-elements are connected to 1-junctions.)
4. Attach inertia 1-port elements to their respective 1-junction.
5. Attach 1-port sources and 1-port sinks to appropriate 1-junctions.
6. Assign a reference direction for the energy flow to each bond (half arrow).
7. Remove all 1-junctions representing a velocity or angular velocity $\equiv 0$ along with all incident bonds and simplify the bond graph.

B.3.2 Non-mechanical Subsystems

1. Identify distinct efforts (potentials of nodes in electrical networks, absolute pressures in hydraulic and acoustic systems, absolute temperatures in thermodynamic systems); represent them by a 0-junction.
2. Insert the non-mechanical power port of a source, energy store, dissipator, transformer or gyrator via a 1-junction between two proper 0-junctions.
   (In bond graphs of electrical systems, the 1-junction represents the voltage drop across the port and, at the same time, the current through the port. In case of an electrical transformer, the 1-junctions at both ports of the associated TF element represent the currents through the coils of the transformer. In bond graphs of hydraulic systems, C elements are inserted via a 1-junction between the 0-junction of an absolute pressure and the 0-junction of the atmospheric pressure. In bond graphs of thermal systems, the thermal port of a C element is attached directly to the 0-junction of an absolute temperature.)
3. Add half arrows to all bonds.
4. Choose a potential as a reference; eliminate its corresponding 0-junction along with all incident bonds. If two sub-circuits of an electrical network are connected via an isolating transformer, a reference potential must be chosen in each sub-circuit.
5. Simplify the bond graph.

For hydraulic subsystems, it is common to choose the atmospheric pressure as reference. After elimination of its associated 0-zero junction along with all incident bonds, 0-junctions represent gage pressures. This results in a simplification of the construction of bond graphs for hydraulic systems. Gage pressures are represented by 0-junction, C elements are attached directly to a proper 0-junction. TF elements in bond graphs of hydraulic systems relate a pressure, p, to its associated mechanical force, F, and a volume flow rate, $\dot{V}$, of incompressible fluid flow to its associated translational velocity v.

$$F = A \cdot p \tag{B.10a}$$

$$\dot{V} = A \cdot v \tag{B.10b}$$

where A is the cross-section area perpendicular to the direction of a one-dimensional fluid flow. That is, the hydraulic port of a TF element is connected to a 0-junction of a gage pressure, while its mechanical port is connected to the 1-junction of a velocity (Fig. B.4).

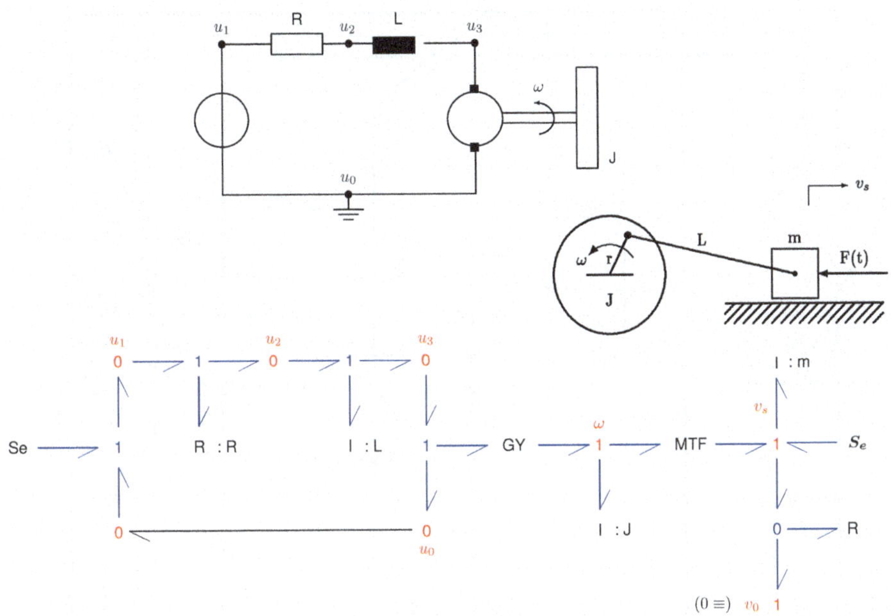

Fig. B.4 Systematic construction of a BG from a schematic of an electromechanical system

B.3.3 Assignment of Power Reference Directions

There are intuitive rules for the assignment of power reference directions to the ports of all types of bond graph elements. For sources, the half arrow points away from the element's port, for storage elements and resistors, the orientation of the incident bonds is towards the element. This is quite intuitive as sources supply energy, energy storage elements store energy temporarily, and resistors transform the absorbed energy irreversibly into heat. TF and GY elements adopt a 'through direction' of the reference direction of the energy flow because energy passes through these elements without storage and without entropy production. In essence, reference directions of the energy flow are from the sources through the junction structure into energy stores, resistors, and sinks (cf. Figs. B.4 and B.5).

There should a *difference* of power variables at 0-junctions in bond graphs of mechanical systems and at 1-junctions in bond graphs of non-mechanical systems because springs and dampers react to a velocity difference and in electrical circuits, it is a voltage drop, viz. a difference of potentials, across a two-terminal element. The through direction of half arrows at the 1-junction corresponds to the reference direction of the current through the two-terminal element.

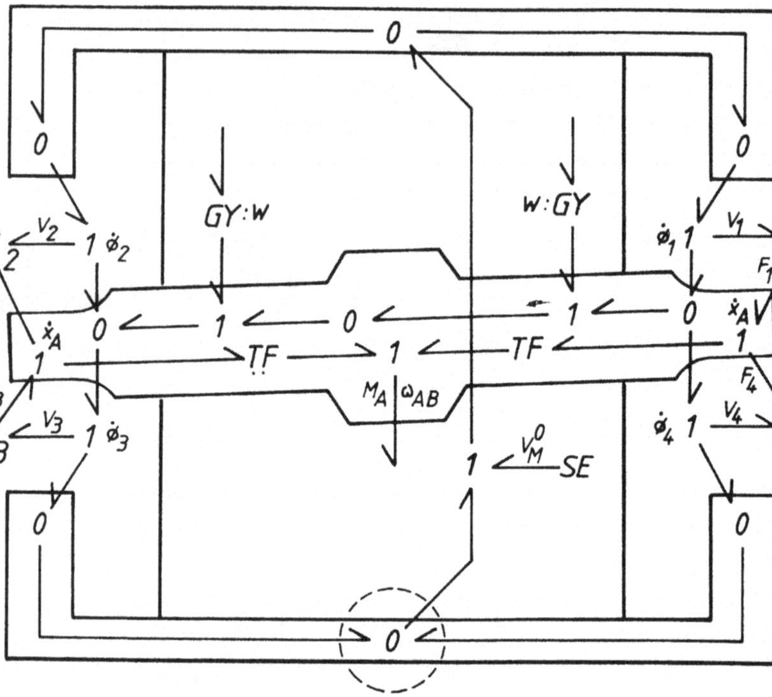

Fig. B.5 Systematic development of a BG model from the schematic of the torque motor of a servovalve

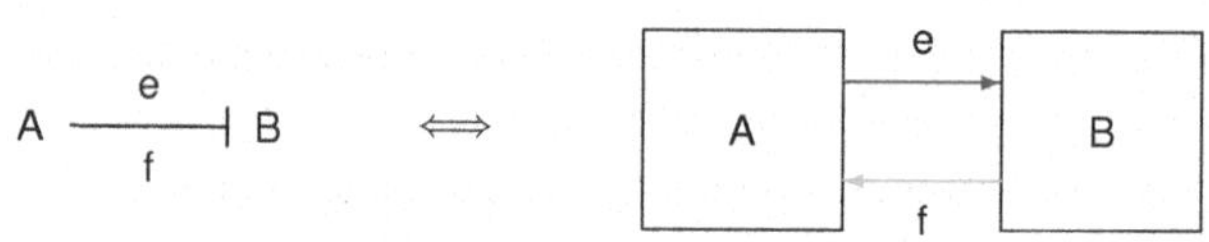

Fig. B.6 Computational causality: indicating the direction of effort and flow

B.4 The Concept of Computational Causality at Power Ports

As each bond connecting two power ports of different nodes A and B carries two power variables, one of the two may be determined by one of the two submodels, while the other is determined by the other model. In other words, from a computational point of view, the effort could be computed by the evaluation of model A, while the flow is computed in model B. It could be the other way around as well. The decision as to local input-output pairs is indicated by a perpendicular stroke attached to the bond and is called the assignment of *computational causality* (Fig. B.6).

The end of a bond without perpendicular stroke indicates the model in which the effort is computed and in which the conjugate flow must be known. Another possible and common view is to consider both power variables of a bond as signals of *opposite* direction. That is, the perpendicular stroke indicates the signal direction of the effort, which implies that the end without the perpendicular stroke displays the direction of the conjugate flow. The perpendicular stroke is called a *causal* stroke. A bond graph is called a *causal* bond graph if a causal stroke has been added to each bond. Note that the half arrow and the causal stroke are orthogonal concepts. That is, there are four possible pattern of half arrow and causal stroke. Assigning causalities in a bond graph means that the bond graph is superimposed with a block diagram representing the *computational* structure of a model.

If sensors and instruments are included in the bond graph modelling of an engineering system, then the power conveyed between two ports of different components can be neglected, if the sensing of signals is of primary concern. This means that one of the two conjugate power variables associated with a bond can be dropped, turning the bond into a so-called *activated* bond or reducing it to a conventional signal arrow. As a result, the ports linked by a bond that has become a signal arrow turn into signal ports.

The hierarchical development of a bond graph model and the connection of component submodels according to the topological structure of a system implies that bond graphs of component models must be non-causal. The decision, which of the two power variables of a power port plays the role of an input signal, forcing the conjugate variable to be an output variable, is determined by the nature of component models. That is, causal strokes cannot be assigned before the hierarchical development of an overall system model is finished and the hierarchy has been 'flattened'. The latter means that each node representing a submodel is to be replaced recursively by a bond graph until the overall system bond graph only contains standard bond graph elements. Causal strokes, or computational causalities, at the ports of one and the same component submodel can be different depending on the component submodels it is connected to. Equations derived from a non-causal bond graph should be expressed initially in implicit form.

B.4.1 Rules for Computational Causalities at Power Ports

At each power port of a component model, it can be decided which one of the two power variables is computed in the component model, or in other words, which of the two power variables is an outgoing signal, or an output variable in one of the constitutive relations. However, these decisions cannot be made completely arbitrarily.

For sources there is no choice. For an effort source, the output is the effort, for a Sf source, it is the flow.

For 2-port TF and GY elements causal patterns must be as displayed in Fig. B.7. A TF relates efforts. That is, if the effort at one port is an input, the effort at the other port must be an output. Since the constitutive equations of GY elements relate the effort of one port to the flow of the other one, both causal strokes must either point to the element or away from it.

At 0-junctions the effort at all incident bonds is the same. Consequently, one causal stroke can point to the junction, while all others must point away from it (cf. Fig. B.7). For the dual 1-junctions, the role of effort and flow is interchanged. That is, at one bond the causal stroke may be pointing away from the junction. At all other bonds it must point towards the 1-junction. This causal pattern reflects the fact that one effort is equal to the sum of all other efforts and simultaneously it indicates that all flows are the same. One flow may be input to the junction, while all others are outputs.

For storage elements the so-called *integral* causality is preferred. The power variable that is integrated with respect to time is the input variable. The conjugate power variable, related to the state of the energy store, is taken as output variable. For a C energy store, this is the effort. Consequently, the causal stroke points away from the port of the C element. For the dual I element, the effort of the power port is taken as an input variable. Consequently, the causal stroke points to the port of the I element. The effort is integrated with respect to time. The flow related to the resulting state of the I energy store is its output. If, for a C energy store, the causal stroke is on the side of the power port, or if the casual stroke points away from the port of an I energy store, then this means that the output is obtained by differentiation

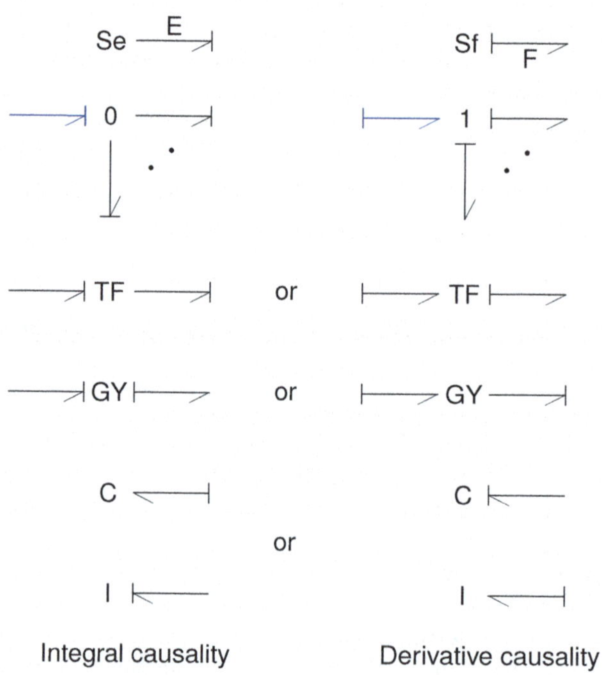

Fig. B.7 Admissible causal patterns at bond graph element ports

of the conjugate power variable. In this case, the so-called *derivative* causality has been assigned to the port.

For resistors with linear constitutive relations, causal strokes may be either on the side of a port (conductance causality) or pointing away from the port (resistance causality). In some cases, however, there is no such choice. For a resistor representing dry friction, only the effort (force) can be the output.

Finally, causal patterns other than the admissible ones are termed *causal conflicts*. They give clear, valuable indications to consequences of modelling assumptions and may give rise to changes of the model.

B.4.2 Sequential Assignment of Computational Causalities

Given the previously discussed rules for the assignment of computational causalities to power ports, the question is in which order are causalities assigned to ports and how is this information added to bond graph propagated through the graph, in other words, how does the choice of computational causality at power port affects the ports of adjacent elements. The step-by-step assignment of causalities follows the Sequential Causality Assignment Procedure (SCAP) introduced by Karnopp and Rosenberg [17, 18]. This procedure has become a standard and is widely used. A modification with regard to the so-called causal bond graph loops has been discussed by van Dijk in [13]. In order to support the derivation of ARRs from hybrid bond graphs that hold for all system modes, Low et. al. [1] recently proposed another modification of the SCAP by introducing preferred causalities for controlled junctions.

Sequential Causality Assignment Procedure (SCAP)

1. Assign causality to one of the sources according to its type and *propagate* this causal information into the bond graph through its junction structure as far as possible by observing causality rules at element ports.
2. Repeat step 1 until all ports of sources are assigned an appropriate causality.
3. Assign preferred integral causality to a port of an energy store and propagate this causal information into the bond graph as far as possible. Propagation of the causality of a storage port may lead to derivative causality at power ports of other energy stores and often entails an assignment of causality at resistor ports. If, by this way, a nonlinear resistor port receives a causal stroke not in accordance with the resistor's constitutive equations, the causality assignment must be reversed at this port. This, however, may result in a causal conflict at the junction the resistor port is attached to. This conflict at the junction must be resolved before preferred integral causality can be assigned to the next storage port. For instance, if an

I element and a 1-port resistor representing Coulomb friction are attached to a 1-junction, then the I element must take derivative causality.

4. If there are any resistor ports left without causality after causality has been assigned to all storage ports, then the procedure continues with assignment of causality to resistors with characteristics that do not have a unique inverse, to ensure their correct formulation.
5. Finally, if there are still resistor ports or internal bonds in the junction structure without causality, one resistor port or an internal bond must be chosen. Causality is arbitrarily assigned and propagated through the junction structure. This step is repeated until no causally unassigned bonds are left.
   Note, if this last step is needed, then algebraic loops, viz. implicit algebraic equations, will be part of the mathematical model. Many of today's software programmes supporting bond graph modelling are able to cope with algebraic loops. They just issue a warning or process the model silently.

B.5 Derivation of Equations from Causal Bond Graphs

Once causal strokes have been added to the bonds of a bond graph, a mathematical model can be derived in a systematic manner. First, it must be decided for which unknowns a set of mathematical relations is to be derived. An obvious choice is the set of states of energy storage elements with integral causality at their ports. These variables determine the energetic state of a system in the sense that they quantify the content of each energy store at all time instances $t \geq 0$. As the output variable of a 1-port energy storage element with integral causality, also called *co-energy* variable, is related to its state, it can be chosen as an alternative unknown. This choice is adopted in this book. Note that energy stores with derivative causalities do not contribute to the system's state. Their output variable algebraically depends on the output variables of energy stores with integral causalities.

Software supporting bond graph modelling such as 20sim®, Symbols™, or CAMPG mentioned at the beginning of this appendix can automatically deduce model equations from a causal bond graph. For small to medium scale models, equations can also be deduced manually in a systematic manner. For that purpose, a procedure is given in the following that uses the notion of a *causal path*.

Definition B.2 (Causal path) A sequence of bonds from one power port of an element to a power port of another element is called a causal path if there is no 2-port gyrator in between and if all bonds have their causal stroke at the same end. A cascade of bonds between two power ports with a gyrator in between is called a causal path if all bonds on one side of the gyrator have their causal stroke at the same end, while all bonds on the other side of the gyrator have their causal stroke on the opposite end. That is, the gyrator switches the direction of efforts on one of its sides. □

Remark B.1 An essential feature of bond graphs is that, once assignment of causalities has been completed, conclusions can be drawn with regard to the form of mathematical models that can be derived from the graph by looking for causal paths in a causal graph. There is no need to know the actual form of possibly nonlinear constitutive relations nor to establish and to reformulate any equations. For instance, if there are no energy storage elements with derivative causality, no causal paths between two ports of different resistors and no closed causal paths in the junction structure, then a mathematical model in the form of an explicit state space model can be derived from the graph. More details may be found, for instance, in [4, 13]. □

One straightforward way towards the formulation of a mathematical model in a modelling language, well suited for automation, is to write the constitutive equations for all nodes of the bond graph and to have all redundancies removed symbolically. If the aim is to perform a simulation, the equations can be sorted and transformed into a programming language. If the equations are linear and if the aim is to come up with the matrices of a linear state space model in symbolic form to be processed by a mathematical program, e.g. the open source software Scilab [24], then, clearly, all algebraic equations must be eliminated. Auxiliary variables can be eliminated when equations are derived from the causal bond graph by walking back causal paths until the input, e.g. into a resistor can be expressed by the outputs of sources and storage elements in integral causality. For bond graphs that are not too large, this can be done manually in a systematic manner. The derivation of an ordered set of equations is guided by the following procedure [2, 27].

B.5.1 Procedure for Manually Deducing Equations from a Causal Bond Graph

1. Write the constitutive equations for all independent sources. Their outputs are given functions of time.
2. In contrast, the output of a controlled source is algebraically related to its input. If the latter is not an output of an independent source or an energy store with integral causality, then it can be expressed by means of such outputs by back propagation of causal paths in the junction structure and by eliminating intermediate variables.
3. The outputs of resistors depend algebraically on their inputs. By back propagation along causal paths through the junction structure, their outputs can be expressed by outputs of sources either independent, or controlled ones and outputs of energy stores. The outputs of dependent sources do not need to be eliminated, since they have already been determined in the previous step.
4. For storage ports, the derivative with respect to time of an output is a function of the input(s). By working back causal paths, the inputs can be expressed by outputs of other energy storage elements, of resistors, or sources.

Note that if there are causal paths between resistor ports, then implicit, algebraic equations will result. This means that the output of the resistor port at one end of the causal path cannot be computed without knowing the output of the resistive port at the other end of the causal path. In this case, intermediate variables cannot be eliminated and expressed by system inputs and state variables by back propagation along causal paths. The mathematical model will be of the form of a DAE system that can be transformed into an ODE system if the system of coupled algebraic equations can be solved symbolically.

Implicit algebraic equations also result, if there are closed causal paths in the junction structure. If there are storage elements that must accept derivative causality, then their output variables are algebraically dependent on the outputs of the storage elements in preferred integral causality and the number of states is smaller than the number of storage elements.

B.5.2 A Circuit with an Operational Amplifier

The systematic manual derivation of equations from a causal bond graph shall be illustrated by considering the simple circuit depicted in Fig. B.8.

The input-output behaviour of the operational amplifier is captured by the simple commonly used model depicted in Fig. B.9. That is, the example circuit contains a dependent controlled source.

Application of the procedure for non-mechanical subsystems (p. 290) results in the causal bond graph of Fig. B.10.

The causally completed BG in Fig. B.10 indicates that the two capacitors C : C_1, C : C_2 are in integral causality and that there are causal paths from R : R_2 to R : R_1, from R : R_2 to R : R_i, and from R : R_2 to R : R_o via the controlled source MSe. These causal paths share bonds. The output V of the controlled source

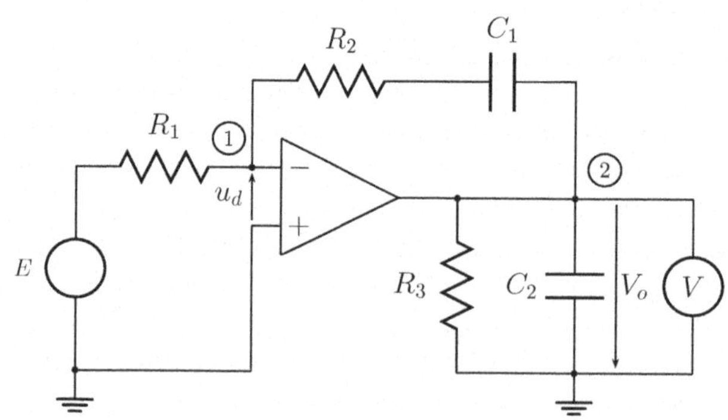

Fig. B.8 Example circuit

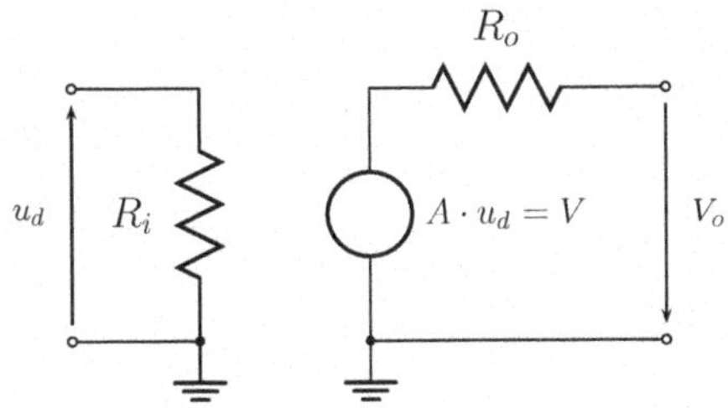

Fig. B.9 Behavioural model of the operational amplifier

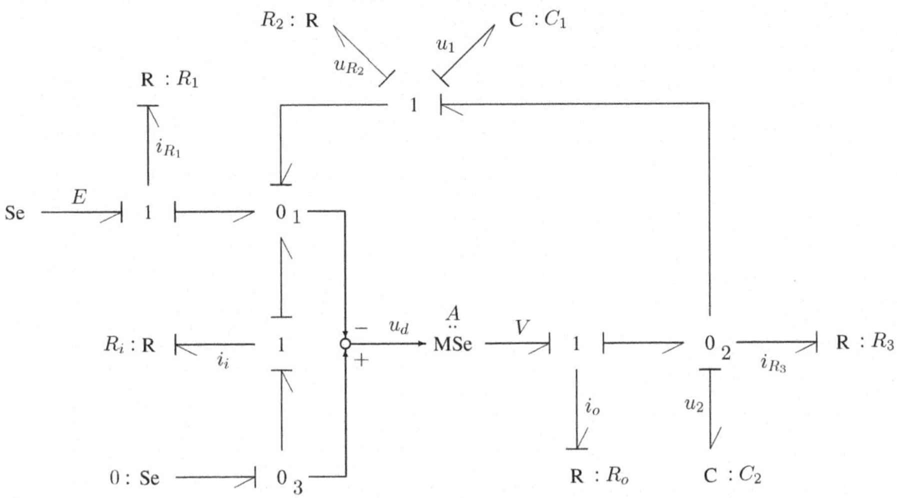

Fig. B.10 Bond graph of the example circuit in Fig. B.8

is algebraically dependent on its input u_d. The latter one algebraically depends on the capacitor voltages u_1, u_2, and the resistor voltage u_{R_2}.

This means that the mathematical model to be deduced will consist of two explicit differential equations including resistor currents and a set of coupled implicit algebraic equations for the outputs of the resistors and the input of the controlled source. As the model is linear, these algebraic equations could be solved symbolically turning the DAE system into an explicit ODE system. Alternatively, the DAE system could be directly formulated in the Scilab script language and evaluated by the DASSL solver.

From the bond graph of the example circuit the following equations can be deduced.

Independent sources:

$$E = f_E(t) \tag{B.11}$$

Dependent sources:

$$u_d = -u_2 + u_1 + u_{R_2} \tag{B.12}$$

$$V = A \cdot u_d \tag{B.13}$$

Resistors:

$$i_{R_1} = \frac{1}{R_1}(E - u_d) \tag{B.14}$$

$$i_i = \frac{1}{R_i} u_d \tag{B.15}$$

$$u_{R_2} = R_2(-i_i - i_{R_1}) \tag{B.16}$$

$$i_o = \frac{1}{R_o}(V - u_2) \tag{B.17}$$

$$i_{R_3} = \frac{1}{R_3} u_2 \tag{B.18}$$

Storage elements:

$$\dot{u}_1 = \frac{1}{C_1}(-i_i - i_{R_1}) \tag{B.19}$$

$$\dot{u}_2 = \frac{1}{C_2}(i_o + i_i + i_{R_1} - i_{R_3}) \tag{B.20}$$

The set of coupled linear algebraic equations (B.12)–(B.18) deduced from the bond graph in Fig. B.10 can be easily solved manually for the input u_d of the modulated source MSe.

$$\begin{aligned} u_d &= -u_2 + u_1 + u_{R_2} \\ &= -u_2 + u_1 - R_2(i_i + i_{R_1}) \\ &= -u_2 + u_1 - R_2\left(\frac{1}{R_i} u_d + \frac{1}{R_1}(E + u_d)\right) \end{aligned}$$

$$\left(1 + \frac{R_2}{R_1} + \frac{R_2}{R_i}\right) u_d = -u_2 + u_1 - \frac{R_2}{R_1} E \tag{B.21}$$

Once u_d is known, the currents of the resistors can be computed so that the right hand side of the two state equations is known. In the case of a larger set of coupled linear algebraic equations, software such as Matlab's Symbolic Math Toolbox™ could be used to solve the subsystem of linear algebraic equations symbolically.

If there were no causal paths between resistor ports, then their outputs could be expressed by the two state variables u_1, u_2 and the input E by back propagation of causal paths. The result would be an ordered set of equations that could be computed in that order. Clearly, if state space matrices are needed, the outputs of the resistors could be inserted into the constitutive equations of the storage elements.

B.5.3 A Switched Circuit

Figure B.11 shows a schematic of an electronic circuit in which a pass transistor has been modelled by means of a switch with two discrete modes. A BG with a generic BG element Sw : m is depicted in Fig. B.12.

In a BG, the switch may be represented by a zero flow source when it is open and accordingly by a zero effort source when the switch is closed. This, however, entails a variable mode-dependent computational causality at the switch port. Moreover, in case the switch is open and fully disconnects inductor I : L_1 from the two right hand side storage elements, i.e. its OFF conductance is assumed to be zero, then the model order is two. In case the switch is closed the model order is three. That is, the number of state variables is mode-dependent. If a non-zero ON resistance of the switch is taken into account and if the switch is modelled as a mode switching resistor with a *fixed* conductance causality then the latter one either causes a causal

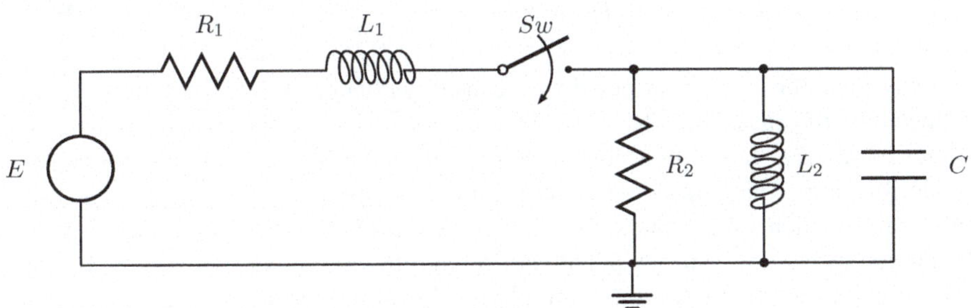

Fig. B.11 Circuit with a switch

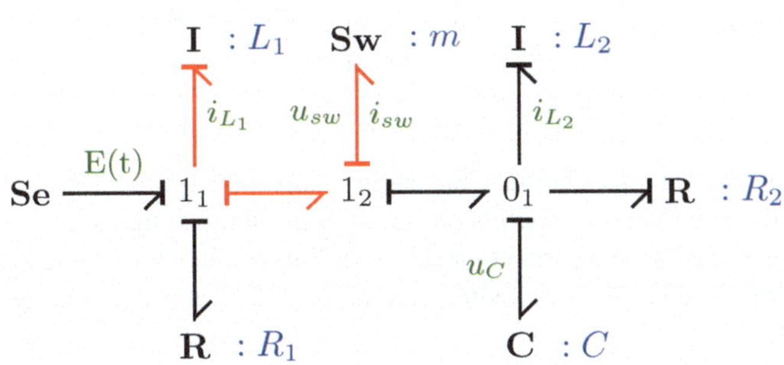

Fig. B.12 BG of the circuit in Fig. B.11 with a generic switch element Sw : m

conflict with the preferred integral causality at the inductor I : L_1 or the inductor would be forced into derivative causality.

Despite the variable causality at a switch port, the standard SCAP has been applied to the BG of the switched circuit. As a result, all three storage elements are in preferred integral causality. By consequence, the switch receives an effort out causality. That is, the causal BG depicted in Fig. B.12 reflects the configuration when the switch is closed. The causal path between the switch and the inductor highlighted in red indicates that their causalities change oppositely, i.e. if the open switch state would be taken into account by a flow out causality, then the inductor would be correctly forced into derivative causality. Nevertheless, a single *implicit* DAE system can be derived from the BG with fixed causalities that holds for both system modes. To that end, the switch is described by an implicit equation regardless of its assigned causality in the BG.

$$Sw: \quad 0 = m u_{\text{sw}} + \bar{m} i_{\text{sw}}\,, \quad m \in \{0, 1\} \tag{B.22}$$

$$L_1: \quad L_1 \frac{d}{dt} i_{L_1} = E - R_1 i_{L_1} - u_{\text{sw}} - u_C \tag{B.23}$$

$$C: \quad C\dot{u}_C = i_{\text{sw}} - i_{L_2} - \frac{u_C}{R_2} \tag{B.24}$$

$$L_2: \quad L_2 \frac{d}{dt} i_{L_2} = u_C \tag{B.25}$$

Only inductor I : L_1 is affected by a causality change at the switch port. This is reflected by the causal path between the inductor I : L_1 and the switch Sw : m. The preferred integral causality at the capacitor port and at the second inductor remain mode-invariant. Their states are collected into a vector $\boldsymbol{x}_{ii} := [u_C, i_{L_2}]^T$. The state of storage elements that change their integral causality into derivative causality due to a mode-change of some switch are collected into a vector $\boldsymbol{x}_{id}$, i.e. $\boldsymbol{x}_{id} := [i_{L_1}]$ in this simple example. In general, an implicit DAE of the form

$$\begin{bmatrix} \mathbf{I} & -\mathbf{M}_{12} \\ \mathbf{0} & \mathbf{M}_{22} \end{bmatrix} \underbrace{\begin{bmatrix} \dot{\boldsymbol{x}}_{ii} \\ \dot{\boldsymbol{x}}_{id} \end{bmatrix}}_{\dot{\boldsymbol{x}}_i} = rhs(\boldsymbol{x}_i, \boldsymbol{u}) \tag{B.26}$$

can be set up where rhs denotes the right hand side of the DAE. The matrix multiplying the state vector depends on the system mode so that some of the ODEs can turn into an algebraic equation, or the right hand side can become zero [7].

Substituting (B.23) into the switch equation (B.22) and accounting for $i_{\text{sw}} = i_{L_1} =: i$ yields

$$\begin{bmatrix} C & 0 & 0 \\ 0 & L_2 & 0 \\ \hline 0 & 0 & m \end{bmatrix} \frac{d}{dt} \begin{bmatrix} u_C \\ i_{L_2} \\ \hline i \end{bmatrix} = \begin{bmatrix} -1/R_2 & -1 & 1 \\ 1 & 0 & 0 \\ \hline -m & 0 & \bar{m} - mR_1 \end{bmatrix} \begin{bmatrix} u_C \\ i_{L_2} \\ \hline i \end{bmatrix} + \begin{bmatrix} 0 \\ 0 \\ \hline m \end{bmatrix} E \tag{B.27}$$

where $\bar{m} = 1 - m$.

In (B.27), the ODE for the inductor current i in the last row turns into the correct algebraic constraint $i = 0$ in case $m = 0$ (open switch). The ideal switch could also be replaced by a non-ideal switch in fixed conductance causality with an ON resistance R_{on} (forcing inductor $I : L_1$ into derivative causality). Replacing u_{sw} in the constitutive switch equation

$$i = m \frac{u_{\text{sw}}}{R_{\text{on}}} \tag{B.28}$$

then gives

$$m\,L \frac{di}{dt} = -(R_{\text{on}} + m\,R_1)i - m\,u_C + m\,E \tag{B.29}$$

which is also correct for both system modes.

B.6 Characteristic Bond Graph Features in a Nutshell

- Lumped parameter models for continuous systems
- Distributed parameter models of subsystems can be approximated by lumped parameter bond graphs
- Starting point: Energy exchange between subsystems
- Small number of basic elements
- Uniform notation for all energy domains
- Bond graph models should comply with the principles of conservation of physical quantities:
  - energy, mass, momentum, charge
- Bond Graph modelling is rule-based. There are rules for
  - the systematic construction of a bond graph from a system schematic
  - the orientation of power bonds (reference directions of energy flows)
  - the assignment of computational causalities at power ports
  - the systematic derivation of a mathematical model from a causally completed bond graph.
- State variables and features of mathematical models to be derived from a causal bond graph are known before writing any equations

- Causal BGs reveal information for control engineering such as structural observability and controllability

B.7 Bond Graphs: A Core Model Representation

Bond graph modelling cannot only support the generation of state space models for the simulation of the dynamic behaviour of a system. Bond graphs can serve as core model representation from which various forms of mathematical models can be directly generated such as

- DAE systems,
- Lagrange equations,
- transfer functions,
- equations of a partial inverse model with respect to a given pair of variables,
- canonical form of state space equations or the standard interconnection form both used for robust control.

Moreover, causal bond graphs can support

- the analysis of structural observability and structural controllability,
- model-based control,
- parameter sensitivity analysis,
- the analysis of a direct LTI model for structural invertibility,
- model-based fault detection and isolation (FDI) for systems represented by a hybrid model
- failure prognosis

B.8 Summary

Bond graph modelling of multidisciplinary systems comprising various energy domains starts from considering energy flows between ports of subsystems or components and is based on physical principles. An essential feature of the methodology is that modellers do not need to start with equations but may focus in a qualitative way on physical phenomena a model should capture. The formulation of equations is postponed to a later stage of the model development process. Moreover, the path down from a conceptional view, from a system schematic that may not be formalised down to a mathematical model is rule based. Rules guide the systematic development of a bond graph, the assignment of causalities, and the generation of equations. A causal bond graph clearly reflects modelling assumptions and displays the topological structure, i.e. the connectivity of components or elements like a circuit diagram and furthermore indicates the computational structure of a model such as a block diagram does.

Bond graph modelling allows for a representation that is uniform over all energy domains, does interface with block diagrams and is formalised so that it can be processed by software. If software supporting bond graph modelling is not available, causal bond graphs can be systematically converted into a block diagram, or the equations of a time domain model deduced from a causal bond graph can be processed by a numerical solver for ODE or DAE systems.

References

1. Low, C. B., Wang, D., Arogeti, S., & Zhang, J. B. (2010). Causality assignment and model approximation for hybrid bond graph: Fault diagnosis perspectives. *IEEE Transactions on Automation Science and Engineering, 7*(3), 570–580.
2. Borutzky, W. (2005). Exchange and reuse of bond graph models based on XML. In J. Granda & F. Cellier (Eds.), *2005 International Conference on Bond Graph Modeling, and Simulation (ICBGM 2005)* (Simulation Series, Vol. 37, No. 1, pp. 129–136). SCS Publishing. ISBN: 1-56555-287-3.
3. Borutzky, W. (2009). Bond graph modelling and simulation of multidisciplinary systems – an introduction. *Simulation Modelling Practice and Theory, 17*(1), 3–21. https://doi.org/10.1016/j.simpat.2007.08.008.
4. Borutzky, W. (2010). *Bond Graph Methodology – Development and Analysis of Multidisciplinary Dynamic System Models*. London: Springer. ISBN: 978-1-84882-881-0.
5. Borutzky, W. (Ed.). (2011). *Bond Graph Modelling of Engineering Systems – Theory, Applications and Software Support*. New York: Springer.
6. Borutzky, W. (Ed.). (2016). *Bond Graphs for Modelling, Control and Fault Diagnosis of Engineering Systems*. Cham: Springer International Publishing.
7. Borutzky, W. (2017). A new approach to the derivation of a single set of implicit state equations from a fixed causality bond graph of a hybrid model. In A. Bruzzone, G. Dauphin-Tanguy, S. Junco, & F. Longo (Eds.), *Proceedings of the 10th International Conference on Integrated Modelling and Analysis in Applied Control and Automation (IMAACA 2017)*, part of the I3M Multiconference, Barcelona.
8. Broenink, J. F. (1999). Introduction to Physical Systems Modelling with Bond Graphs. Technical Report, University of Twente, Enschede. https://web.mat.upc.edu/carles.batlle/MOSS/broenink.pdf.
9. Brown, F. T. (2001). *Engineering System Dynamics: A Unified Graph-Centered Approach*. New York: Marcel Dekker. ISBN: 0-8247-0616-1.
10. Controllab Products. (2020). 20-sim Software for Modeling Complex Physics. http://www.20sim.com.
11. Damic, V., & Montgomery, J. (2016). *Mechatronics by Bond Graphs* (2nd ed.). Berlin/Heidelberg: Springer.
12. Das, S. (2009). *Mechatronic Modeling and Simulation Using Bond Graphs*. Abingdon: CRC Press.
13. van Dijk, J. (1994). On the role of bond graph causality in modelling mechatronic systems. Ph.D. Thesis, University of Twente, Enschede.
14. Gawthrop, P., & Bevan, G. (2007). Bond graph modeling. *IEEE Control Systems Magazine, 27*, 24–45.
15. Gawthrop, P., & Smith, L. (1996). *Metamodelling: Bond Graphs and Dynamic Systems*. Hemel Hempstead: Prentice Hall International (UK) Limited. ISBN: 0-13-489824-9.
16. Granda, J. (2007). CAMP-G – User's Manual. Cadsim Engineering, P. O. Box 4083, Davis, Ca 95617. https://www.bondgraph.com.

17. Karnopp, D., & Rosenberg, R. (1968). *Analysis and Simulation of Multiport Systems – The Bond Graph Approach to Physical System Dynamics*. Cambridge: MIT Press.
18. Karnopp, D. C., Margolis, D. L., & Rosenberg, R. C. (2012). *System Dynamics: Modeling, Simulation, and Control of Mechatronic Systems* (5th ed.). Hoboken: Wiley.
19. Merzouki, R., Samantaray, A., & Pathak, P., & Ould Bouamama, B. (2013). *Intelligent Mechatronic Systems*. Springer: London.
20. Mukherjee, A., Karmakar, R., & Samantaray, A. (2006). *Bond Graph in Modeling, Simulation and Fault Identification*. New Delhi: I.K. International Publishing House. ISBN: 81-88237-96-5.
21. Paynter, H. (1961). *Analysis and Design of Engineering Systems*. Cambridge: M.I.T. Press.
22. Samantaray, A. (2017). SYMBOLS 6 Software Installation (Windows XP to Windows 10). https://www.researchgate.net/publication/317277532_SYMBOLS_6_Software_Installation_Windows_XP_to_Windows_10.
23. Samantaray, A., & Ould Bouamama, B. (2008). Model-based process supervision – a bond graph approach. In *Advances in Industrial Control*. London: Springer.
24. Scilab Team, ESI Group. (2020). Scilab. https://www.scilab.org/.
25. The Mathworks. (2020). MATLAB®/Simulink®. http://www.mathworks.com/products/simulink/.
26. Thoma, J. (1975). Entropy and mass flow for energy conversion. *Journal of The Franklin Institute*, 299(2), 89–96.
27. Wellstead, P. (1979). *Introduction to Physical System Modelling*. London: Academic.

Appendix C
Some Mathematical Background

C.1 A Lyapunov Function

Finding a Lyapunov function with a negative definite time derivative may be difficult in the case of a nonlinear time-variant system $\dot{x} = \dot{\mathbf{f}}(x(t))$, $x(0) = x_0$. For an LTI system $\dot{x} = \dot{\mathbf{A}}(x(t))$, one way is to choose a quadratic candidate Lyapunov function $V : \mathbb{R}^n \rightarrow \mathbb{R}$, $V(x) = x^T \mathbf{P} x^T$ and a symmetric matrix $\mathbf{Q} > 0$ and to solve the Lyapunov equation

$$-\mathbf{Q} = \mathbf{A}^T \mathbf{P} + \mathbf{A}\mathbf{P} \tag{C.1}$$

for the unknown matrix **P**. If it is positive definite, the eigenvalues of the system matrix **A** do have negative real part. That is, the system is globally asymptotically stable and $V(x) = x^T \mathbf{P} x^T$ is a Lyapunov function. Its time derivative reads

$$\begin{aligned} \dot{V}(x(t)) &= \dot{x}^T \mathbf{P} x + x^T \mathbf{P} \dot{x} \\ &= (\mathbf{A}x)^T \mathbf{P} x + x^T \mathbf{P} \mathbf{A} x \\ &= x^T (\mathbf{A}^T \mathbf{P}\, , + \mathbf{P}\mathbf{A}) x \end{aligned} \tag{C.2}$$

If there is a positive definite matrix **P** such that

$$-\mathbf{Q} = \mathbf{A}^T \mathbf{P} + \mathbf{A}\mathbf{P} \tag{C.3}$$

for any chosen positive definite matrix **Q**, then $\dot{V} = -x^T \mathbf{Q} x < 0$. That is, the time derivative of V along the trajectory $x(t)$ negative definite. The system $\dot{x}(t) = \dot{\mathbf{A}}(x(t))$ is asymptotically stable.

W. Borutzky, *Bond Graph Modelling for Control, Fault Diagnosis and Failure Prognosis*, https://doi.org/10.1007/978-3-030-60967-2

Remark C.1

1. The Lyapunov equation (C.1) may be solved by means of the Octave function `lyap()` in the control package.
2. The eigenvalues of **A** can be computed by the Octave function `eig()`.

C.2 LaSalle's Invariance Principle

Theorem C.1 (LaSalle (1960)) *Let $\mathbf{x}_e = \mathbf{0}$ be an equilibrium point of $\dot{\mathbf{x}} = \mathbf{f}(\mathbf{x}(t))$, $\mathbf{x}(0) = \mathbf{x}_0$ and $V : \mathscr{D} \subseteq \mathbb{R}^n \rightarrow \mathbb{R}$ a continuously differentiable function that is positive definite with respect to $\mathbf{x}_e$ for which $\dot{V}(\mathbf{x}) \leq 0$ holds in $\mathscr{D}$.*

Let S denote the manifold $S := \{\mathbf{x} \in \mathscr{D} : \dot{V}\mathbf{x} = 0$ and suppose that no solution can stay identically in S, except the trivial solution $x(t) \equiv \mathbf{x}_e$. Then the equilibrium point $\mathbf{x}_e$ is asymptotically stable.

If moreover, $\mathscr{D} = \mathbb{R}^n$ and V is radially unbounded, $\mathbf{x}_e$ is even globally asymptotically stable.

C.3 Implicit Function Theorem

In the context of ARRs it is sufficient to consider a simplified version of the implicit function theorem.

Theorem C.2 (Implicit function theorem) [1] *Let $D \subset \mathbb{R}^n$ be a domain, ϕ a real valued function $\mathbb{R}^n \rightarrow \mathbb{R}$ that is continuously differentiable on an open set $D_1 \subset D$, $\mathbf{x}^0 = (x_1^0, x_2^0, \ldots, x_n^0) \in D_1$ and $\phi(\mathbf{x}^0) = 0$. Suppose that $\partial\phi(\mathbf{x}^0)/\partial x_1 \neq 0$.*

Then there exists a neighbourhood $U(x_2^0, x_3^0, \ldots, x_n^0) \subset D_1$, an open set $V \subset \mathbb{R}$ containing x_1^0, and a real valued function $\psi_1 : U \rightarrow V$ such that

$$x_1^0 = \psi(x_2^0, x_3^0, \ldots, x_n^0) \tag{C.4a}$$

$$0 \equiv \phi(\psi(x_2^0, x_3^0, \ldots, x_n^0), x_2^0, x_3^0, \ldots, x_n^0) \tag{C.4b}$$

C.4 Inverse Model of Non-reduced Order

Classical algebraic construction of an inverse model reported in the literature usually takes place in two steps. First, an inverse model is constructed by successive differentiation of the observation equation of a linear state model. The obtained inverse model is of the same order as the forward model. In a second step, a state transformation is then used to reduce the order of the inverse model [2]. For demonstration, in following, the equations of the inverse model of non-reduced order are derived for the case of a MIMO LTI system.

Let

$$\dot{\boldsymbol{x}}(t) = \mathbf{A}\boldsymbol{x}(t) + \mathbf{B}\boldsymbol{u}(t) \tag{C.5a}$$

$$\boldsymbol{y}(t) = \mathbf{C}\boldsymbol{x}(t) \tag{C.5b}$$

be a state space model with $\boldsymbol{x}(t) \in \mathbb{R}^n$, $\boldsymbol{u}(t) \in \mathbb{R}^m$, $\boldsymbol{y}(t) \in \mathbb{R}^p$, and constant coefficient matrices $\mathbf{A} \in \mathbb{R}^{n\times n}$, $\mathbf{B} \in \mathbb{R}^{n\times m}$, $\mathbf{C} \in \mathbb{R}^{p\times n}$.

Furthermore, let α be the smallest integer so that $\mathbf{CA}^k\mathbf{B} = \mathbf{0} \quad \forall\, k < \alpha - 1$ and $\mathbf{CA}^{\alpha-1}\mathbf{B} \neq \mathbf{0}$. Then successive differentiation of $\boldsymbol{y}(t)$ with respect to time yields

$$\boldsymbol{y}^{(k)}(t) = \begin{cases} \mathbf{CA}^k\boldsymbol{x}(t) & k < \alpha \\ \mathbf{CA}^k\boldsymbol{x}(t) + \mathbf{CA}^{k-1}\mathbf{B}\boldsymbol{u}(t) & k = \alpha \end{cases} \tag{C.6}$$

Left multiplication of $\boldsymbol{y}^{(\alpha)}(t)$ with the inverse of matrix $\mathbf{CA}^{\alpha-1}\mathbf{B}$ gives

$$\boldsymbol{u}(t) = -(\mathbf{CA}^{\alpha-1}\mathbf{B})^{-1}\mathbf{CA}^{\alpha}\boldsymbol{x}(t) + (\mathbf{CA}^{\alpha-1}\mathbf{B})^{-1}\boldsymbol{y}^{(\alpha)}(t) \tag{C.7}$$

Substituting the result (C.7) into (C.5a), one obtains the state equation of the inverse model

$$\dot{\boldsymbol{x}}(t) = [\mathbf{A} - \mathbf{B}(\mathbf{CA}^{\alpha-1}\mathbf{B})^{-1}\mathbf{CA}^{\alpha}]\,\boldsymbol{x}(t) + \mathbf{B}(\mathbf{CA}^{\alpha-1}\mathbf{B})^{-1}\,\boldsymbol{y}^{(\alpha)}(t) \tag{C.8}$$

The inverse model (C.8) and (C.7) is of the same order as the forward model (C.5). Note that the input is not $\boldsymbol{y}$ but the time derivative of order α. That is, the function y must be sufficiently smooth, i.e., a member of the class C^{α} at least.

Reduction of the order of the inverse model makes use of the fact that the output $y(t)$ and its time derivatives up to the order α are known.

References

1. Krantz, S. G., & Parks, H. R. (2013). *The Implicit Function Theorem – History, Theory, and Applications*. Springer. https://doi.org/10.1007/978-1-4614-5981-1, (Modern Birkhäuser Classics, Boston, 2003, second printing).
2. Silverman, L. M. (1969). Inversion of multivariable linear systems. *IEEE Transactions on Automatic Control, AC-14*(3), 270–276.

Glossary

In this glossary, some key notions used in the text are listed in alphabetical order along with short explanations in accordance with a list of definitions compiled by the Safeprocess Technical Committee of IFAC, the International Federation of Automatic Control [4].

Analytical redundancy relations (ARRs) are mathematical equations that relate known system inputs, known parameters and quantities obtained by measurements from a real system. Their evaluation results in the so-called ARR residuals that are identical to zero or close to zero in narrow limits as long as the system is healthy. Residuals that deviate distinguishably from zero serve as fault indicators. If nonlinear constitutive element equations do not permit to eliminate unknown variables in a candidate for an ARR in closed symbolic form, then residuals are given implicitly and can be determined by numerically solving a set of equations. As inputs into ARRs may be time derivatives of measured quantities, measurement noise is to be filtered appropriately. The differentiation is carried out in discrete time. p. 71

Diagnostic bond graph A system bond graph model with storage elements in derivative causality and sensors in inverted causality from which ARRs can be systematically deduced. In online model-based FDI, initial conditions are difficult to obtain. Therefore, storage elements are in derivative causality. With regard to FDI, measured quantities provided by sensors are known inputs into a diagnostic bond graph model. Its purpose is to provide ARR residuals as fault indicators to a diagnosis module. p. 83

Disturbance is an unknown and uncontrollable system input. p. 72

Failure A failure is a permanent interruption of a system's ability to perform a required function. It can only be accommodated by a reconfiguration of the system. p. 1

W. Borutzky, *Bond Graph Modelling for Control, Fault Diagnosis and Failure Prognosis*, https://doi.org/10.1007/978-3-030-60967-2

Fault A system fault is a deviation of the system structure or the system parameters from the nominal conditions [1]. Appropriate actions may enable to recover from a component fault without replacing the component. The fault may be accommodated through fault tolerant control. p. 1

Fault accommodation means to assess the severity of a fault and to decide if and what actions can be taken. If a fault cannot be accommodated, a controlled shutdown may become necessary. If it can be accommodated, possible actions may be a change of controller parameters or a controller redesign, or a system reconfiguration, i.e. a malfunctioning component is replaced by one with similar functionality. p. 178

Fault detection means to constantly monitor the behaviour of system and to determine if it has deviated from its normal operation beyond permissible limits. p. 82

Fault diagnosis means to detect and to isolate faults and to determine their type and magnitude. p. 1

Fault isolation If values of fault indicators are beyond acceptable limits and an alarm has been raised then fault isolation means to locate possibly faulty components. p. 82

Fault quantification (identification) means to determine the type of a fault and its severity. If the fault is due to a parametric change that is not too severe, then parameter estimation can be used to determine an approximation of the actual parameter value. p. 231

Fault tolerant control (FTC) is the ability to automatically accommodate system component faults so that overall stability and acceptable performance of the faulty system are maintained without replacing hardware. p. 177

Feature extraction is the process of processing collected raw sensor data so that their amount is reduced, transformed, and presented in a form that can be used in data-driven fault diagnosis and failure prognosis. The preserved data are called features. It should be possible to classify features by assigning them to classes that are sufficiently distant from each other. Features should be uncorrelated from each other. The effort to obtain them should be small. It should be possible to associate features with a physical meaning and a mathematical definition. p. 68

Feature selection means to choose a subset of an available pool of features. Feature selection is application dependent. p. 135

Hybrid system model A hybrid system model makes use of the abstraction of instantaneous state changes and captures the dynamic behaviour in various system modes as well as discrete events. The latter ones are either controlled by local automata or take place autonomously and cause the system to instantaneously change from one mode into another. p. 165

Hybrid system A System is called a hybrid system for short if its dynamic behaviour is appropriately described by a hybrid model. p. 165

Malfunction is an intermittent, i.e. a temporary irregularity in the fulfilment of a system's desired function. A malfunction is due to one or more faults [3]. p. 51

Modelling uncertainties denote all kinds of discrepancies between a mathematical model and the actual faultless system due to imperfect modelling [2] p. 166

Prognosis Failure prognosis means the ability of an early detection and isolation of incipient faults that may lead to a component failure, to determine the progression of the fault and to predict the remaining useful life (RUL), i.e. the time to failure given the current state of a system. p. 1,131

Supervision means monitoring a physical system and taking appropriate actions to maintain the system's operation in the case of faults [1]. p. 179

References

1. Blanke, M., Kinnaert, M., Lunze, J., & Staroswiecki, M. (2006). *Diagnosis and Fault-Tolerant Control*. Berlin: Springer.
2. Frank, P. M. (2004). Control systems, robotics and automation. In *Fault Diagnosis for Linear Systems* (Vol. XVI) Encyclopedia of Life Support Systems (EOLLS) edn, Oxford: EOLLS Publishers. http://www.eolss.net.
3. Isermann, R. (2006). *Fault-Diagnosis Systems An Introduction from Fault Detection to Fault Tolerance*. Berlin/Heidelberg: Springer.
4. Isermann, R., & Ballé, P. (1997). Trends in the application of model-based fault detection and diagnosis of technical processes. *Control Engineering Practice, 5*(5), 709–719.

Index

W. Borutzky, *Bond Graph Modelling for Control, Fault Diagnosis and Failure Prognosis*, https://doi.org/10.1007/978-3-030-60967-2

The manufacturer's authorised representative in the EU is Springer Nature Customer Service Centre GmbH, Europaplatz 3, 69115 Heidelberg, Germany. If you have any concerns regarding our products, please contact ProductSafety@springernature.com

Printed and bound by CPI Group (UK) Ltd, Croydon, CR0 4YY
22/09/2026
02221062-0017